Prehistoric Stress in Australian Aborigines

A Palaeopathological Study of a Hunter-Gatherer Population

Stephen Webb

BAR International Series 490
1989

B.A.R.

122 Banbury Rd, Oxford, OX2 7BP, UK

BAR -S490, 1989: 'Prehistoric Stress in Australian Aborigines'

ISBN 9780860546276 paperback
ISBN 9781407347790 e-book
DOI https://doi.org/10.30861/9780860546276
A catalogue record for this book is available from the British Library

This book is available at www.barpublishing.com

CONTENTS

LIST OF CAPTIONS
(Short Titles)

Note:-
All Figures and Tables have been placed at the back of the volume.

FIGURES

TABLES

PLATES

MAPS

ACKNOWLEDGEMENTS

I received help from many people during this study. It is not possible to name all of them but I hope they will know who they are. Nevertheless, Professor Graeme Ryan, Geoffrey Kenny, Ron Vanderwal, Mans Lofgren, Julia Clark, Alan Wallace, Wally Wood, Mike Quinnell, Ron Lampert and Tasman Brown all played a vital role in my quest for data. Alan Thorne provided much needed argument and rethinking and Professor Jack Golsen taught me never to believe what you write until you have rewritten it. Others who offered advise with diagnosis were Emeritus Professor Frank Fenner, Don Ortner, Michael Goodman, Cecil Hackett, Neville White, Lindsay Richards and Len Freedman.

All illustrations are the work of Winifred Mumford and Betsy-Jane Osmond and photography was carried out by Dragi Markovich. Without the patience of Yvonne Ptelkow and Bill Craig management of the vast data base would have been impossible.

Above all I want to thank my wife, Marilyn, without whose years of support none of this would have been possible.

INTRODUCTION

More than a decade ago Macintosh and his colleagues wrote that:

> *It is not too much to hope that Australian archaeology will shed light on disease introduction to this continent and how far back in time that first occurred.* (Macintosh *et al.* 1970:94)

In effect, this statement highlights the little that was known then about the palaeopathology of Australian Aborigines. Since then little data concerning patterns of precontact Aboriginal health and disease have emerged, let alone what kinds of disease were introduced during the last 40,000 years that humans have inhabited this continent. In fact, no systematic palaeopathological study has been undertaken in Australia, until now.

Apart from Hackett's work (1933a, b, c, 1963, 1968, 1974, 1976), only one paper dealing directly with disease in Aboriginal skeletal material had been published before 1970 (MacKay 1938). From then until 1984 only six others were published (Sandison 1973a,b, 1980; Zaino and Zaino 1975; Prokopec 1976; Jurisich and Davies 1976; Webb 1982). It is worth noting that Sandison's 1973a and 1980 publications are almost exactly the same. Most of these papers were written in the form of pathological description and with the exception of Zaino and Zaino's contribution, there is little concern for quantitative assessment of the palaeopathology and what this approach might tell us about the human biology and ecology of the original Australians. The first attempt to address these themes, albeit limited, was my own publication of 1982. It was, therefore, the singular lack of interest in precontact Aboriginal health and paucity of this important form of biological information about traditional Aboriginal society that provided the main stimulus for this work. The study was undertaken for a doctoral thesis. It was completed in 1984 and no major revision of the literature has taken place since then.

The 1960's marked a turning point in palaeopathology around the world, with an upsurge of fresh interest and renewed dynamism derived from improved methods of analysis, systematic assessment and a view to understanding patterns of disease as a cultural marker and an indicator of cultural change and human activity. The rather staid approach taken in previous decades had followed the birth and concomitant enthusiasm for the subject, when major advances in medical science at the turn of the century were immediately applied to the diagnosis of past diseases. The publication of Jarcho's (1966) and Brothwell and Sandison's (1967) edited volumes, however, inspired new approaches and spawned many scholars. Palaeopathology began to experience a rebirth which took the form of an exploration of new developments in the fields of physics, chemistry and medicine. With increasing use of various 'non-specific' bone markers, such as Harris lines and dental hypoplasia, as well as detailed analysis of bone composition, the aims of palaeopathology have expanded so that in the last 10 years its philosophy and approach have changed radically from its Virchowian beginnings and, with this, its emphasis. Rather than the major theme being the discovery of disease in the past (which is by definition self-limiting) or the publication of 'interesting' or 'unusual' medical conditions or 'bad' injuries, the accent of palaeopathological study has turned to a wider, if more general, interpretation of the data. Used in a mutually supportive way, non-specific stress indicators, together with studies of bone remodelling dynamics and trace element analysis, now enable researchers to come to grips with past societies as living populations reflecting the vagaries of their particular environments.

A number of palaeopathological studies have shown that the overall patterns of stress in various prehistoric communities is a factor of their respective lifestyles. Temporal changes and spatial differences in the frequency of various forms of stress, have been correlated with subsistence and dietary changes, population growth and, in particular, sedentism. The development of these interpretations has led to a greater understanding of the evolution and variation among prehistoric societies, together with their adaptive capabilities, human ecology, epidemiology, palaeobiology and, in a wider sense, the breadth and diversity of human society. This is the spirit in which this study has been undertaken.

This palaeopathological survey is concerned with the Australian mainland and Tasmania. The main aim is to study a selection of pathologies which provide some measure of the stress that Aboriginal people suffered in the past. Originally I intended that it would deal with Aboriginal disease generally. Unfortunately this aim was far too ambitious. While on the one hand not all diseases leave their mark on the skeleton, the range of pathological data that could be gathered from bone was too great for the time available for the study. I therefore selected a series of pathologies that might tell us most about the general health of the people; produce a biological picture of traditional Aboriginal lifeways in a way not possible using other forms of investigation and that would be most useful in comparing and contrasting widely separated Aboriginal groups across the continent. Integral to this exercise was the comparison of skeletal samples from various parts of an ecologically highly variable continent. Naturally, with such diversity, it became logical to ask: 'what were the worst areas for disease and stress in Australia'? In other words, who was the sickest? In the end this became a primary question to resolve.

To answer this question there was a need to present a broad, generalized study rather than concentrating either on one particular area of Australia or one pathology. Moreover, in a primary study it is essential to be able to compare the data obtained. Of course, comparison with data from other parts of the world is possible and, at one level, this is essential. Equally important, however, is the need to provide an inter-regional comparison within Australia, so that the health profiles of a number of groups, inhabiting different environmental settings and, perhaps, following different lifestyles can be assessed. For instance, a 'healthy' group within a continent the size of Australia cannot be determined as such until it has been compared with many others from around the country. For these reasons a continental survey was deemed the most appropriate way to establish whether there was any continental variation in the palaeopathological picture and if so, what were the extremes, where were they and why.

Similarly, by surveying only one or two pathologies little can be said about the general continental picture. At that level all that becomes known is that one area has a higher frequency of the particular pathology than another. This is not very constructive by itself. Using a number of pathologies in an integrated approach, however, it is possible to gauge the overall health pattern in an indirect but substantial manner. This method also provides us with a greater understanding of what the data is reflecting, as well as defining subtle differences in what sort of stress is affecting a particular group. Not all stress indicators mean the same thing: some reflect long-term, persistent or chronic forms, others short-term or acute stress. The application of these two categories in interpreting the skeletal biology of a hunter-gatherer society is extremely valuable. For instance, short-term stress may be caused by seasonal factors in food resources, whereas long-term or chronic stress may reflect population size, or variation of lifestyle and settlement patterns.

I have used three types of pathology as stress indicators: cribra orbitalia, dental hypoplasia and Harris lines. These reflect stresses that originate from a nutritional, metabolic and/or physiological disorder. Cribra orbitalia is the only one of these believed to have a singular cause - anaemia. The other two are more idiopathic, although recent research has shown that dental hypoplasia, like cribra orbitalia, probably indicates a longer stress episode than Harris lines, which are an indication of something more acute or short-term.

Chapter Outline

Chapter 1 is a general methodology chapter. It outlines the sexing and ageing methods used on the skeletal samples, together with recording techniques and the layout of recording sheets. The division of Australia into two sampling levels and the composition of those samples are described. I have also included a discussion of the usefulness of early historical observations for the reconstruction of pre-contact Aboriginal health and disease.

In Chapter 2, 3 and 4 cribra orbitalia (anaemia), dental hypoplasia and Harris lines are discussed, respectively. These chapters deal with the skeletal stress markers. It has been shown that the incidence of these conditions is often related to the socio-economic status of the society in which they are observed. In general higher frequencies are found in those people living under some demographic or environmental stress. These chapters are set out in the manner of a scientific publication: Introduction, Methodology, Results, Discussion and Summary. It was decided that this would be the best way to handle and describe a large amount of data, as well as being the most appropriate manner for biological data in a survey of this nature to be presented.

The next three chapters deal with pathologies which can be termed 'secondary' stress indicators. These are Non-specific and Treponemal Infection, Osteoarthritis, and Trauma. It was hoped that because they are often products of behavioural and cultural factors, they would provide complementary information to that derived from the stress indicators regarding demography, work patterns and social tension. In this way a series of aetiologically variable data is built up to provide the widest approach to the palaeopathology of stress among Australian Aborigines.

Chapter 5 discusses bone infection both of a non-specific origin and that resulting from treponemal disease. The importance of this lies in the fact that more infection occurs in those societies which are under demographic pressures such as sedentism, population aggregation and general population growth. In the absence of a diagnostic technique which accurately separates non-specific bone infection from that caused by treponemal disease, I have defined and used periostitis, osteitis and osteomyelitis of the long bones as an indication of non-specific infection. Conversely, the diagnostic criteria for treponemal disease of the cranium, set down by Cecil Hackett (1976), have been used to determine the frequency of treponemal disease. Tibial bowing is also discussed in this chapter because of its past association with treponemal disease, its fairly frequent appearance among Aboriginal groups living traditionally and other arguments concerning its aetiology.

Chapter 6 examines the incidence of osteoarthritis of the knee, elbow and temporomandibular joint. This provides information about day-to-day activity and on work stress on both jaws and major joints. Therefore, analysis of the patterns of osteoarthritis offers some direct evidence of biomechanical stress in a particular society, information regarding everyday work patterns and, thus, the extractive effort expended.

Chapter 7 investigates, separately, trauma to the long bones and to the head. By analysing the position of head injuries some ideas concerning their origin can be suggested. From the frequency of the use of violence some assessment of belligerence and social tension within and between groups can be made, together with differences between male and female individuals. Trephination was described in the original work as a surgical technique previously unknown among Australian Aborigines, but this has now been published separately (Webb 1988a).

Finally, Chapter 8 is a discussion which draws together conclusions concerning individual pathologies which are then discussed in terms of each major area of the continent. The implications of the results are examined in the light of recent claims concerning Aboriginal subsistence intensification and population destruction in the face of introduced diseases. Some comments regarding the basic demographic makeup of Aboriginal society in southeastern Australia is presented.

I want to point out here that I see this study as only a first step in our understanding of past Aboriginal health, physiology and adaptation. Nevertheless, it does provide fresh insights into Australian prehistory by presenting an extensive palaeopathological data base which deals with the continent as a whole. Moreover, it shows, once again, that traditional Aboriginal society was divers and more complicated than the blanket 'hunter-gatherer' tag imposed many years ago, would have us believe.

Chapter 1

GENERAL METHODOLOGY

The Survey

Palaeopathological work is largely associated with the identification of diseases suffered by peoples who have no written records. Some diseases are identified in the course of investigations in mummified remains, but frequently the palaeopathologist seeks to identify and interpret past diseases only from their remnant marks left on skeletal material. Naturally, this restricts what can be said about the range of prehistoric diseases to those affecting the skeleton only and, moreover, those which can be readily identified. Hence, a number of cautionary statements have been raised by researchers concerning the equivocal nature of skeletal pathology, the difficulties of accurate identification of lesions and the classification of disease (Sandison 1968; Dastugue 1980). Dastugue remarks:

> *At every step of his work, the paleopathologist must be cautious and aware of his own limitations.....In short.... many paleopathological diagnoses are mere approximations..... Each of us must be aware that his duty is not to give a solution at all costs but to assemble honestly the data of the problem.* (1980:5-6)

The possibility arises also that osteological lesions may not be readily classifiable. It might be because they derive from diseases that were unique to a particular era, or they no longer have a modern counterpart or that they are ancestral to a contemporary disease but manifest different symptoms (Klepinger 1983).

The use, therefore, of a set of skeletal markers (pathologies) caused by certain environmental, pathological, nutritional, biomechanical and demographic circumstances is particularly appropriate in a preliminary study of this sort. These markers, used separately or in concert, have the capacity to tell us more about prehistoric lifestyles than most other archaeological evidence remains or analytical approach. They are, increasingly being used in a mutually supportive role to verify or refute hypotheses arrived at using them separately. This is how they are used here.

The skeletal markers and pathologies used in this survey are:

Cribra Orbitalia, Dental Hypoplasia, Harris Lines, Non-specific Infection, Bowing, Treponemal Infection, Osteoarthritis and Trauma.

Most of these have been used in surveys elsewhere and have proven most useful in the interpretation of human ecology and lifestyle in past populations. Further, they are, in most cases, readily identifiable and generally accepted as being products of known pathological and epidemiological processes.

Osteology

As this study was designed to deal with the Australian continent as a whole, the large amount of skeletal material available dictated constraints on how much was used. The survey was restricted, therefore, to the major long bones (humerus, radius, ulna, femur, tibia and fibula), the cranium and associated mandible. These bones were deemed the most useful for diagnosing the pathologies selected. Long bones had to be complete. Those badly fractured and/or displaying periostial erosion were rejected (Tables 1-1 and 1-2). Bones displaying several post-mortem breaks were included, provided they constituted a complete bone. These were included particularly if they originated from an area where human remains were underepresented. Some long bones had been painted by museum staff at some time in the past, these were rejected, as well as material whose poor condition made diagnosis difficult.

Crania were selected in a similar way so that those fractured and incomplete or with surface erosion were rejected (Table 1-2 and 1-3). Unfortunately these standards effectively reduced sample size in areas which were not so well represented in collections. Skeletal remains from the desert region, where rapid surface erosion is closely followed by fragmentation in dry, arid conditions, posed a particular problem in this regard.

Post-mortem tooth loss was not thought to detract from the overall value of a cranium, neither were broken zygomatic arches. Basi-cranial destruction was ignored because none of the pathologies surveyed were sought in this region. If these factors had not been ignored, severe limitations would have been imposed on the sample size from each region which would have possibly resulted in the elimination of many locations.

Any crania suspected of mixed genetic origin were not included, together with those of dubious provenance, known individuals and those who died in hospital, prison or from European type diseases.

Sample numbers vary, depending on the type of pathology under analysis. If the portion of bone necessary for the detection of a particular pathology is partially or wholly missing, the whole specimen was excluded from that sub-survey. During analysis of a second pathology, however, this same specimen was included if the respective portion of bone needed for the analysis was present. A typical example of this is a cranium with damaged or missing temporomandibular joints. While being excluded from the osteoarthritis sub-survey, it could be used in assessing cribra orbitalia, trauma and other pathologies.

Pleistocene and early Holocene remains were excepted from these criteria because of their obvious scientific value and included in the survey if at all possible. Cremated, highly fragmented, badly eroded and calcified Pleistocene bone, however, had to be rejected. Only obvious pathological processes were noted and these will be discussed individually (Table 1-4).

All the material used for this survey was stored in museum collections in capital cities throughout Australia.

Age Assessment

Sub-adult crania were divided into three basic age groups: 0-5 years, 6-11 years and 12-20 years, alternatively termed groups 1, 2 and 3 (see Appendix 1 and Table 1-2). Tooth eruption sequences were used to define these three categories (Barrett and Brown 1966; Scott and Symonds 1977) and fully erupted third molars and fused spheno-occipital synchondroses to define adults. Initially, post-cranial remains were to be divided into sub-adult and adult material, respectively. So little sub-adult post-cranial material was available for study, however, that these divisions abandoned, leaving only adult post-crania.

Adult crania were divided into two generalized age groups: young adults and old adults, coded 6 and 7 respectively. The variability in Australian Aboriginal toothwear, both regionally and sexually, precludes its use for determining adult age with any accuracy although some gross categorization is possible. Therefore it was was used in combination with ectocranial suture fusion as a basic form of demarcation between young and old adults. While it is recognised that there is some variation in the latter (Krogman 1962), at least half the coronal and sagittal sutures had to be be closed to qualify the individual as an 'old adult'. No exact age marks the boundary between young and old adults, but 35-40 years can be used as a generalised division (Abbie 1950).

Sexing

The sexing method developed by Larnach and Freedman (1964) for coastal New South Wales Aboriginal crania was used on all adult crania, except those from the Murray River. The predominant robustness of the latter and those from elsewhere prompted Brown (1981a) to develop a separate multivariate sexing technique for crania from the Murray Valley and this was used where necessary. Any crania having associated pelves were sexed using these (Davivongs 1963).

As yet no techniques are available for independently sexing the Australian Aboriginal tibia, fibula, ulna or radius. These were sexed only when they accompanied crania, pelves, or when large rugose or small, very gracile bones were encountered which obviously represented males and females, respectively. Femora were sexed using Davivongs' technique (1963), but van Dongen's method (1963) for sexing humeri was found to be unacceptable. The reasons for this were that such minimal amounts of material could be satisfactorily sexed using this technique that the time invested in the exercise could not be justified. Moreover, van Dongen's samples were so small and of such narrow provenance that the value of the technique was thought to be minimal, particularly when one takes into account the morphological variety of post-cranial Aboriginal remains across the continent.

Coding and the Recording Sheet

I want now to describe the method I developed to record and manage the mass of data involved in this work. The study necessitated a computerized method of coding, recording and collation to be developed. All data, therefore, was recorded on two formated sheets. Data Collection Sheet 1 (Fig 1-1) was used for a pathological bone or cranium and Data Collection Sheet 2 (Fig 1-2) for a maximum of five non-pathological samples.

Sheet 1 is divided into three records of 80, 51 and 115 bits of information. Record 1 contains three forms of data: (1) basic description of the specimen such as collection number, collection, provenance, age and sex; (2) metrical information, used for sexing Murray Valley crania; and (3) further pathological data. The latter includes Harris Lines, Cribra orbitalia (C.O.), Dental Hypoplasia and Dental Pathology. Dental pathology includes caries, alveolar disease and pre-mortem tooth loss (edentulism) for both the maxilla (Max) and the mandible (Mand). For a complete breakdown of the codes used for each category of pathological and non-pathological data I draw your attention to Appendix 1. Definitions of individual pathologies will be discussed in their appropriate chapters.

With the exception of the Collection Series Number (CSN), Records 2 and 3 contain pathological information only. Record 2 deals with ostearthritic data for the major joints (shoulder, elbow, hip and knee) and the temporomandibular joint. Trauma, Non-specific disease, Treponemal disease (CR) and two Comment boxes are contained in Record 3. The Comment boxes were rarely used.

Data Collection Sheet 2 is divided into five identical records each containing 46 bits of information. These 46 bits correspond to the first 46 boxes on Sheet 1 with the exception of a permanent '1' in the last box (46), which show that the bone or cranium is non-pathological. On Sheet 1 there is a permanent '2' in box 46 indicating a pathological specimen. With by far the majority of bones being non-pathological, there was a need to reduce the actual number of data sheets and prevent massive wastage of space on a sheet. Therefore, compaction of non-pathological information was achieved by entering five bones or crania on each sheet 2. This method also meant less recording sheets needed to be taken into the field and handled during data processing. Analysis and collation was undertaken using Statistical Package for the Social Sciences softwear (S.P.S.S.) (Nie *et al.* 1975).

Skeletal Samples

At the time of this study there were only three large collections of archaeologically recovered skeletal remains in Australia. These were: Broadbeach (Haglund-Calley 1968a, 1968b; Wood 1968; Haglund 1976), Roonka (Pretty 1977; Prokopec 1979) and Kow Swamp (Thorne 1969, 1971a, b, 1972, 1975, 1976, 1981) (Maps 1-1 and 1-2). A fourth, and much older collection, the Willandra Lakes Hominid series, has been the focus of another separate study to this (Webb 1989). Other remains have been found in dated context during archaeological investigations, such as those reported by Mulvaney (1960), Lampert (1971), McBryde (1974), Jurisich and Davies (1976) and Schrire (1982), but those discussed in respect of physical anthropology and/or palaeopathology constitute only a minute portion of the material held in collections. It is felt, therefore, that no further discussion concerning these individuals is necessary. All other remains are undated, some without provenance, and these form the bulk of materials dealt with in this study.

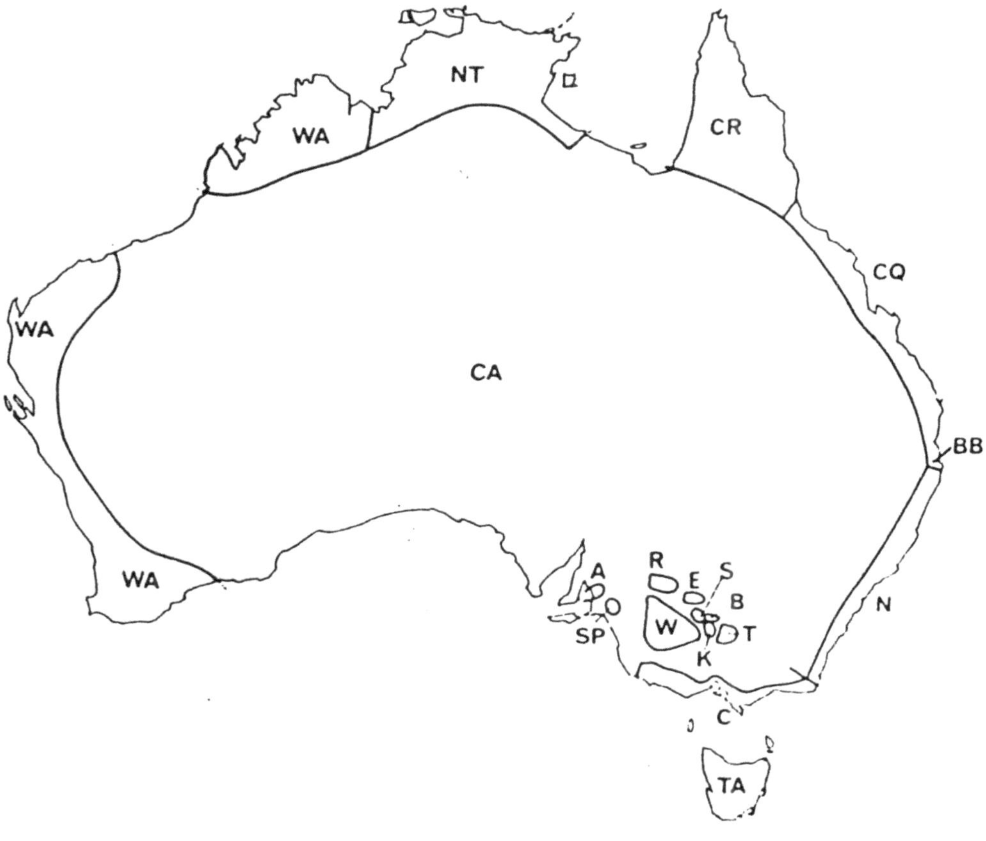

T - Tocumwal
K - Kow Swamp
B - Baratta
S - Swan Hill
E - Euston
R - Rufus River

SP - Swanport
A - Adelaide
C - Coastal Victoria
W - West central Victoria
N - Coastal New South Wales
CA - Central Australia

TA - Tasmania
NT - Northern Territory
CR - Cape York/Cairns
CQ - Coastal Queensland
BB - Broadbeach

Map 1-1 'LOCATIONS' used in this survey

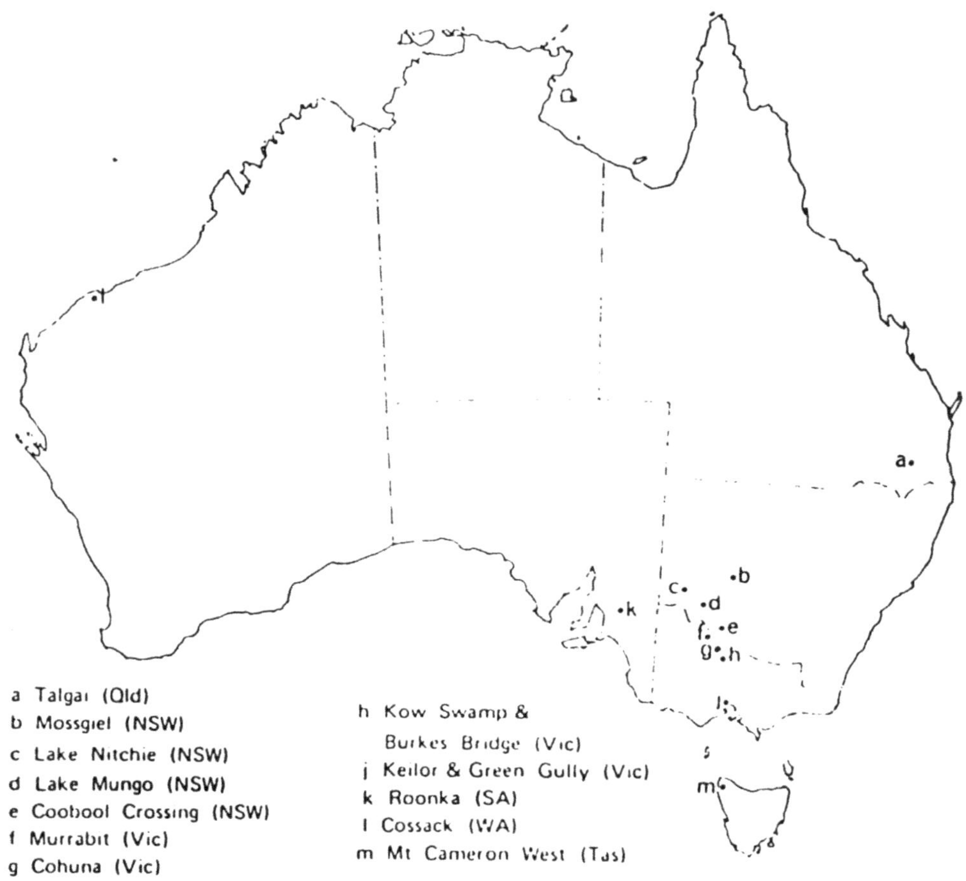

Map 1-2 Australian Pleistocene Sites

Australia's two largest collections of skeletal material are the Murray Black collection from the Murray Valley and the Swanport collection. These have been the mainstay of osteological and physical anthropological research in Australia for a number of years, but with a few exceptions none of this material has been dated. Since this study took place, the Murray Black collection has become the main focus of Aboriginal concerns over the storage, curation and

continued study of their ancestral skeletal remains (Webb 1987a). Moves by the Aboriginal community to have the Murray Black collection reburied have been supported by both State and Federal political and legal agencies. This has resulted in the return of the collection to communities from which it was gathered, for reburial. In these terms, this study now becomes unique and the information more valuable.

The samples respresenting the Murray River locations in this study come from the Murray Black Collection, named after the late George Murray Black, who, with others, retrieved them from the banks and floodplains of the Murray River over a period of about 20 years during the 1930's and 1940's (Sunderland and Ray 1959). While archaeological methods were not employed in the retrieval of these remains, certain inferences can be made about their age. The nature of the burials suggest a traditional interment in the fashion of discrete cemeteries. When land along the Murray was divided into 'runs' by European settlers, its continued use by Aborigines became restricted and, no doubt, Aboriginal people themselves probably would not have wished to continue burying their dead near European settlement. If we accept the recent proposition by Butlin (1983) that the Murray Valley Aboriginal population was devastated by smallpox and other introduced diseases ahead of actual European settlement, it is likely that some of these remains originate from those times (?1828-31). Both these factors put the vast majority of individuals in the Murray Black Collection into a pre-contact context, even if some only by a few years. Moreover, many of the pathologies dealt with in this study began during childhood, 20-50 years before death. This puts almost all individuals in these samples well into the pre-European context.

The Swanport collection from the lower Murray River, South Australia, is not dated but it comes from a discrete burial complex accidently exposed during levee building at the turn of the century (Stirling 1911). One advantage of this group is that it represents a single unit of people, originating from one tribe and confined to a common territory. By the uniformity of preservation of the Swanport material and its good condition it is likely that it is not very old, spanning perhaps two or four hundred years prior to European arrival.

The Roonka collection from the lower Murray River north of Swanport is dated but is very fragmentary, poorly preserved, severely eroded and much material is yet to be properly cleaned and freed entirely from adherent soils. Further, post-mortem deformation of crania have made their reconstruction difficult and in many cases where this has been attempted it is inaccurate. For these reasons the collection was not included in the study. For reports concerning the physical anthropology and palaeopathology of the Roonka group, see Prokopec 1975 and 1979.

The Adelaide sample (A) consists of a series of individual and multiple burials uncovered during the past 150 years by public works projects in various suburbs and places within the city limits.

All samples from Western Australia, the south and east coasts, Central Australia and much of the Northern Territory consist of individuals from widely scattered origins within those areas.

The Broadbeach collection has similar problems to those of Roonka so far as erosion, fragmentation and preservation are concerned. The limitations these factors impose on the use of such material for palaeopathological study reduced the Broadbeach population to a small sample only. Acceding to the wishes of the descedant Aboriginal community, the collection is now under scientific moratorium but remains in the Anatomy Dpeartment of Queensland Museum.

Locations and Areas

It was thought that the most appropriate way to tackle this type of continental study was by using two levels of sample analysis. The smaller the region the more precise the results for that group, but in those places where sample sizes became a problem an extended view, based on the environmental similarities between adjascent populations, was deemed appropriate. These levels have been termed 'locations' and 'areas' and exemplify pathological trends in both a fine and coarse manner.

Locations

In most cases individual specimens that are identified geographically lack precise origin. Rather they are located to the nearest large town or city, i.e. Tocumwal, Swan Hill and Rockhampton. Others are located to a range of hills or the nearest landmark along a coastline. Such circumstances prevent specimens from being connected with any particular linguistic, cultural or tribal group (Tindale 1974), even if these were now believed to be accurate. As a result I have had to put material into units for study by reference to the geographical information supplied. I have called these units or groupings 'Locations'. The criteria of their formation were that they should comprise adequate samples (25 is a generally accepted minimum) and come from a relatively small, discrete locality.

For some regions substantial amounts of material can be located to particular towns or geographical features. This is especially true for much of the Murray Valley: e.g. Euston (E) or Rufus River (R) (Sunderland and Ray 1959; Pietrusewsky 1979; Brown 1981a, b; Webb 1981, 1982; Green 1982). The Rufus River sample consists of material recovered from Lake Victoria, Lindsay Creek, Chowilla and Rufus River. This forms a discrete area around the junction of the Murray and Darling Rivers and represents an intermediate zone between the central Murray area and Swanport at the mouth of the Murray River. Because of this and the substantial size of the sample, Rufus River will be referred to as both a 'location' and an 'area'. It is the only example of this double title.

The Pleistocene (P) sample forms a location because all the specimens contributing to it come from a small area of the Murray Valley (Map 1-2). This consists of the combined crania from Kow Swamp, Coobool Crossing (Brown 1982) and the individual from Cohuna (Macintosh 1952a, b, 1953, 1965; Macumber and Thorne 1975). The material from Kow Swamp is dated 14,000-9,500 B.P. (Thorne 1975, 1981). Only one individual from the Coobool Crossing series has been dated (Brown 1987). The 12,500 b.p. date makes it contemporary with those from Kow Swamp with whom, on grounds of morphological similarity, the Coobool population had already been associated, as had Cohuna. Some examples of Pleistocene and early Holocene remains have not been referred to if pathology was not found (see Table 1-4).

In some cases, the paucity of material from restricted areas required merging those areas to form a location. This has been done, for example, with the location called 'T'. Here, skeletal material from Tocumwal, Echuca and Deniliquin has been analysed as a whole unit because as individual samples they were too small. Also, they are part of an area, the upper Murray, that environmentally allows them to be treated as a representative group from this part of the river.

Elsewhere, the requirement of a minimum sample size of 25 often needed the formation of a location covering a larger geographical area while maintaining a general ecological or geographical distinction. An example of this is the location 'coastal Victoria' (C). Initially this coastline was divided into eastern and western sections at Melbourne. After sample numbers had been calculated for these divisions, however, it was deemed necessary to combine them to form a southern coastal location. Other examples of extensive locations are the New South Wales coast (N), Cape York and Cairns, forming CR, and western (WNT) and eastern (ENT) Northern Territory forming NT.

Yet another type of combination is that of AT, which is composed of samples from eastern Northern Territory (ENT) and western Northern Territory (WNT) plus Northern Territory (NT) remains lacking exact provenance. The Western Australian location (WA) is similarly constructed because the three subdivisions of northwest Western Austrlia (NWA), west central Western Australia (CWA) and southwestern Western Australia (SWA) are too small to stand alone.

The problem of deciding how material was best combined to form locations was a difficult one. Superficially, the composition of one or two locations seems unreal. An example of this is the WA sample. This is composed of material extending from 14 to 35 degrees latitude, a distance of over 2000 kilometers. It brings together individuals from totally different ecological zones which, presumably, include some social differences also. I recognise this sort of anomaly but do not think it precludes the formation of a location representing Western Australia. Moreover, the small sample size of each location would give little credence to any findings resulting therefrom. Even as a sample representing Western Australia, contrasts can be made with other parts of Australia on a very general level. The alternative, ignore Western Australia, seems too drastic when contemplating the sort of restrictions for study that might be imposed on the Western Australian material in the future.

The composition of all locations is set out in Appendix 2. In the figures which accompany analyses they appear in a fixed order starts with the Murray people and finishes with the Pleistocene sample, as follows: T, K, B, S, E, R, M, SP, A, C, W, N, CA, WA, NT, AT, TA, CQ, CR, BB, QB and P (Maps 1-1 and 1-3).

The offshore islands of Bathurst, Melville, Groote and the Keppells were initially included in this study, but due to very small or almost non-existent samples they will not be generally referred to.

Areas

The use of locations allows a close comparative analysis of adjacent groups, which helps define differences of pathology and stress between outwardly very similar people. The drawbacks are sample size is barely adequate and, secondly, differences in disease and stress among geographically disparate populations can be masked by splitting these into smaller groups or sub-populations (i.e. locations). Therefore, it is appropriate to examine groups of locations, based upon larger geographical and/or environmental considerations. This wider level of analysis

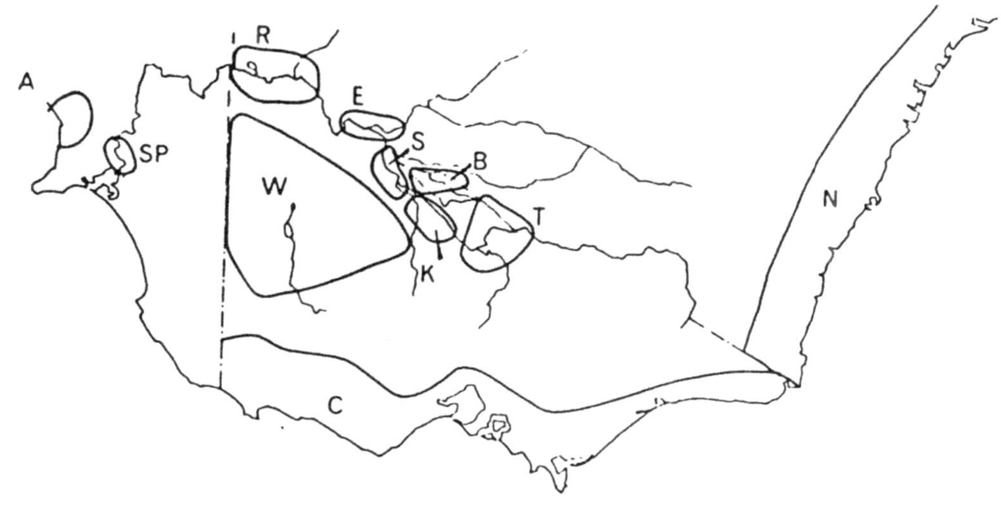

Map 1-3 'LOCATIONS' in southeastern Australia and the Murray Valley

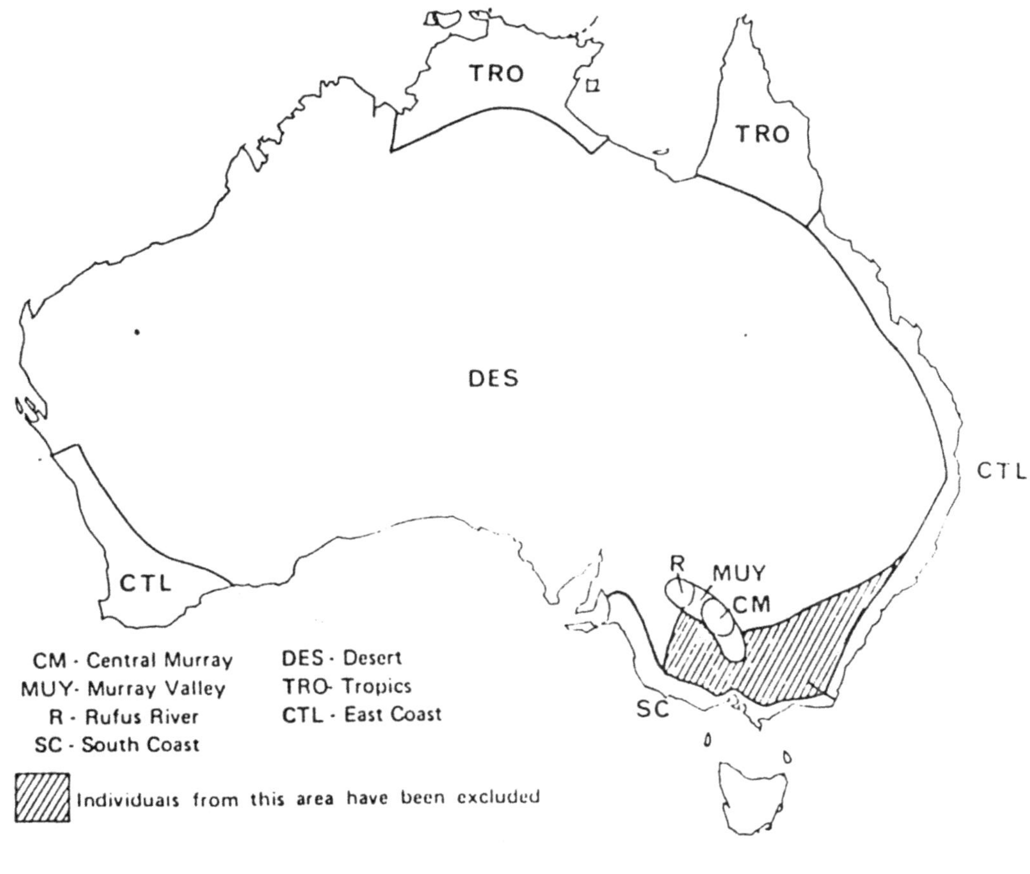

Map 1-4 'AREAS' used in this survey

method of comparing and contrasting aspects of stress and disease across the continent. Therefore, by merging many of the 21 locations seven large groups, or **'Areas'**, have been formed, covering all the basic ecological zones in Australia. These seven are: the central Murray (CM), Murray Valley (or River) (MUY), Rufus River (R), south coast (SC), desert (DES), the tropics (TRO) and the eastern coast (CTL) (Map 1-4). The composition of each area is set out in Appendix 2 and will follow each other in the same order on all figures (CM, MUY, R, SC, DES, TRO and CTL) (Map 1-4).

These two levels of analysis (location and area) will be used separately or concurrently in all the following chapters.

Assessment of Historical Source Material

There is a vast amount of historical source material relating to every aspect of the life of Australian Aborigines. Whatever its value to archaeologists and anthropologists attempting to reconstruct Aboriginal lifestyles at the time of European contact, on the whole it is usually not very enlightening for those wishing to know more about traditional health and disease.

The use of this body of literature to corroborate, support or refute the findings and hypotheses arrived at in this study depends on a number of factors. The closer in time to the first contact, the more reliable any information is likely to be, even though contact times vary greatly from one part of the continent to another. One exception is the more remote desert regions where there are reports of indigenous disease that remain fairly reliable up to the late 1950's and early 1960's, when medically trained personnel were able to examine Aboriginal people still living there. Therefore, those with more knowledge of medicine and the Aboriginal way of life were able to report more accurately on these matters than early observers.

The earlier literature was generally not written by those with medical training, so any description of disease in these accounts is likely to be very general, misleading or plain inaccurate. A common theme in these reports is the description of Aboriginal men being 'muscular', 'well built', 'strong' and 'fit looking' (Bennett 1832:119; Eyre 1845:206; Leichardt 1847:166; Sturt 1849:137; Bonwick 1857:67; Lang 1861:50). This presents basic evidence for the well-being of males, although quite often the opposite picture is presented for females and descriptions of them being ill-treated and inferior in stature and appearance are common (Bennett 1832:250; Mitchell 1839:259; Eyre 1845:206; Sturt 1849:77; Westgarth 1848:78; Lang 1861:51; Chauncy cited in Smyth 1878 II:259). At this level such reports can be of value, albeit that this type of assessment was subject to European cultural standards of the eighteenth and nineteenth centuries. These reports also contain distortion, rumour, anecdotes (Smyth 1878), misdiagnosis (Howitt cited in Smyth 1878 II:302; Mackillop 1892) euphemisms, as a form of Victorian moral censorship, (Eyre 1845 II:379; Sturt 1833 II:96) and bias (Cunningham 1827 II:45). Moreover, most accounts date to well after the first European contact, thus rendering many of them useless for our purposes. Moreover, by the time many ethnohistorians began writing, the health of Aborigines living in southeastern and southwestern Australia had deteriorated drastically following the spread to the Aboriginal community of a whole range of diseases which brought to this continent by white settlers.

By the early 1830's the intervention of highly infectious diseases such as smallpox had drastically altered Aboriginal society, including its demography, particularly in southeastern Australia (Butlin 1983; Webb 1987b) and possibly elsewhere (Campbell 1983). Accounts of smallpox visitations both for the centre of Australia (Giles 1889; Schultze 1891) and around almost the whole continent (Green 1981; Campbell 1983) suggest that Aboriginal people were affected ahead of spreading white settlement and even explorers. Consequently, the original condition of Aboriginal society in some areas can only be known through the auspices of archaeology and physical anthropology. Early European accounts are, in many respects, unintentionally distorted because of the these processes. Following this, surviving Aborigines were increasingly enmeshed in a European pattern of health and disease, and societal collapse. This situation has helped also confounded attempts by those trying, for example, to differentiate between the endemic forms of treponemal disease (yaws and treponarid or irkintja) and European introduced venereal syphilis (Hackett 1936, 1976) (see Chapter 5).

All the above factors call for extreme caution when using medical observations contained in most historical literary sources in reconstructing the state of precontact Aboriginal health or even to support independent palaeopathological evidence. On the other hand, not all historical accounts of a medical nature can be discounted. The assessment of the accuracy or value of a particular report used in this study has remained, therefore, somewhat subjective. It has been based, however, on the known competence of the observer, the type of description given, the region involved, the amount of previous European influence and/or contact experience of the Aboriginal people described and the likelihood of introduced disease being present. Since there is no universal rule for this type of assessment each source must be judged by these criteria and those peculiar to the source itself. Experience, however, makes one a little wiser judge.

Chapter 2

CRIBRA ORBITALIA (ANAEMIA)

Introduction

Cribra orbitalia (henceforth termed 'cribra') consists of one or several clusters of foramina located on, but not necessarily confined to the orbital plate, of the frontal bone. Since Welcker (1885, 1887, 1888) first noted this condition, it has only rarely been referred to in anatomy (Wolff 1954; Duke-Elder *et al.* 1961) and pathology texts (Hogan and Zimmerman 1962). Rather it has been researchers dealing with skeletal populations, in particular the palaeopathologist and physical anthropologist, who have noted its frequency in spatially and temporally disparate people.

Originally it was suggested that cribra was a racial characteristic (Toldt 1886; Welcker 1887, 1888; Adachi 1904; Oetteking 1909) originating as an embryological phenomenon (Toldt 1886). Others thought a pathological aetiology more likely (Koganei 1912) and nobody has seriously challenged this idea since. Some of the pathologies believed to cause cribra are compression or infection of the lacrimal gland (Blumberg and Kerley 1966), specifically leprotic infection of this gland (Moller-Christensen 1961; Hogan and Zimmerman 1962), mumps (Moller-Christensen 1978), trachoma or localized periostitis (Blumberg and Kerley 1966), poor nutrition (Henschen 1961; Nathan and Haas 1966), infantile scurvy (Still 1915) and genetically determined blood dyscrasias (Angel 1964, 1967). The latter, which includes thalassemia and sickle cell anaemia, form distinctive lesions on the outer cranial table due to hyperplastic activity by haemopoetic tissue in the diploe (Britton et al. 1960; Aksoy *et al.* 1966). The lesions, called symmetrical osteoporosis or cribra cranii, can appear also as the result of severe iron deficiency anaemia (Henschen 1961; Steinbock 1976) and are almost always associated with the most extensive form of cribra. I have also observed this association in crania from Papua New Guinea where severe iron deficiency anaemia is common and thalassemia is endemic.

It is this link between cribra and anaemia that has prompted a consensus of opinion with regard to the aetiology of the pathology (Mosely 1965; Hengen 1971; Saul 1972; Carlson *et al.* 1974; El-Najjar *et al.* 1976; Steinbock 1976; Cybulski 1977; Lallo *et al.* 1977; Trinkaus 1977; Hartney 1978; Mensforth *et al.* 1978; Huss-Ashmore 1982; Goodman *et al.* 1983; Sandford *et al.* 1983). Further, it has been suggested that anaemia could be a common link between all those people in whom cribra has been observed (Hengen 1970; Carlson *et al.* 1974; El-Najjar 1976; Cockburn 1977; Cybulski 1977; Lallo *et al.* 1977; Mensforth *et al.* 1978). Certainly, indirect support is lent to this hypothesis by the absence of cribra in modern European populations. Improvements in nutrition and community health standards have been cited as important factors in the relief of chronic anaemia (Henschen 1961). I too have accepted the aetiology of anaemia as responsible for the cribra seen in this survey.

In Australia cribra has almost always been assumed (mistakenly) to indicate a genetically based non-metrical trait rather than a pathology (Brothwell 1963; Larnach and Macintosh 1966, 1970; Milne 1982). Brothwell mentions a frequency of 5.26% for Australia and thought cribra might be of 'some value as an environmental 'indicator'' (1963:95). Larnach and Macintosh (1966) found none in a collection of New South Wales crania and only 1.1% in crania from Queensland. Milne's (1982) figures, derived from a Western Australian sample, are not comparable with those obtained in this survey, because his determination of what constitutes cribra and the fact that I use a grading system prevents it.

Sandison acknowledged cribra as a pathological indicator but thought it 'extremely rare except in children' in Australia (1980:46). Unfortunately he gave no further details concerning his findings. In a survey of 544 crania purported to be from New South Wales, Zaino and Zaino (1975) found 26.6% in sub-adults and 2.4% in adults. No other subdivisions for age, sex or provenance were made. Most puzzling, however, is that there are no extant records of 544 crania from New South Wales alone.

Methodology

The orbits of all crania were inspected for signs of cribra. Dirt or other material preventing observation was removed from the orbit and any other pathology was noted. Variation in the form of the cribrous lesion was recorded using the three categories defined by Nathan and Hass(1966) as a basic guide, plus a fourth which I have previously interpreted as a 'recovery' or scar category (Webb 1982). Subsequently I found this category described elsewhere as a 'remodelled lesion' (Mensforth *et al.* 1978:23). I feel, however, there are some differences between those scars described by Mensforth and colleagues and the ones I have recorded with respect to their appearance and occurrence (see below). Descriptions of the classification used here are as follows.

> Category 1. **Porotic.** This is the 'mildest' form, least noticeable and most easy to confuse. It consists of a minimum of eight foramina forming a round or elongated cluster covering not less than 25 square millimeters (Plate 2-1). This distinguishes it from other small, random foramina that often punctuate the orbital plate.

> Category 2. **Cribrotic.** In this category both area affected and foraminal diameter are generally larger than those of category 1 (Plate 2-2). Although Nathan and Hass (1966) suggest that foramina in this category are discrete, I have included shallow channels in the outer bone table that often permeate the affected area and run in random fashion from one foramen to another and in the areas between foramina. The presence of this shallow channelling helps define this category.

> Category 3. **Trabecular.** In this most extreme form of cribra large foramina become confluent and the channelling produces deeply etched furrows and/or complete erosion of the outer table. This process leaves behind trabeculated networks of bone in the form of islets and strands (Plate 2-3). These trabeculae vary in size and thickness and normally occupy a larger area than that seen in category 2.

> Category 4. **Recovery.** This category is thought to indicate a reversal in the expression of anaemia in the individual. Here, foramina generally contain newly deposited bone which fills them. This is particularly so for those of larger diameter, but a few very small foramina may persist, scattered across the old lesion. Bone remodelling is thought responsible for this change as the orbital bone recovers its natural surface form in response to a termination of diploeic hyperplasticity caused by relief of the anaemia. Large foramina are eliminated during this process and an irregularly deposited, smooth lamellar bone remains on the periosteum and remnant channelling (Plate 2-4). Again, there is some variation in this process.

It should be noted that while all categories are sufficiently distinct to allow their separation, intermediate forms can sometimes be difficult to classify. This is particularly true with the first three forms. Often various categories of cribra were found, either mixed together in one patch or occupying separate areas of the

Plate 2-1 Category 1, POROTIC cribra orbitalia. A small cluster of narrow diameter foramina lies in the roof of the orbit.

Plate 2-2 Category 2, CRIBROTIC cribra orbitalia. This consists of larger diameter foramina (some confluent) and shallow channelling of the orbit roof.

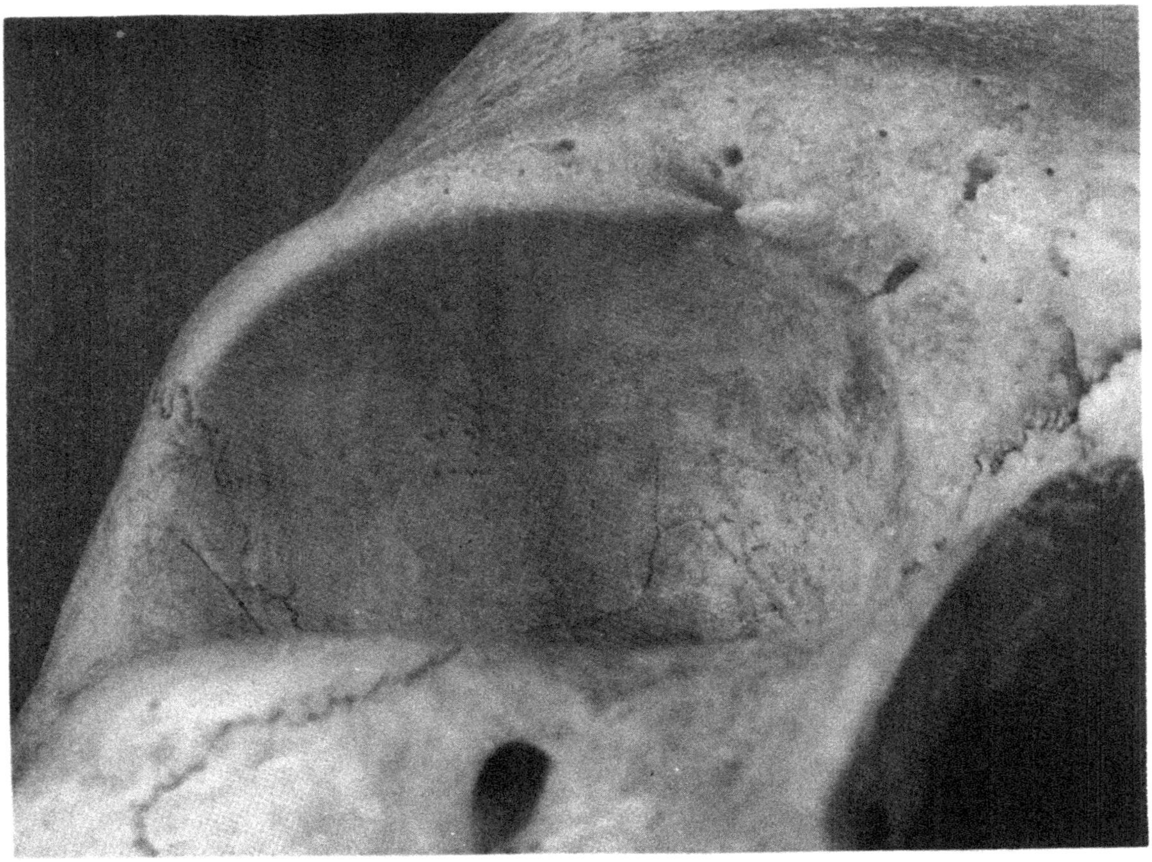

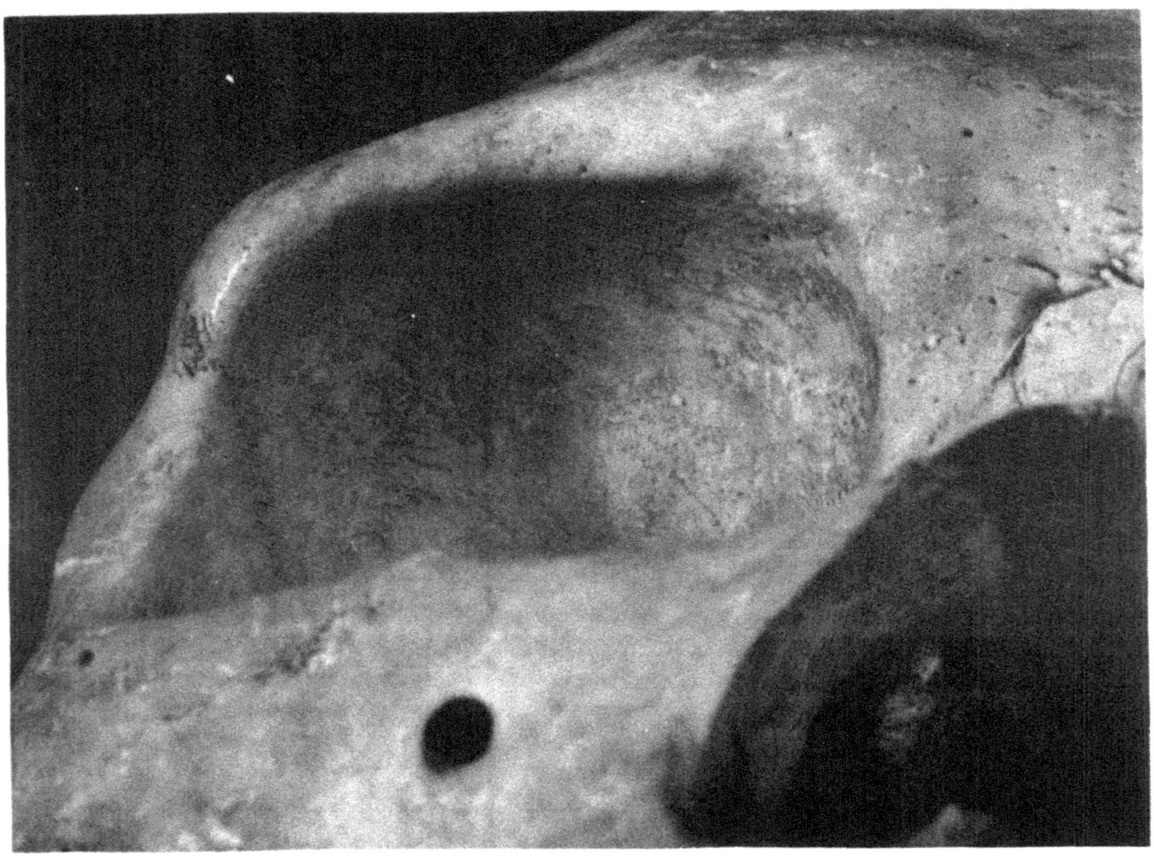

PLate 2-3 Category 3, TRABECULAR cribra orbitalia. Confluent foramina, some of large diameter, produce erosion of the outer bone table. In other areas there is a network of trabeculated bone which gives the characteristic appearance of this extreme form of the condition.

Plate 2-4 Category 4, RECOVERY phase of cribra orbitalia. Formaina are filled with new bone, although a few very small holes are still present. Remnant shallow channelling persists but large foramina have disappeared.

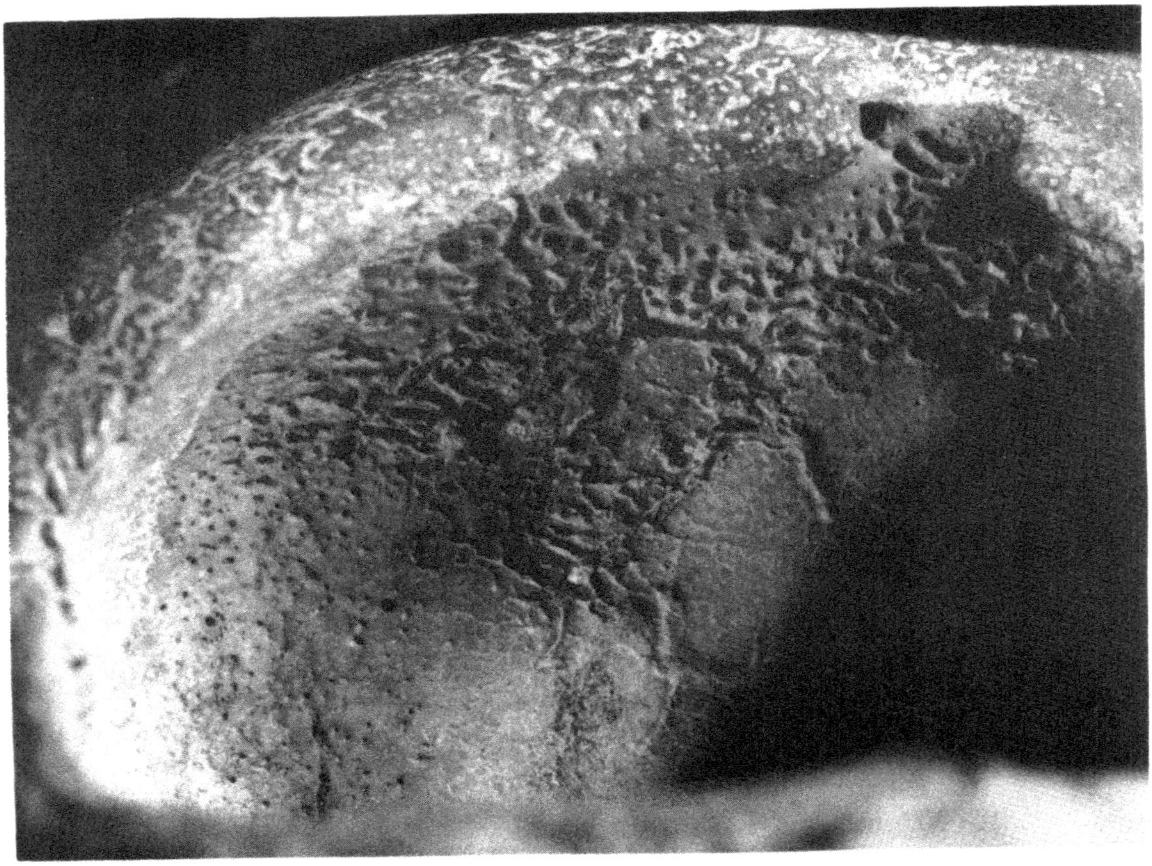

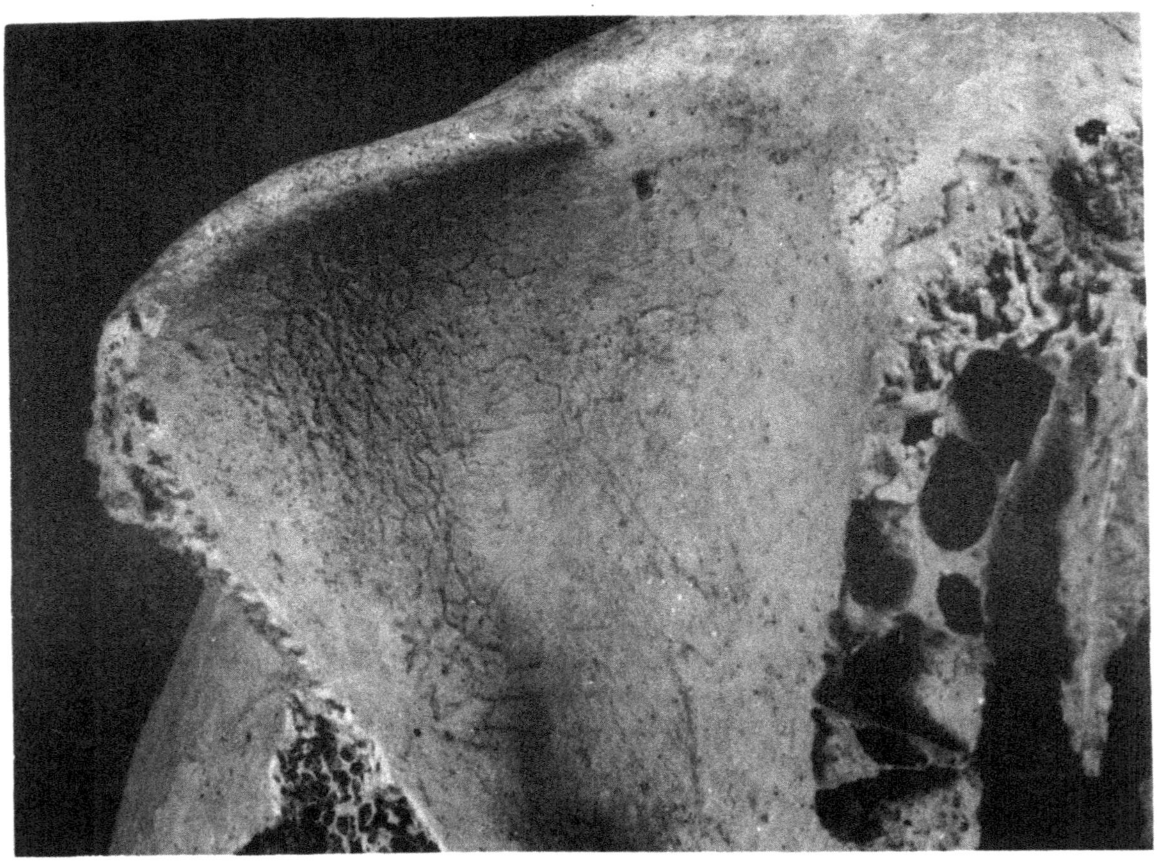

orbit. In both situations the orbit was categorised by the most extreme form present and the individual was categorised by the most extreme form noted in either orbit. Unless otherwise stated, however, all frequencies referred to below represent categories 1-3 combined.

The first three categories provide some measure of the overall severity of cribra in a particular population. Comparisons can then be made between three sub-adult age grades; between sub-adults and adults, and between males and females, either by area or location depending on sample sizes (see Chapter 1). Throughout this study I have tried to maintain a mimimum sample size of 25. Unfortunately this is not always possible when dealing with individuals on a location basis. Therefore, while recognising the drawbacks of using small sample sizes, I include a smaller sample where palaeopathological information would otherwise be lost. Further, in cases of small sample size, some comparisons can only be made at the level of population total or frequency (adult males and females combined). Only by taking the Murray as a whole, from Tocumwal to Rufus River (Map 1-1), is it possible to examine each of the three age grades for each category of cribra in sub-adults. Elsewhere, either the first two or all three age grades have been merged to maintain sub-adult sample size. For these same reasons sub-adult analysis has been confined to area. The Murray (MUY) sample, however, allows analysis of the association between age and each category of cribra. The five age grades used in this chapter are: 0-5, 6-11, >11, young adults and old adults.

It is worth reiterating here that 'location' refers to the origin of a sample e.g. B = Baratta, SP = Swanport or W = west central Victoria, and an 'area' indicates a larger geographical portion of the continent comprising a number of locations. Codes both for locations and areas will follow the respective location or area name in this chapter only. It is hoped that this will provide some familiarity with the codes used on all histograms, figures, tables and maps. A 'population' or 'area total' refers to an adult frequency irrespective of sex.

Results

The first observation that can be made from location and area totals is the great variation in numbers of affected adults across the continent (S2-1). These totals range from a frequency of 33.7% in the central Murray (CM) to 8.3% in the desert areas; statistically, a highly significant difference ($p=<0.001$) (Tables 2-1 and 2-2). Secondly, Murray people have more cribra than any other area and this reaches a peak in the central Murray. The difference between the central and other parts of the Murray, however, is only significant at the $p=<0.05$ level. A third major feature is that the desert has a significantly lower amount of anaemia than any other area of the continent ($p=<0.01$).

Females in all areas (S2-2) and locations (S2-3) have higher frequencies of cribra than males. This is highly significant for the Murray River (p=<0.01) (Tables 2-1 and 2-3). The individual locations in S2-3 show almost the same characteristics as those revealed in the coarser area analysis. The highest incidence of anaemia in Australia is in the central Murray locations of Kerang, Baratta, Swan Hill and Euston (K,B,S and E), followed by the coastal groups of New South Wales (N) and Victoria (C) (Table 2-4).

Central Australia (CA) has one of the lowest frequencies of anaemia, together withp eople around Adelaide (A), which shows also the only real reversal of male and female frequencies. This reversal can be seen in the Pleistocene (P) sample, where males have 19.2% and females 13.3%, but the small female component (n=15) may be too small to reflect real frequencies of this condition among this population. It is worth noting, however, that this group has a far lower frequency for both sexes than their modern counterparts from the same location (K), who have 50.0% and 46.6% for males and females respectively.

The highest frequency of cribra in sub-adults is seen in the Murray (MUY) sample (S2-1). Here 63.8% are affected (categories 1, 2 and 3 combined) which significantly different from the 39.6% for all other areas combined (p=<0.001) (Table 2-5). The overall sub-adult pattern follows adult trends, but sub-adult frequencies from both Rufus River (R) and the Tropics (TRO) display proportionally higher frequencies than adults from the same areas. This small deviation from the adult picture could, however, be a function of the sub-adult Rufus River sample size (n=16). Because sexing of sub-adults was not possible (see Chapter 1), male/female analysis is precluded. Three age groups have, however, been used to trace trends in anaemia throughout childhood. By using age grades in combination with the three cribra categories, some idea of the dynamics of the condition in different places can be shown. Moreover, by combining all three cribra categories in the five sub-adult and adult age groups (males and females combined) a reconstruction of the progression of anaemia in a particular area by age is possible. In the central Murray (CM) and on the south coast (SC) sample size has forced a combination of age groups 1 and 2. In other areas all three sub-adult age groups have been combined (Table 2-2 and S2-4).

In the Murray 83.3% of the 0-5 year olds and 78.1% of the 6-11 year group are affected, while the oldest age group (12-20 years) fares substantially better (45.5%) (S2-4). Further frequency reductions occur in young (34.9%) and old adults (26.9%). This pattern is repeated in the central Murray (CM), except that young and old adult frequencies are very similar (34.2% and 33.5% respectively) and somewhat higher than those for Rufus River (R) (24.2% and 20.7% respectively). Elsewhere, frequencies are, age for age, much lower than those of the central Murray region, with adult groups from outside areas significantly lower (Table 2-2). In the eastern coastal group (CTL) there is a marked increase in frequency from the under five's to the 6-11 age group, contrary to patterns elsewhere. Both coastal groups (SC and CTL) show dramatic reductions in cribra from young to old adults, which have a high statistical significance $p = < 0.01$ and $p = < 0.001$ respectively, while the Rufus River (R), Euston (E) and Swan Hill (S) locations show a distinctive upswing of this kind of stress in old adults.

The dynamics of anaemia in sub-adults can be seen in Figs 2-4 and 2-5. S2-4 highlights the dramatic drop in the incidence of cribra after the age of 11, while S2-5 shows the prevalence of category 1 in the youngest age group and a change of emphasis to category 2 in the 6-11's. Those affected by the severest category seem to remain constant for both these age groups.

Discussion

To date no genetically based blood disease has been found in Aboriginal Australians (Horsfall and Lehmann 1953, 1956; Moodie 1971; Kirk 1981). These aetiologies can, therefore, be discounted as a cause of the anaemia recorded in this survey and the cause of this stress among Aboriginal people sought elsewhere.

Sub-adults

From what we have seen above some observations can be made concerning cribra in sub-adults. Using these observations the dynamics of cribra become more apparent and, in turn, the questions of why and how anaemia occured in pre-contact Aboriginal children come closer to being answered.

There is a major difference in the incidence of cribra between children under 11 years and those over 11 years. The youngest children suffer the most and after 11 years there is a substantial decrease. Most of the youngest group have the mildest form of the lesion with fewer individuals having the intermediate and severe forms. Chronicity of anaemia through the first two age groups is demonstrated by the shift of greatest frequency from category 1 to category 2. This is caused by lesion enlargement and hence a change in its category type. It has been shown that infants in their first six months do not have this condition, but acquire it increasingly as they get older (Mensforth *et al.* 1978; Goodman *et al.* 1983). One likely reason for this is that after six months of age there is rapid loss of natural iron reserves, leaving the child less and less to fall back on during weaning and later development.

In the 12-20 year olds there is a reduction in all categories of cribra, although chronicity remains in some (note category 3). These patterns suggest that while nutritional or pathological stresses are maintained throughout childhood, the capacity for recovery is increased with age. As stated previously, sample size prevents me confirming whether the Murray pattern holds true for children from other areas.

Palaeopathological studies have shown that children generally suffer more anaemic stress (cribra) than adults (Nathan and Haas 1966; Carlson *et al.* 1974; El-Najjar *et al.* 1976; Cybulski 1977; Lallo *et al.* 1977; Mensforth *et al.* 1978) (Table 2-6). These individuals are, of course, the first to suffer the effects of social and environmental pressures and change and, thus, reflect these to a greater degree than adults. One example of these processes has been studied in skeletal remains from the Dickson Mound sites in Illinois, U.S.A.. Three sub-adult groups representing hunter-gatherer (Late Woodland, A.D. 900-1050), transitional (Mississippi Acculturated Late Woodland, A.D. 1050-1150) and agricultural (Middle Mississippian, A.D. 1150-1350) phases of social and economic change show cribra frequencies of 13.6%, 31.2% and 51.5% respectively (Table 2-6) (Goodman *et al.* 1983). Moreover, this work supports the idea that this stress begins in children at an early age. It is suggested here that a similar pattern applies to Aboriginal children, with anaemia beginning in earnest at about the time of weaning. A number of conditions can arise at this point in the child's development which, separately and synergistically, tend to weaken the constitution of the child. Weaning nearly always means a drop in nutritional intake, the loss of natural immunity received from breast milk, greater vulnerability to gastric infections (weanling diarrhoea), the loss of adequate, hygienic, regular or frequent feeding, and nutritional stress from the introduction of unsuitable weaning foods (Britton 1963; Scrimshaw 1964; Scrimshaw *et al.* 1968; Stini 1971; Smith 1976; Hamilton 1981). All of this can result in poor health, undernourishment and iron-deficiency anaemia (Cook 1970). Although these factors could affect all the locations in this study to some degree, there is probably some inter-populational variation. Such variation probably occurs with respect to weaning age and weaning processes also and it is important to bear this in mind in any discussion which compares groups with different environmental and/or demographic backgrounds (Berndt 1981; Hamilton 1981).

Consideration must be given also to parasite infestation, a natural hazard for all societies living physically close to their environment. Parasitism can lower survival chances and depress the natural growth rates of those infected (Anderson and May 1979). With the introduction of adult food the need for the child to fend more for itself and explore its surroundings leads to a greater likelihood of contracting parasites in a traditional setting. What may constitute a substantial parasite load for a toddler, however, may hardly be bothersome for older children or adults. So the degree of infestation and the individual's age are important when considering how infestation affects sub-adults. Moreover, parasitism can negate the benefit of an adequate diet because parasite activity in the gastrointestinal tract prevents nutrients passing across the gut wall. The resultant malabsorption causes nutritional inadequacy, diarrhoea and loss of body fluids which in turn removes electrolytes, salts and minerals. These effects often result in iron deficiency anaemia and have other, more serious, consequences for the child, typically affecting growth, later development and lowering survival potential. The Murray valley and eastern coastal areas provide a natural environment for the support of various species of intestinal parasites, such as *Trichuris trichuris*, *Ascaris lumbricoides*, *Strongyloides stercoralis* and *Enterobius vermicularis* (Sweet 1924; Johnston *et al.* 1937).

General developmental processes among the 6-11 year olds impose stresses which can only be combated by good standards of health and nutrition. From 4-8 years the child has a chance to gain some of the ground lost during the first four. After the age of eight the growth spurt begins. This is a critical time for long bone development and the growth of cerebral, optical and other soft tissues (Black *et al.* 1938; Harrison *et al.* 1977:320). These processes aggravate any long-standing pathological or nutritional inadequacies and tend to run down the healthiest of constitutions. It is likely that these influences played a part in the maintenance of anaemia in Aboriginal children of this age group and were responsible for the observed frequency of cribra. Some reduction in stress, compared with the previous age group, is probably due to the increasing ability of some individuals to fend for themselves independently of adult providers, together with a greater physical capacity of the older child to combat stress.

Further improvement can be seen in the oldest age group. This might be expected to result from a decrease in growth processes, increasing participation in the community's food quest, the lifting of any food taboos and, in the case of many young men, full acceptance into the adult group. On the other hand, with the onset of menstruation and early childbearing in females together with the possible imposition of food taboos, which may have been in place from an early age, any extant anaemia resulting from causes described above would be compounded (Eyre 1845 II:283; Berndt 1981:155). Therefore, allowance must be made for the female component in the 12-20 age group, as it is in the adult group discussed below. With these considerations in mind it is suggested that the majority of those having category 3 cribra are more likely to be females.

Although the sample size of sub-adults from the Northern Territory is small, the 50% frequency of anaemia in these children probably reflects, in a general way, that toddlers here are as susceptible to anaemia as they are in other parts of Australia. Hookworm (*Ankylostoma duodenale*) is one of the main causes of anaemia in tropical regions today, where up to 40% of young children can be anaemic (Hennessy 1979). Hookworm is thought by some, however, not to have been a problem for precontact Aboriginal people (Cook 1970), but 'precontact' with whom? The first introduction of hookworm into Australia was probably through contact with Macassan fishermen or people from countries to the north (Abbie 1960). Certainly, hookworm is part of contemporary Aboriginal parasitism in tropical Australia (Billington 1960; Jose *et al.* 1970; Moodie 1971; Walker 1975). The evidence for precontact hookworm is lacking even if some was introduced later by Chinese goldminers (*Necator americanus*) and Melanesian kanakas (*A. duodenale*) (Cilento 1942). It may, however, be not unrealistic to posit hookworm as part of the cause for high rates of anaemia among children in the tropics.

Adults

Before discussing the active scar in adults, it is worth noting the frequency pattern of category 4. Looking at the areas in S2-4 it can be seen that the frequency of this category grows with age, except at Rufus River. Exactly the same pattern can be seen in two individual locations from the Murray (Swan Hill and Euston). It is interesting to note that in adults, category 4 follows a trend completely opposite to that for the other three categories. It is proposed, therefore, that this, together with the physical evidence of a reversal in the processes causing cribra, represents a concomitant reversal of the mechanism responsible for the other categories of cribra. If this is so, does it tell us something about the potential of individuals from various areas to recover from this form of stress?

When male and female frequencies are examined on an area basis, it can be seen that, with the exception of the desert, old adult males have a greater potential for recovery than females (S2-6). At some time during their life females from all but two areas show an appreciable recovery, whether it be in early life (TRO), later life (CTL) or throughout life (R and SC). Two areas show minimal female recovery. These are the central Murray (CM) and the desert (DES). In the desert group we must refer to the frequency of active cribra (S2-2) which shows the smallest amount of this type of stress for any area studied. It is only reasonable to assume, therefore, that where there is little active cribra there will be minimal recovery. Further, those areas having minimal amounts of anaemic stress must also have the greatest potential for recovery. In the central Murray area it is a different story. Here, there is greater active cribra stress than anywhere else but little evidence for recovery in females. As already noted, men, and old men in particular, do recover. This evidence seems to suggest that cribra stress in females from the central Murray is continuous (with little relief), while the high recovery frequency in old men points to a real differential between the sexes for relief of the condition and, possibly, cultural bias towards males.

Adults show a substantial drop in the amount of active cribra (categories 1 to 3) in all areas and locations (Figs 2-1 and 2-2). Presumably this is a generalized osteological reaction to some relief of anaemic stress and probably stems from the benefits of an adult constitution and the advantages of attaining adulthood in a tribal society. With the termination of most growth processes and the ability of the adult constitution to cope better with parasite infestation and other pathological processes, as well as any nutritional imbalances, some recovery from sub-adult iron deficiency anaemia is likely. This situation would be improved further with relaxation of any food taboos and full participation in food gathering and distribution processes.

It is interesting to note that adult frequencies reflect those of sub-adults from the same area. Some areas are more prone to anaemia than others and it is precisely those areas which show less relief in the form of category 4. The best example of this is in the Murray (MUR), particularly the central Murray (CM), while the best example of recuperation is the tropics (TRO).

There is a world wide trend for women to have a higher incidence of cribra than men and this is echoed in Aboriginal females (Figs 2-2 and 2-3 and Table 2-6). Child bearing, lactation, menstruation and the imposition of food taboos are probably the primary causes. Frequencies of cribra akin to those recorded for the central Murray and, to some extent, coastal New South Wales are rarely, if ever, found in hunter-gatherer groups that are not facing extremes of hardship. They resemble more those found in people leading a more sedentary lifestyle and/or with higher population numbers and densities than found in hunter-gatherer societies (Lallo *et al.* 1977; Goodman *et al.* 1983). Both young and old have frequencies similar to those found in Eskimos (Nathan and Haas 1966), Peublo and Illinoian agriculturalists (El-Najjar 1978) and Mayan Indians (Saul 1972). These people are known also to have been under considerable nutritional, pathological, social, demographic and environmental pressures. It is clear that high levels of cribra occur in association with increasing sedentism and farming and, thus, they appear to be a good indicator of population increase and aggregation (Lallo *et al.* 1977; Trinkaus 1977; Mensforth *et al.* 1978; Huss-Ashmore 1982). High frequencies of cribra among some North American agriculturalists, however, are partly to be explained by the fact that maize prevents proper metabolism of dietary iron (El-Najjar 1976).

The move from hunter-gathering to agriculture often brought about substantial changes in demography and the lowering of community sanitary standards. This encouraged high levels of parasitism, infectious disease and, as frequently happens with population increase and the inroduction of new forms of subsistence, reduced nutritional circumstances. The synergistic effect of all these factors produces skeletal indicators, such as cribra, which have become useful in the interpretation of prehistoric societal change (see Chapters 3, 4 and 5).

Australian Aborigines have various degrees of mobility during their quest for food. In the Murray Valley and on the east coast food was plentiful and varied. It included crustaceans, frogs, reptiles, fish, fowl, terrestrial fauna, vegetable foods and iron-rich shellfish (Bickford 1966; Lawrence 1967; Allen 1972; Thorne 1975). Some of these foods were subject to seasonal factors and, along the Murray, environmental disturbances including localized flooding and drought (Bickford 1966). The implications of the palaeopathological data presented for our view of Aboriginal life, both in the Murray Valley and some parts of the east coast, will, however, be taken up in Chapter 8.

Generally, coastal areas (SC and CTL) have significantly lower frequencies of cribra in men and women than the central Murray ($p=<0.005$). With abundant, year round resources and a normally stable environment this is to be expected. When, however, the constituent locations of the SC area are looked at, a different picture emerges (S2-3). Females from the Victorian coast (C) have rates of anaemia approaching those of locations in the central Murray (34.2%) and over twice as much as males (16.7%). Similarly, when the coastal New South Wales (N) group is separated from the CTL area, it reveals higher frequencies (22.1% and 30.0% for men and women respectively) than those on an area basis (Table 2-1). This is obscured when the small coastal populations are merged into areas.

It seems, then, that at least some of those factors responsible for the high rates of cribra in the Murray may apply to Australia's southeastern coastal strip. Overall, adults along the coasts show a distinct similarity to frequencies of anaemic stress in the central Murray but children have substantially less. In sum, the southeastern coastal region comprises a second region in Australia where anaemic stress is great but not, generally speaking, as high as it is in the central Murray.

There are other pathological indications that suggest these stresses in coastal people may have a somewhat different origin than they do among Murray people. These will be discussed in Chapter 4. Further discussion of these differences also appears in Chapter 8.

The relatively low rates of anaemia in men (10.9%) and women (18.0%) from the Northern Territory are very different from those in southeastern Australia. This indicates a capacity (or societal circumstances) to overcome the stresses of childhood (50% in sub-adults), whatever the population size. These figures are similar to those of other hunter-gatherer groups world-wide and probably reflect different life circumstances for these people as opposed to people living in the southeast of the continent.

When Elphinstone (1971) examined 57 women living tribally in the Western Desert, all apparently with any previous association with Europeans, he found only 10.5% had sub-clinical anaemia (similar to the 9.8% revealed in this survey, Table 2-1), even though he believed them to be last on the food distribution scale. This minimal amount of anaemia he attributed to infrequent menstruation. Although infrequent or irregular menstruation occurs among nomadic Aboriginal women, it is something which is less typical of more settled or sedentary groups (Roth 1903:24; Roheim 1933:233; Abbie 1940:142). This physiological adaptation to a momadic way of life acts to reduce the chances of childbirth and, therefore, creates greater spacing between births, which helps maintain minimal population pressure. It also helps directly and indirectly reduce the chances of anaemia in females by relaxing metabolic and nutritional demands on them.

The similarity between frequencies of cribra in desert men and women show that conditions causing anaemia operate equally for both sexes. In other regions this seems not to be the case. Elphinstone (1971) found also that there was little sign of parasite infestation in the desert women. This ties in with the ecology of these people who regularly move camp, live in widely separated, small family groups in a relatively sterile environment and often bury their excreta to prevent its use by enemies for magic puposes (Dunn 1968; Abbie 1969; Berndt 1981). Some parasitism must have been derived from eating undercooked game and from close association with dogs, though these sources of infection were no doubt common to most Aboriginal communities in prehistoric Australia (Mackerras 1958). Our understanding of desert dweller ecology seems to be strengthened by the low rates of anaemia seen in this survey together with the small amount of medical evidence available to us (Elphinstone 1971). Also, there is some argument for believing that, in an area notorious for inflicting nutritional and physical hardship, mineral deficiencies may be the norm and human physiological/metabolic adaptation to this has taken place in at least 22,000 years of adaptation (Smith 1987). Indeed, hardship and deficiency in these arid areas were thought by some early ethnographers to be the reason for the diminutive stature and slight build seen in some of the Western Desert groups (Helms 1889).

Summary

(1) Aboriginal children have some of the highest frequencies of anaemic stress found anywhere in the world among traditionally living people.

(2) children from the Murray region have appreciably higher amounts of cribra than those from other areas of Australia.

(3) children always have more cribra than adults from the same area.

(4) Aboriginal adults from the Murray region have one of the highest incidence of cribra in the world.

(5) women everywhere have more cribra than men.

(6) with only minor exceptions, cribra incidence falls with age.

(7) the age/frequency relationship of category 4 supports its identification as a form of recovery scar.

As with most other pathologies in this study, there is no uniform pattern to the prevalence of cribra orbitalia. While fully accepting that there are differences in incidence between people around the world who share the same general mode of existence, the degree of stress revealed in the central Murray communities finds no suitable explanation in terms of an ordinary hunter-gatherer lifestyle. Not only did the Murray people live in, supposedly, one of the finest environments, from the point of view of food choice and nutrition, but there is no evidence for the absence of mineral iron in soils or food eaten by them.

Socio-demographic factors would appear to be the only suitable explanation for the evidence presented above but a full discussion of all findings will be pursued in Chapter 8.

Chapter 3

DENTAL HYPOPLASIA

Introduction

It has been recognised for over 40 years that dental enamel records nutritional and/or pathological stresses imposed on the individual (Sarnat and Schour 1941). These stresses disrupt the body's growth processes and leave their marks as permanent or semipermanent scars on teeth in the form of dental hypoplasia (Acheson 1960; Park 1964). Since dental hypoplasia has been recognised as one of the permanent skeletal reminders of a stressful episode, its use by palaeopathologists to assess stress in past populations has become widely accepted (Sarnat and Schour 1941; Giro 1947; Pindborg 1970; El-Najjar *et al.* 1978; Goodman et al. 1980).

Dental or enamel hypoplasia is the result of a metabolic upset to normal developmental cell (ameloblast) function during enamel formation. The result is a visible defect on the tooth surface caused by a deficiency in enamel thickness which appears as single or multiple grooves, lines or pits (Sarnat and Schour 1941). Because these lesions correspond to a particular time of enamel deposition, in both deciduous and permanent dentitions, they provide a longitudinal record of stress events which can be used to pinpoint the age of the individual at the time they occurred. Several lines or grooves on one particular tooth indicate that a number of stressful events have taken place. If one of these events takes place during enamel formation on several teeth, because of different rates and starting times for enamel laydown, the line will appear at different levels on different teeth.

Tooth enamel forms initially along the occlusal or incisal surface and is progressively laid down toward the cemento-enamel junction (Scott and Symonds 1977). Although there is some individual variation, the time taken for enamel deposition is known for each tooth. Therefore, the distance of the hypoplasia from the occlusal or incisal surface can be measured, thus giving a relatively precise age at which the stress occurred (Sarnat and Schour 1941; Goodman *et al.* 1980; Green 1982). This technique was not employed in this survey but is mentioned because of its importance to the discussion below.

There are three basic causes of dental hypoplasia: localised injury or infection (Clement 1963; Scott and Symonds 1977), hereditary factors (Weinmann *et al.* 1945; Darling 1956; Rushton 1964) and systemic stress. The first two of these are rare but distinctive and thus easily separated from those of concern here. Only dental hypoplasia caused by systemic stress is examined in this study.

During the first half of this century a range of nutritional and pathological causes was cited as responsible for dental hypoplasia (Giro 1947). These included syphilus, rickets, tetany, measles, whooping cough, pneumonia, gastro-intestinal infection, highly fluoridated drinking water and vitamin and mineral deficiencies, to mention only a few. Sarnat and Schour (1941), among others, found little correlation between hypoplastic events and exanthemous diseases and in 50 per cent of cases the cause of hypoplasia was unknown. More recently, tetracyclin has been implicated as a causative agent (El-Najjar et al. 1978), but this, if anything, widens the number of possible aetiologies of dental hypoplasia and further entrenches the present belief that it has no specific aetiology and it remains idiopathic. That is, it appears in different people as a function of different nutritional and/or pathological stresses, singly or in combination.

As well as being artificially produced in experimental animals (Molnar and Ward 1975), dental hypoplasia has been identified in wild and captive primate populations (Schuman and Sognnaes 1956; Molnar and Ward 1975), australopithecines (Robinson 1956; Tobias 1967; White 1978), Homo erectus (Brothwell 1963), Neanderthals (Sognnaes 1956) and prehistoric populations from around the world (Sognnaes 1956; Brothwell 1963; Moller-Christensen cited in Alexandersen 1967; Wells 1967; Molnar and Ward 1975; Schulz and McHenry 1975; McHenry and Schultz 1976; Rose 1977; Cook and Buikstra 1979; Sutton 1979; Clarke 1980; Goodman *et al.* 1980, 1983; Green 1982). The study of more modern groups has shown that this lesion is tied to socio-economic circumstance and is more prevalent in those people facing economic and nutritional hardship (Giro 1947; Moorees 1957; Falin 1961; Sweeney *et al.* 1969; Cohen and Diner 1970; Jelliffe and Jelliffe 1971; Sweeney *et al.* 1971; Enwonwu 1973; Infante 1974; Infante and Gillespie 1974, 1977; El-Najjar *et al.* 1978). In a prehistoric context greater frequencies of dental hypoplasia have been seen in those populations undergoing socio-economic and demographic changes which produce nutritional and, thus, pathological stresses (Cook and Buikstra 1979; Clarke 1980; Goodman *et al.* 1980, 1983). This correspondence between stress and the incidence of dental hypoplasia in contemporary and prehistoric people has reinforced the importance of hypoplasia as a general gauge of overall stress suffered by societies in the past as a result of environmental and/or socio-economic change.

Methodology

For this survey the canine and third molar of the permanent maxillary dentition were inspected for hypoplasia. The decision to inspect these teeth alone was dictated by their respective times of enamel formation. These form two fairly distinct age categories, with the canine enamel laid down between 4 months and 6-7 years and that of the third molar between 7-9 and 12-16 years (Barrett 1957; Moorees *et al.* 1963; Scott and Symonds 1977). The individual variation in these times is not important; what is important is that two distinct sub-adult age groups can be monitored for the defect, and consequently any variation that may occur between them.

If both maxillary canines or third molars were missing, their mandibular counterparts were used. All teeth were cleaned with ethanol using one of three toothbrushes, a child's soft bristle and one each of an adult's medium and hard bristle variety. Inspection was made using a 10x hand-held magnifying glass. Any lines, grooves and pits conforming to continuous or discontinuous depressions laterally placed across the enamel surface were noted. These defects were classified following Sarnat and Schour, but that study's definitions of 'single' (acute), 'wide' (chronic) and 'multiple' events were not used (1941:1997). Lines and grooves were recorded as the same form of lesion, with grooves the more accentuated, and were given the code '1'. (Plates 3-1 and 3-2). Pits were given a separate code '2' in the belief that they might derive from a separate aetiology, have a sexual bias or be more prominent in certain regions (Plate 3-3). This was thought to be a worthwhile exercise that required no extra work. Dental attrition is often excessive among Aboriginal groups where the hight of the crown is often reduced to the level of the gum. This caused sample sizes to be reduced and for this reason they are smaller than might othwise be expected by looking at the overall numbers of crania available and used for other pathologies.

The frequencies given below consist of codes 1 and 2 combined (lines, grooves and pits), unless otherwise stated. All results will be discussed using the location and area format outlined in Chapter 1.

Results

Canine

The canine survey revealed a substantial incidence of dental hypoplasia for many groups (Fig.3-1 and 3-2). There is frequency variability between males and females from the same location and/or area and for males and females from different locations and areas separately. With a few exceptions, female frequencies follow a pattern similar to

Plate 3-1 Dental hypoplasia. Lines (category 1) are visible on the right canine and first molar of this individual.

Plate 3-2 Dental hypoplasia. Grooves (category 1) occur on both central and lateral incisors.

Plate 3-3 Dental hypoplasia. Pits (category 2) on the right lateral incisor, canine and first premolar. In this particular case the condition appears bilaterally.

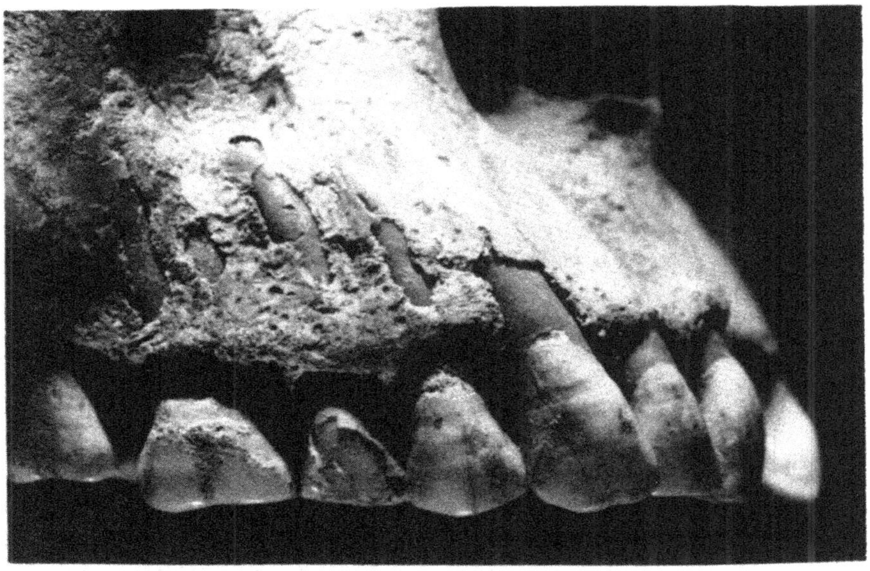

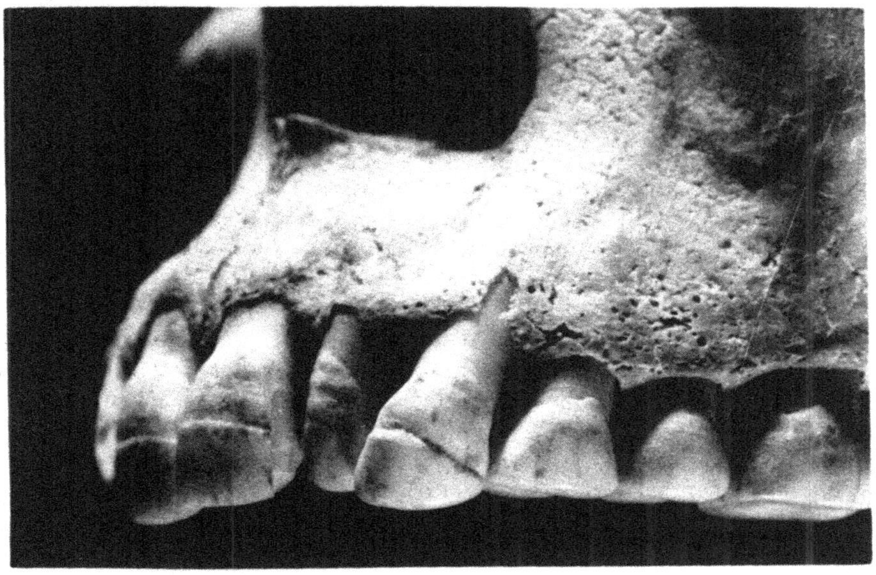

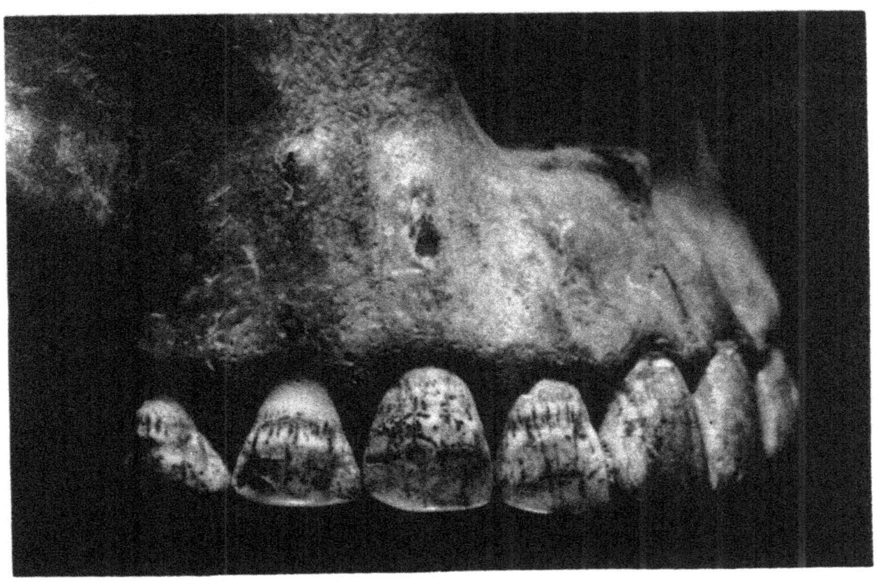

males. For example, the difference between males from Euston and Rufus River shows up in the respective female samples (Fig.3-1). Likewise, the male frequency gradient for coastal New South Wales, central Australia and Western Australia is reflected in the females from those places. In five populations, Barrata, coastal Victoria, west central Victoria, coastal Queensland and the Pleistocene group, females display more dental hypoplasia than males. It should be noted, however, that female sample size in all these groups is below 25 and the west central Victoria (n=12) and Pleistocene (n=5) populations are particularly small (Table 3-1). Similarly, the Broadbeach group reveals the highest frequencies of dental hypoplasia for males and females anywhere in Australia, but the sample sizes of 17 and four respectively are too small to permit meaningful comparison with other locations.

Male frequencies range from zero in a sample of 10 individuals from west central Victoria to 82.4% among Broadbeach people. Sample size probably plays a large part in these results, however. When only locations consisting of more than 23 individuals are included, the male frequency range moves to a low of 23.7% in Western Australia, and a high of 55.1% at Euston, with an average of 40.0% (Table 3-2). Female frequencies range from 6.7% in Western Australia to 50% in the Euston and coastal Queensland locations, but again, if only those samples with more than 23 are included, the range is from 15.3% at Rufus River to 50.0% at Euston, with an average for all locations of 28.5% (Table 3-2).

Using population totals (males plus females) (Fig.3-3), but excluding west central Victoria (n=20), Broadbeach (n=21) and coastal Queensland locations, frequencies range between 18.9% (Western Australia) and 53.4% (Euston). On an area basis (Fig.3-4, Table 3-3) this narrows to a low of 20.7% at Rufus River and a high of 45.2% along the eastern coast. When the groups from arid areas (DES plus R) are compared to the rest (CM, SC, TRO plus CTL), they average 23.9% as against 41.1% respectively, a high statistically significant difference ($p = < 0.001$). The statistical difference between males and females was not explored on a location basis because many sample sizes were too small. The only statistically significant difference between males and females from any area occurs in the tropics ($p = < 0.001$) (Table 3-3).

Molar Three

Less dental hypoplasia occurs on the third molar than the canine (Figs 3-3 to 3-6). The average frequency for males and females, combining all samples of >23 individuals, is 13.7% and 12.2% respectively (Table 3-2). The only exceptions to this general picture are Rufus River, where the female frequency is slightly higher for the molar three (15.7%) than the canine (15.3%), and west central Victoria, which is 25% on the third molar as opposed to 10% for the canine. The sample sizes for both, however, makes these frequencies suspect (n=15 and n=11, respectively). Third molar hypoplasia was not observed among Northern Territory people and females from Western Australia, while the lowest registered frequencies for a location were 2.1% and 2.5% for Swanport, males and females respectively.

There is a statistically significant difference between the high frequency of dental hypoplasia on the canine and the low frequency on third molars among males for all areas of Australia (p=<0.005). This difference is less significant for females, however (Table 3-3). The greatest difference between the frequencies of canine and molar three hypoplasia is 49.1% and 8.8% respectively in males from the tropics (p=<0.001). The central Murray displays the highest overall frequency of dental hypoplasia both for males and females. The Tocumwal, Kow Swamp, Barrata and Euston locations all display frequencies far above the average.

Discussion

From these results a number of things is clear. For the younger age group the evidence from the canine shows that males nearly always suffer greater stress than females; males and females from a particular location reflect, proportionally, the stress occuring at that location and, with few exceptions, stress on young children is of a somewhat higher frequency nationwide (normally at or around 25%) than those in the older age group. The evidence from the third molar shows that there is a substantial fall in the incidence of dental hypoplasia for all locations and areas relative to canine frequencies. The lowest falls are recorded in the central Murray and on the east coast. In these areas there is an absolute and relative maintenance of the stress or stresses causing dental hyploplasia that begin in early childhood as opposed to their decline in in older children in other parts of the continent.

Although several studies of dental hypoplasia have been carried out among contemporary Aboriginal people (Cran 1952; Moody 1960; Schamschula *et al.* 1980), these do not reflect the precontact incidence of the condition. Rather, they are the product of the poor living conditions that the Aboriginal people who were surveyed endured. They are, therefore, of little value for predicting the precontact situation or comparing with the results obtained in this survey. They do, however, provide an insight into changes in the frequency of this condition consequent upon European contact.

Canine

The only study of dental hypoplasia in precontact Aboriginal remains is that of Green (1982). He surveyed the permanent incisors and canines of 100 individuals from the Murray River and found a frequency of 85.3% and 75.8% for males and females respectively, compared to 38.4% and 31.4% in this survey. It is notable that the proportional male/female differential occurs both these surveys and another undertaken by Green subsequently. The substantial differences between Green's frequencies and those obtained here may be caused by a number of factors. As he points out, his figures may unintentionally include individuals with hypocalcification, probably because he used higher magnification during inspection and spent a greater time inspecting each tooth. Further, dental hypoplasia can occur on the incisors without appearing on the associated canines. Green would have counted these, whereas they would not be included here. Also, this survey extended to other collections and includes nearly six times as many individuals as Green's sample, which was derived from one collection only. There are, then, two conclusions to be drawn. Firstly, the frequencies presented in this study reflect minimum numbers of those affected by dental hypoplasia and underestimate rather than overestimate this form of stress. Secondly, direct comparions between these frequencies and those of Green's cannot be made, primarily because they are derived from surveys designed for different purposes. The value of Green's work, with respect to this study, is that he pinpoints the age at which early dental hypoplasia occurs in Aboriginal children. The greatest incidence peaks at about four years in both males and females, after which there is a dramatic fall (Fig.3-7). This roughly coincides with peaks found by other workers studying North American Indian skeletal populations (Shulze and McHenry 1975; Rose 1977; Rose *et al.* 1978; Clarke 1980; Huss-Ashmore *et al.* 1982; Goodman and Armelagos cited in Huss-Ashmore *et al.* 1982). Peak frequencies do, however, vary among different groups. In Californian Indians the highest incidence of 39.5% occurs between four and five years (Schulze and McHenry 1975), the 53.8% peak in Illinoian Dickson Mound children occurs around the 3-3.5 year mark (Rose 1977) and a Anasazi/Dickson Mound sample shows about 50% affected at 3 years (Clarke 1980). The 62.5% peak for dental hypoplasia in the younger of the two Aboriginal age groups (canine) occurs at about the time of weaning, as it probably does in other groups around the world (Saul 1972; Clark 1980).

The weaning 'shock', brought about by the negative effect of a rather sudden and nutritionally inadequate change in diet and any resulting illness, is the most positive aetiology for the widespread occurrence of dental hypoplasia among this age group. Weaning best explains the age at which the greatest frequency occurs and the rapid rise and decline of values on either side, which broadly correspond with the weaning age in tribal societies. Although the above peak frequencies are not large, their amplitude depends upon the age spread of the affected group. Two thirds of the dental hypoplasia in a Chicago group (Sarnat and Schour 1942) occurred during the first 12 months of life, then tapered to almost nothing in the next 20 months. This peak is large because most weaning in western society occurs towards the end of the first 12 months of the infant's life. Among Aboriginal and Indian people weaning took place later and over a greater time period. Hence, the wider spread of incidence and lower peak frequency compared to modern populations (Fig.3-7). It is worth reiterating that although no single aetiology can be blamed entirely for hypoplastic events, weaning may be an exception as it is a term used (particularly here) to encompass a variety of symptoms, such as gastro-enteritis, that often accompany the weaning process all of which can contribute to the formation of hypoplastic lines and pits.

As mentioned above socio-economic circumstance has been shown to play an important role in the incidence of dental hypoplasia. Frequencies of canine lesions in a group of poor blacks and whites from the turn of the century and modern black and white Cleveland schoolchildren revealed 64.4%, 94.4%, 26.0% and 12.0% respectively for males and 56.2%, 62.8%, 33.0% and 6.0% for females (El-Najjar *et al.* 1978). Three points are of interest here: one, there is the overall reduction in stress for both blacks and whites in the last 70 years; two, contemporary blacks suffer greater stress than their white counterparts; and three, females have lower frequencies than males in all categories. Excluding the Pleistocene, Broadbeach, west central Victoria and coastal Queensland locations because of sample size, in 15 of the 18 remaining locations males have greater frequencies of dental hypoplasia than females and this is so for all areas. These results follow a pattern revealed in studies elsewhere; that male toddlers seem to have a greater susceptibility to stressful events in their formative years. This is particularly evident in the group from the tropics.

Rufus River, Western Australia and central Australia seem to have the lowest incidences of canine hypoplasia. The highly significant difference ($p=<0.001$) between the arid area groups and those inhabiting the rest of Australia could be a reflection of the similarity of child rearing practices in these arid areas (Fig.3-4, Table 3-3). While this is open to speculation a limitation on births and greater spacing between them would give a mother more time to wean and care for a toddler, thus reducing nutritional pressure on the infant from competing peers.

Molar Three

The most obvious feature of this part of the study is that dental hypoplasia occurs much less on this tooth than the canine. This indicates one of two possibilities: either there has been a reduction in the amount of stress on children in the older age group, or the older child is physically more able to withstand the unrelieved stress of its infancy. The latter argument would imply that weaning is not the major cause of the high rates in the younger age group because the stress is persistent. If this were the case the clustering of the susceptibles around the 2.5-4.0 year age period is inexplicable. Indeed, this whole argument flies in the face of the rather strong evidence for weaning stress presented above and cannot be accepted as a valid explanation. Furthermore, if the reduction in incidence is due to physical growth, it should be equal in all parts of Australia.

These arguments bring us back to the first possibility: that there is a genuine reduction of stress as the child grows. The susceptibles curves produced by Green and others demonstrate a link between weaning and dental hypoplasia. So, when weaning is over, we should expect at least some reduction in the frequency of dental hypoplasia. Generally, that reduction does take place, but in some locations substantial frequencies of dental hypoplasia still occur. It is shown quite clearly in Fig.3-5 that in the older age group there is a persistence of stress among people living along the Murray River, in west central Victoria, Tasmania and the Cape York/Cairns region. If those locations having less than 25 males and females are ignored, the Cape York/Cairns males are the only individuals outside the Murray with a frequency greater than 20%. In an area comparison the central Murray presents significantly more stress persistence of this type than any other ($p=<0.05$) (Table 3-4).

Although the Broadbeach sample is small, it may, nevertheless, give us some indication of the prevalence of hypoplasia in a discrete, precontact coastal population. At first glance the canine frequency appears to be anomalous compared to other locations, but if we remember that Green has shown similar frequencies in people from the Murray River, it may not be regarded as extraordinary.

There are only five females in the Pleistocene group, a sample too small for any interpretation. The 26.4% of affected males in the toddler category is lower than that for any modern sample from the Murray region, except Rufus River, and well below the modern frequency for males from the same area. Whether this can be interpreted as an increase in the amount of stress over the last 8-10,000 years, however, is open to speculation. A much larger sample of modern and Pleistocene individuals is needed from this area before more accurate comparisons can be made. Nevertheless, it is significant that no molar hypoplasia was recorded for the Pleistocene sample, perhaps adding some support to a general interpretation as follows. Weaning has always been a physiologically stressful time for hunter-gatherer children in this part of the Murray as elsewhere, we can see this in the canine hypoplasia frequency. As certain environmental and demographic pressures built up during the late Holocene, however, other causes of hypoplasia followed weaning, prolonging stress into later childhood and, thus, have become reflected in the substantial incidence of molar three hypoplasia in late pre-contact Aboriginal people from this region.

As in the case of anaemic stress, there is great continental variation in the degree and frequency of hypoplastic stress suffered by Australian Aborigines. Everywhere, weaning seems to be a particularly stressful time with comparable, if not higher, incidences of dental hypoplasia to those found in North American Indians. A commensurate increase in the frequency of dental hypoplasia and Wilson bands, another dental indicator of metabolic insult, has been noted among archaeological Indian populations known to have changed, or were in the process of changing, their socio-economic base from hunter-gathering to agriculture (Rose 1978; Goodman *et al.* 1980, 1983). Sciulli says that:

> *Severe linear enamel hypoplasia....... is found in every Amerindian group and is present at a significantly higher frequency in focal agricultural groups.* (1978:193).

The reasons for this are understood and stated clearly by Cook and Buikstra as:

> *....increased population pressure on food resources, increased reliance on carbohydrates for weaning diets, and environmental deterioration. It seems plausible to suggest that all of these factors were at work, and that enamel defects have provided us with an indirect measure of the pressures associated with increased reliance on agriculture.* (1979:658)

Goodman and his colleagues (1980, 1983) have demonstrated that both the number of hypoplastic events per individual and the total incidence of dental hypoplasia rise with cultural change. They found also that as Indians moved from a traditional hunter-gathering lifestyle (Late Woodland) through a transitional phase (Mississippian Acculturated) to farming (Middle Mississippian), so the incidence of hypoplasia rose from 45% to 60% to 80%, respectively. It is interesting that the male/female frequencies for the latter horizon are 81% and 76% respectively, similar to Green's results for the Murray people. Rose *et al.* (1978), using Wilson bands as a pathological stress marker, found similar changes of 10.3%, 21.4% and 40.0% respectively in the Late Woodland, Mississippian Acculturated and Middle Mississippian groups respectively. Cultural changes in Ohio Valley people between 2,000 B.C. and 1600 A.D. seem to have produced similar increases in dental hypoplasia, although sample size is small and consists of immature individuals (Sciulli 1977). Greater nutritional stress in the later agricultural people (post-Hopewell) has been cited as the most likely cause of the increase. Sciulli's results, however, may represent those sub-adults selected from the population because of other health problems, thus causing a bias in results.

Elsewhere, a significant relationship (p=<0.05) has been shown to exist between dental hypoplasia and cribra orbitalia (Cook and Buikstra 1979). In this study, however, the only significant association between dental hypoplasia and cribra orbitalia was in canines of east coast females (p=<0.005) and the molar three in desert males (p=<0.001) (see Table 3-5). Because only two of the 28 possible groupings tested showed any association, it is thought that these results are fortuitous.

Pits

This brief discussion of hypoplastic pitting is set apart from the main discussion in order to show its particular pattern and to see how that pattern differs from that for lines and grooves.

Little research, experimental or otherwise, has been undertaken into why pits form in tooth enamel instead of lines or grooves. To date it has been accepted that these three lesions are all expressions of the same process, but, even to the untrained eye, pits look like the product of a far greater disturbance of ameloblastic activity than that represented by the other two forms. Curious also is the fact that this phenomenon almost always occurs as discrete patches around the tooth which are etched more deeply into the enamel than either lines or grooves. This too would seem to justify this separate discussion.

The overall incidence of hypoplastic pitting on the canine is not great, with a number of locations showing little or none (Fig.3-8). It is obvious that the bulk of hypoplasia is made up of lines and grooves and there is no location where pits contribute greatly to its overall incidence. Evidence from the molar three presents a different picture for the older age group, accounting for proportionally more hypoplasias. This is particularly so in Murray River people, where it increases somewhat in comparison with the younger age group, and coastal New South Wales its frequency is 9.4% and 11.1% in males and females respectively. With the exception of these groups and males from the Cape York/Cairns region, almost all locations have a negligible incidence of hypoplastic pitting.

While accepting that no histological or other form of experimental study accompanies these findings, the results show that locations with more hypoplastic pits are also those that have the highest frequencies of lines and grooves (compare Figs 3-5 and 3-8). It is suggested, however, that this association is not due to the likelihood that populations with more hypoplasia will, by chance, have more pits. This conclusion is supported by evidence from Kow Swamp, the Murray and New South Wales locations, which shows increased frequencies of pits between the two age groups together with effective reduction in the other forms of dental hypoplasia. If there was a genetic component predisposing some individuals to the formation of pits rather than lines or grooves, it might be expected that: one, there would be an even reduction in the incidence of pits from the younger age group to the older because the latter has less hypoplasia; two, they might favour one sex over the another; and three, individuals with an inherited tooth defect would, particularly in a society suffering considerable dental attrition, have been selected out or reduced to a minimal frequency over thousands of years. Another possible explanation is that molar teeth are more susceptible to this form of hypoplasia. If this was the case, however, surely a more even distribution of pitting in all locations might be expected.

I propose, therefore, that hypoplastic pits are a result of ongoing stress which, because of extended chronicity, acts upon the tooth enamel in a more dramatic manner than that of which lines and grooves are an expression. Such chronic, unrelieved stress can, therefore, be seen to a greater extent in Murray River people than in any other.

Summary

Dental hypoplasia occurs in almost all areas of Australia, generally affecting more young males than young females. Dividing the continent into areas shows that while the period of canine growth is a stressful time for all infants, it is less so for those inhabiting the more arid regions.

The value of a canine/molar three survey is made clear by the results obtained here. The vast differences between the frequency patterns of hypoplasia in young and in old sub-adults add further support to the belief that weaning causes the large frequencies seen in the canine. The Murray people, however, serve to remind us that not all hypoplastic stress is a result of the events of early childhood. The persistence of stress in the central Murray and to some degree on the east coast, highlights these areas, once again, as those in which the inhabitants are subject to a stressful existence on a continuous basis.

Chapter 4

HARRIS LINES

> *....by far the greatest technical advance was made when radiology began to be used in the examination of anthropological and paleontological materials* (Sigerist 1955:41).

Introduction

This is the last of three chapters dealing with skeletal stress markers resulting from a direct episode of pathological and/or nutritional stress. It discusses lines of increased bone density which were first observed in dry bone before the discovery of X-rays (Wegner 1874). Initially, they were not thought to be the result of a pathological process, a discovery which waited for another 40 years (MacConaill 1969). It was more than 50 years, however, before experimental work showed this association conclusively (Asada 1924; Harris 1926, 1931, 1933).

Harris (or Harris's) lines are transverse networks of trabecular bone that completely or partially span the medullary cavity of tubular bones or trace the outline of irregular or flat bones through their cancellous tissue. They range from 1 mm to 1 cm in thickness and are found most commonly in the distal tibia and, decreasingly, in the proximal tibia, distal femur, distal radius and metacarpals (Steinbock 1976). The mechanism of their formation is thoroughly understood but there is still some doubt concerning their exact aetiology (Park and Richter 1953; Park 1954, 1964). Broadly speaking, they appear after bouts of illness, whether chronic (Acheson 1959) or acute (MacConaill 1969), or even temporary and relatively mild (Follis and Park 1955). Causes cited include both malnutritional, starvation in particular, and systemic pathology of one kind or another (Wells 1961; Steinbock 1976). Experimentally induced protein calorie malnutrition has produced Harris lines in pigs (Stewart and Platt 1958; Platt and Stewart 1962), while 84% of boys and 74% of girls suffering kwashiorkor had them, compared to 22% and 10%

respectively in healthy children (Jones and Dean 1959). Other causes are the ingestion of heavy metals (Caffey 1931; Park *et al.* 1933), testosterone treatment and repeated blood transfusions (Garn *et al.* 1968), surgery, chicken pox, smallpox immunization and whooping cough (Gindhart 1969).

The formation process begins with a reduction in the thickness of the epiphyseal cartilage plate beneath which osteoblasts form a thin bone layer called the 'primary stratum'. With recovery from the stress, osteoblasts immediately start laying down bone on the inferior surface of the primary stratum, thickening it and producing the thin line of bone visible in radiographic examination (Steinbock 1976). It may be said, therefore, that these lines are not only indicators of a stress event but show also recovery from that event.

The majority of Harris lines seems to form in the first five years of life (Dreizen *et al.* 1964). Peaks occur at about 2 to 3 years coinciding, it is thought, with the prevalence of childhood diseases (Garn *et al.* 1968; Gindhart 1969). Boys have more lines than girls, but because of greater rates of resorption in males the lines do not persist as long. Bone undergoes greater restructuring in males due to greater physical activity, muscular function on the bone and overall body size, so that by adulthood males display far fewer lines than females, making the latter look more susceptible to stress (Garn and Schwager 1967; Garn *et al.* 1968). Some redress in this differential between males and females takes place the longer people live and many will have disappeared among the elderly. Garn and Schwager (1967) have shown that line frequencies drop from 14% in young males to 8% in old males, and 30% in young females to 14% in old females. Similarly, Garn and colleagues (1968) found a difference of 53% to 28% in young and old adults, respectively. This is mainly due to age-associated sub-periosteal apposition and endosteal resorption processes which are at their maximum in the 50th decade. Two points can be made here with regard to this evidence. Firstly, the age of the skeletal population is important to any conclusions drawn from a Harris line survey and, secondly, the incidence of lines between sample groups may be radically different. These factors are important to this work and will be discussed further.

Because Harris lines only form during the time of bone formation, they have become accepted as natural bone markers, showing not only periods of pathological or nutritional stress in the sub-adult but the age at which it occurred. They provide the palaeopathologist, therefore, with a longitudinal record of an individual's health during childhood and can contribute to the picture of general morbidity within a particular community. Wells (1961, 1963, 1967) seems to have been the first to realise the value of these lines for assessing, comparing and constrasting stress in past populations. Since then, Harris lines have been noted in skeletal material from Europe (Kuhl 1980), North America (McHenry 1968; Garn and Baby 1969; Clark 1981), Central America (Woodall 1968), South America (Allison *et al.* 1974), New Zealand (Sutton 1979; Houghton 1980) and Australia (Gill 1968).

The negative correlation between Harris lines and other pathological indicators noted by some workers has, however, somewhat dampened enthusiasm for their use as a mutually supportive stress indicator for use in concert with others such as cribra orbitalia and dental hypoplasia (Schultz and McHenry 1975; Clarke 1982; Goodman *et al.* 1983). This has led to their partial removal from the general run of skeletal stress markers because of an increasingly held belief that they may be an osteological reaction to stress similar to cribra orbitalia and dental hypoplasia but appear for entirely different reasons. This will be taken up in the discussion section later in this chapter. The main drawbacks to using Harris lines seems to be, however, the differential resorption rates among adults and their appearance in some individuals who have apparently not undergone any form of stress or trauma. In the light of these points, an exact assessment of what they mean is difficult, which can limit their value. Added to this is the fact the they have no exact aetiology. This latter point matters little because, firstly, few other stress indicators have an exact aetiology, secondly, it is not imperative to have a known cause, and, finally, Harris lines do represent an episode of stress, which is the main aim of using them in this study. Further, as obvious as it might seem, it needs to be emphasized that although a number of causes for Harris lines seem likely, in the context of traditional Aboriginal society many of those mentioned above can confidently be ruled out.

Methodology

The bulk of the 1002 right tibiae examined for Harris lines were radiographed using a Pecker-Andrex 3002 FT X-ray apparatus in the Department of Prehistory, Australian National University. Ilford Ilflex film was used for about half this material until its unavailablility, when Agfa-Gevaert Osray RP1 was substituted. A general setting of 45 kV, 8 mA for 7.5 minutes, at a distance of 1.5 metres, was used for most specimens. These settings were varied to accommodate bones of differing cortical density such as those that were pathologically altered in some way, large or rugose specimens and Pleistocene specimens which were often heavily mineralised and impregnated with calcium carbonate.

Damaged tibia were rejected if the damage interfered with internal bone structures, whence the left tibia was substituted. The left was used if the right was unavailable. A bone with a complete single break was used if there was no damage to internal structures. Occasionally, the distal articular surfaces were found to be damaged or partially missing, particularly the medial malleolus, allowing soil into the medulla. If loose, as much of this was removed as possible so that internal structures would not be obscured. All tibia were X-rayed in an antero-posterior aspect. Exceptions to this were pathological and fossil specimens, those with thick cortices and others which, due to their shape (i.e. 'boomerang' tibia, see Chapter 5) could not be supported in that position.

The definition of a Harris line used here is a transverse or oblique radiopaque line that extends into the medullary space by a minimum of a quarter of its radius. Because of the fragmentary and rare nature of Pleistocene and early Holocene remains, any substantial portion of a long bone was examined if internal structures were present. Some bones showed U-shaped and long, wispy lines. These were not considered Harris lines but part of the pattern of their internal trabecular bone.

Two levels of analysis have been used in this chapter: one on the basis of three age groups and the other on all individuals from a particular location or area, regardless of age. The distal end of the tibia was divided into three age groups, basically following the method used by Hunt and Hatch (1981) (Fig.4-1). These age grades are: less than 5 years, 5-10 years and greater than 11 years. They are henceforth referred to as H1, H2 and H3 respectively. Considering the possibility of differential growth rates between the American white children examined by Hunt and Hatch and Australian Aboriginal children, between males and females and because of the geographical and temporal spread of people across this continent (not to mention variations in activity and nutrition levels), I maintain some scepticism as to the exactitude of these age group boundaries. Such divisions do provide, however, a basic insight into the longitudunal occurrence of andbasic age at which stressful events were most common in sub-adult Aboriginal people and the variability of that stress throughout their childhood.

The Index of Morbidity (IM) developed by Wells (1967:392) has been used in this study because it is thought to be the most convenient, easiest and expressive way to measure Harris lines and compare different populations. It has been applied both to individual age categories (H1, H2 and H3) and the total number of lines per individual regardless of age, the 'H' total. The IM was adopted also because of its compatibility with other studies around the world. The index (average number of lines per individual) is arrived at by dividing the total number of lines by the total number of tibiae (individuals). An additional location has been added to the list of locations and is used in this chapter only. This is a sample of 16 tibiae from Goolwa (G), on the southeastern coast of South Australia. It is analysed as a location and used to bolster the sample from the south coast area. Due to the total lack of post-cranial material from Tasmania and the very small samples from the Northern Territory (n=5) and Cape York/Cairns (n=9), these locations do not appear on the histograms. Finally, because of the minimal numbers of sexed tibia, it is possible only to analyse males and females separately at Baratta, Swan Hill, Rufus River, Swanport, Adelaide and coastal New South Wales. Some caution is necessary with interpretations of the Adelaide group as the female sample size is particularly small (n=4).

Results

The main discovery of this particular survey is the number of stress episodes per individual on the south coast. Here the IM is nearly twice that of the next highest, the central Murray (Table 4-1, Fig.4-2). The distribution of lines, however, shows that a similar number of individuals in both these areas displays one or more periods of growth arrest (Fig.4-3). Although the central Murray people have the second highest IM, they form a group with Rufus River and the Desert areas, between whom there is no statistically significant difference (Table 4-2). There is, however, a highly significant difference between the east coast and all other areas. The few Harris lines that appear in these people do so in less than half the population (Fig.4-3).

When the south coast is divided into locations, Swanport, with an IM of 3.54, and Goolwa (G) (2.88) are outstanding for stress events, although all locations along the south coast have an IM of greater than 2.0 (Tables 4-1, 4-3 and Fig.4-4). Further, it is interesting to note that this area, as indeed over the continent at large, the locations with the highest IM are those having the greatest frequency stress occurring during H2 (5-10 years) (Figs 4-5 and 4-6). Fig.4-7 shows the wide distribution of lines in each location (SP, A, G and C) making up the south coast area and Fig.4-8 shows clearly that males from Swanport and Adelaide have inordinately more lines than females from those locations do (H). This is further emphasised by the fact that more males than females have greater than four lines in all locations able to be tested in this way (Table 4-4). The male propensity for lines can be seen again in the three age group analysis, where they have more lines than females in the H1 and H2 grades, a trend basically reversed in H3 (Fig.4-8).

It is interesting to note that males and females in the central Murray locations of Baratta and Swan Hill have less lines in each age category than Swanport and Adelaide, while the IM for males and females sit close to each other, except H3 females from Swan Hill (Fig.4-8). This central Murray pattern is completely different from that of the south coast, particularly Swanport. Also, one individual at Swanport has 20 lines and over one third of the population (34.4%) has more than four lines (Plates 4-1, 4-2 and Fig.4-9A). In the central Murray the frequency is far lower, with 8.5% at Baratta and 13.3% at Swan Hill. Rufus River is also low with 8.8% (Fig.4-9B).

Plate 4-1 Approximately 20 Harris lines have developed in this tibia from Swanport (SP) on the lower Murray River.

Plate 4-2 Four individuals from the south coast (SC) area show the 'step ladder' configuration of lines indicating repetative stress events.

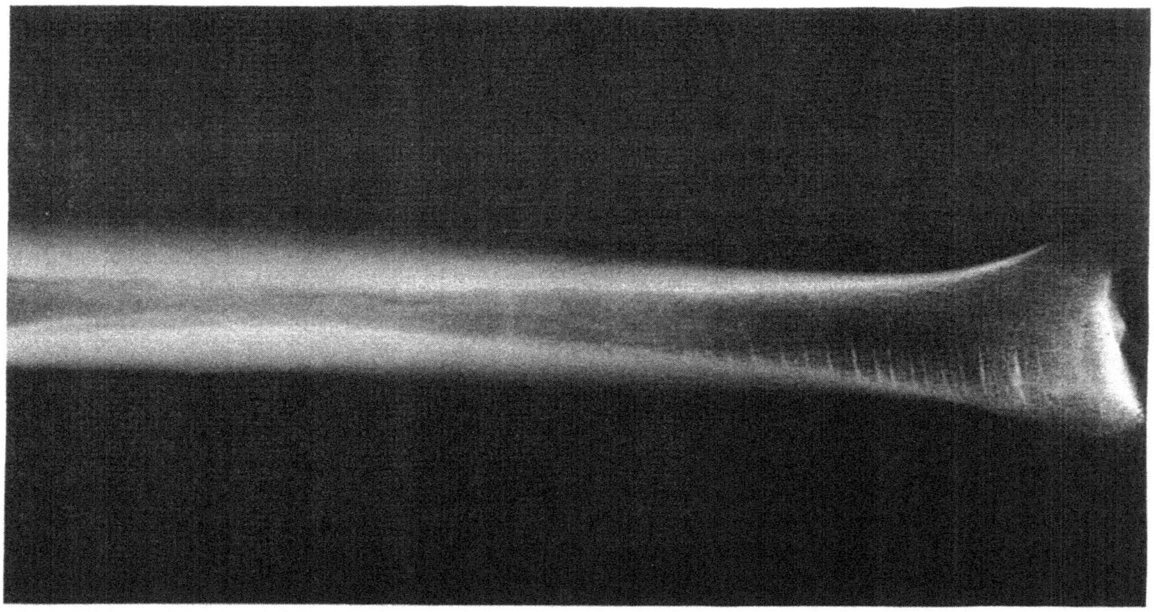

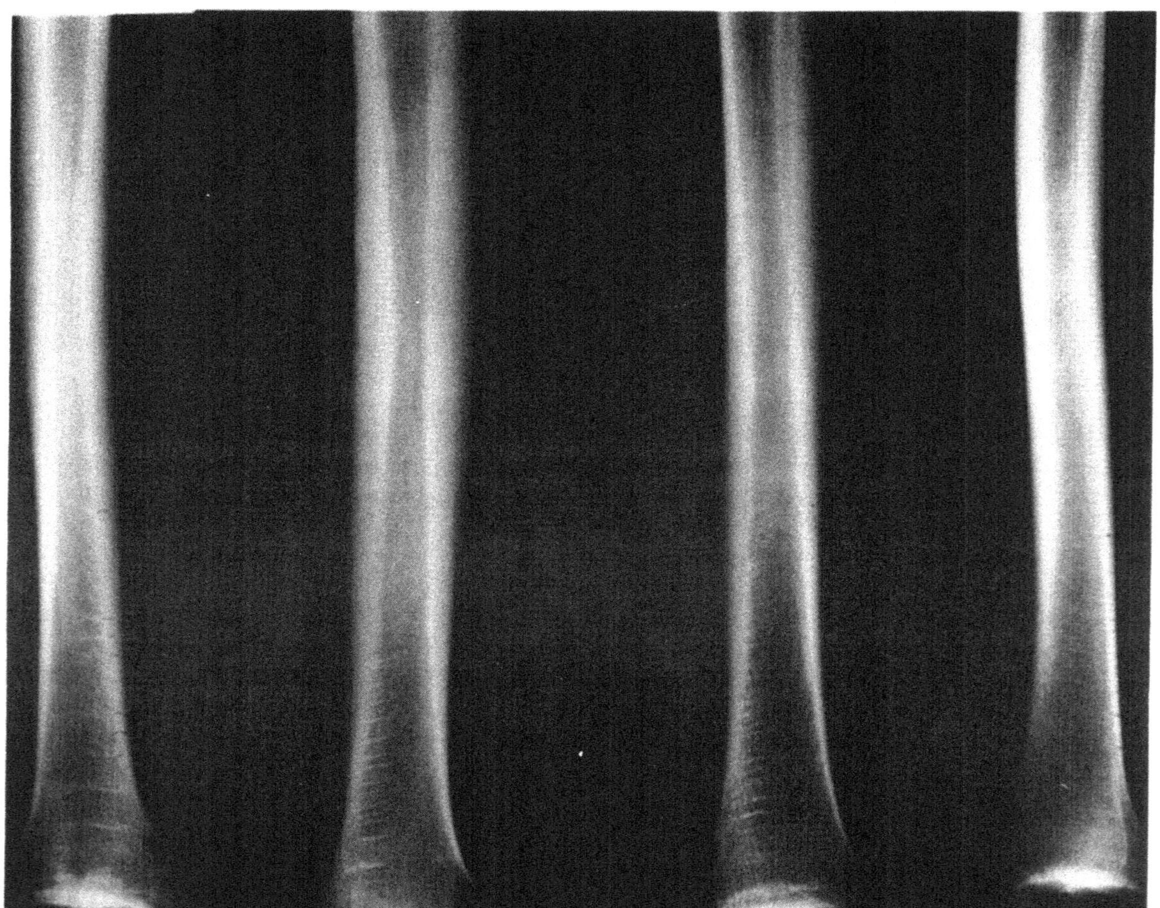

The coastal New South Wales group is different again. These individuals display very few if any lines after about five years, but the susceptibility of males is still visible (Fig.4-7). Further, the IM is one of the lowest recorded: less than 50% of these people have lines, while Fig.4-9C shows that no individual displays more than four. This picture is characteristic of all people living on the eastern seaboard (see also Broadbeach and coastal Queensland).

Before moving on to the discussion it is appropriate to introduce the only other observations concerning Harris lines in Australian Aboriginal bones.

The only study of Harris Lines in Aboriginal Australians is that of Gill, who used three recent skeletons and one from the mid-Holocene. His findings, however, are not clear, as he simply says '...the fossil skeleton showed no significant variation from the control skeletons'. (1968:215).

Hackett notes lines in a number of Aboriginal tibia he inspected during his work on 'boomerang' tibia, but when talking of one particular tibia appears to refer to lines of arrested growth and transverse lines as two separate entities:

> *There are no lines of arrested growth, but in the lower part of the middle of the shaft are two transverse lines.* (1936:50).

This confusion seems to be compounded when he tries to explain their origin:

> *It would appear that the transverse lines in the shafts of these bones are not directly related to* [boomerang leg]. *It was found that these lines.... were due to transverse cribriform plates.... The origin of these plates is unknown. The lines of arrested growth are the result of some interference with the general well-being of the individual (Harris, 1933), but it is not possible to attribute them definitely to the early stages of [boomerang leg] or to dietetic deficiencies.* (1936:50).

Discussion

As in the two previous chapters, I want to concentrate the discussion on the salient features arising from this Harris line survey rather than discuss every detail, which would become both confusing and meaningless in a primary survey of this nature. Another reason for a more general discussion is simply that because some samples are small and very roughly provenanced, only general statements can be made. Specific treatment of some parts of Australia must be left for a time when both sample size and better provenance allow it. Some locations such as Baratta, Swan Hill, Rufus River and Swanport are exceptions, however, (see Table 4-3) and these will be discussed more fully than others.

So far as numbers of stress episodes are concerned, three distinct groups emerge: the south coast, the Murray River and the desert plus east coast. If the number of individuals affected is considered, group clustering changes to central Murray with south coast, the desert and the east coast, plus Rufus River. South coast people have , therefore, not only the greatest number of lines but the highest percentage of individuals having them, while those living along the east coast have the least in terms of both these categories.

Because a survey of Harris lines in adult bones essentially looks at the childhood stress of the individual, it is noted that some line resorption can be expected to have taken place since that time. The figures presented above, therefore, are likely to represent minimum numbers of stress episodes. These results suggest that line resorption has been minimal in those places where large numbers are found, such as the south coast. There is an apparent contradiction, however, between these data and those of some other studies with reference to overall numbers of lines found and their distribution between males and females (Garn *et al.* 1968). Rather than fewer Harris lines in males, which seems to be the general pattern elsewhere, in this study the results are opposite, with the central Murray the sole exception (Table 4-4). In the light of what is known about line resorption, these are contradictory results, but the following would seem to be the most likely reasons for them:

1) Aboriginal people have different resorption processes from other people around the world;

2) females are not subject to the same stresses that cause lines in males;

3) males have more stress episodes than females;

4) both males and females undergo similar degrees of stress but resorption in females is greater;

5) males are less active than females, thus, line resorption is less and lines are retained;

6) women are living longer than men, thereby increasing the chances of line resorption, or

7) males and females undergo similar amounts of stress but males have a better chance of recovery.

In the absence of direct evidence to support notion one, it must be rejected. As regards the second and third, it is clear that Aboriginal females, as we have seen in the previous two chapters, are subject to more stress than males. However, these may not be the same stresses that cause Harris lines. If, however, Harris lines are caused by a different form of stress then evidence from previous chapters may not be valid for refuting reason three. We may, then, be dealing with something that affects males and not females. Reasons four, five and six are essentially linked in that the loss of lines in females might be caused by them experiencing greater levels of activity than males and/or living longer. The latter, as we shall see later, is clearly not the case. This leaves us with reason seven, which arises from the fact that Harris lines are indicators not only of a stress event but also of recovery from it.

South Coast

Many, if not all, individuals from the south coast with a lot of Harris lines have them equally spaced in a 'step ladder' pattern (Plate 4-2). Also, the periodicity between lines closely follows known estimates of annual growth increments of the tibia. The overall effect is one of decreasing width in line spacing with age (see also McHenry 1968). This pattern is produced by a regular sequence of stress episodes suggesting annual or biannual nutritional shortages rather than annual bouts of acute illness. At Swanport and Adelaide, particularly, males show this pattern more clearly than females, which might indicate the former's greater potential for recovery from these events (see above). It could be suggested that age plays a large part in male/female differentials and that if lines are lost with age, then perhaps men were dying earlier than women. Fig.A1 shows this not to be the case, however. Overall, women at both Swanport and Adelaide die earlier (less than 35 years) than men.

It is reasonable to conclude, therefore, that annual or biannual periods of nutritional shortage took place on the south coast. Further, the incidence of lines in males points to them being not more susceptible to stress but more capable of recovery, which could reflect a cultural bias towards male children in this region. It should be noted also that, unlike the central Murray, neither sex shows undue morbidity trends as reflected by other pathological markers or early mortality. It does seem significant that both sexes were affected in the second trimester of childhood, perhaps indicating an ongoing stress affecting all but getting worse as natural growth processes increased and exacerbated any extant stress. Obviously, breast feeding plays no part in the amelioration of nutritional stress experienced by the H2 group as, no doubt, it does for the majority of the younger H1 group. Finally, the processes of resorption seem to play a smaller part in the elimination of lines in south coast people than are claimed for other skeletal populations. Early death and inactivity are obvious limitations on line resorption. The former cannot be said to be a feature of this area, but what of the latter? This question will be raised again in Chapter 8.

Central Murray

There seems to be a different type of stress acting on the people of the south coast than on individuals from other areas. Numbers of affected people living in the central Murray are similar to frequencies observed for the south coast but male/female indices are closer matched. This gives the impression that stress is affecting both sexes equally and, unlike the south coast, there are equal chances for men and women to recover. However, recovery is not as readily achieved, this is reflected in the lower IM of this region. Therefore, although 'equal chances' exist, they are depressed for everyone. Alternatively, it could be argued that the lower IM of the central Murray is real and reflects a situation of less stress. From evidence presented elsewhere in this study, however, this seems unlikely, but the question will be taken up again in Chapter 8. A third alternative concerns, again, the age factor: how many lines have been resorbed? No estimate of this can be made, but by looking at Fig.A2 it can be seen that chances of resorption are less because the central Murray has the highest death rates among both young men and women anywhere. We might expect from this that women would apparently have more lines than men, but they do not. At Swan Hill, where 80% of females die as young adults, this should be particularly obvious but there is no significant difference between the male and female IM, although males live longer. This situation is repeated for Baratta. Although most stress in these locations takes place during the H2 period, it is not as great as in the same age group on the south coast. Two of six locations along the Murray display substantial stress in children of this age and they, as elsewhere, are also those places along the Murray River with the highest incidence of lines.

Whatever the reasons are for the line patterns in the central Murray, they seem to be different from those of the south coast. Further, if the IM is a true reflection of stress events in this area, then, as shown in other work, they do not correlate with the other stress indicators used in this study. On the other hand, if recovery from stress is non-existant, impeded or slow Harris line indices cannot be used as a direct measure of stress in this area.

East Coast

The east coast provides a third distinctive pattern of Harris lines. This area has the lowest indices; the least number of people affected; no individual with more than four lines and most stress events taking place in the H1 age group. The latter was found to be the case in modern populations also (Dreizen *et al.* 1964; Garn *et al.* 1968). If for the moment we accept the H1 figures as being 'normal' or 'expected' for everyday living, the IM for the east coast drops from 1.00 to 0.36. As shown in other chapters, there are good reasons why the primary age group, from 0-5 years, would be more vulnerable to stress events but, in this respect, the east coast presents a different picture from that of the south coast and some parts of the Murray.

It could be argued, as it has been above, that the absence of lines in eastern coastal groups is no proof that stressful periods did not occur and that they are absent because no recovery phase took place. The fact, however, that less than half the people are affected; nobody has the 'step ladder' pattern and few lines form after the H1 growth stage suggests three possibilities: one, there is a real absence of the stress which causes growth arrest lines; two, that a cycle of feast and famine was not part of the east coast life, as it seems to have been along the south coast; and three, that rates of resorption are far stronger than they are elsewhere. Furthermore it seems that other evidence of stress on this coast which has been presented in the previous chapters points, once again, to the fact that these lines do not correlate with other skeletal stress markers.

Rufus River

Rufus River presents a mix of several characters that have been seen already. Less than half the population is affected (45.6%), which is similar to the east coast but unlike other parts of the Murray. The overall IM is the second lowest of any area, while peak line formation takes place in the H2 group, as it does on the south coast. So, like most other areas, Rufus River presents its own line pattern. Most importantly, this pattern separates it from areas further up the Murray River and those at the mouth of the river. There is some indication that seasonal stress, perhaps in the form of food shortage, may have played a part in the formation of lines, with up to 11 in some individuals, but this is not as distinctive as it is for the south coast. Perhaps, relief from these periods came less often to the Rufus River people than it did for those living on the south coast.

Pleistocene and Early mid-Holocene Material

Because of the incomplete and fragmentary nature of Australian Pleistocene and early Holocene remains, very few individuals were suitable condition for inclusion in this survey. Of these only two have lines. These include an unpublished male from Bourke's Bridge, believed to be about 15-17,000 years old (Thorne pers. comm.), and an adolescent female from the Kow Swamp series (K.S.16) (Thorne 1975). There may be some faint lines in the distal tibia of K.S.5 but these are too indistinct to be positively identified. Furthermore, this individual is of advanced age which may have caused any resorption of lines which formed in youth.

There are at least two Harris lines in the distal left femur of Bourke's Bridge. As with K.S.5, other faint lines were present but were too indistinct to be positively identified as Harris lines. K.S.16 displays a number of lines in various long bones, 14 in the distal left tibia, four in the right distal tibia and seven in the distal left femur. Once again, the regular step ladder line pattern is more likely to be explained by seasonal shortage than annual disease. No individual of later date in this part of the Murray has a similar number of lines, but, as I have intimated above, this may not mean that nutritional stress has not been a continual problem for people living in this area for thousands of years.

A Possible Meaning For Harris Lines In Aboriginal Australians

An exact interpretation of what the occurrence of Harris lines mean in precontact Australian Aboriginal populations is difficult to determine. This is mainly because of the shortage of comparative data from elsewhere. An IM of 3.18 for the south coast cannot, for example, be readily compared to the 3.4 obtained by Wells (1961, 1967) for an early Saxon population. Clearly, the respective aetiologies behind the formation of lines in these groups are different. The incidence of infectious disease which largely contributed to the frequency of Harris lines among the early Saxon population, sprang from its overall lifestyle which differed enormously from that of Aboriginal people. Though it is illogical to compare these two widely differing populations, Wells' study nevertheless leads to a conclusion of some interest. Line frequency increases with population growth, rising social complexity and the formation of large towns and villages (see Wells 1967:392). It is hard to know what to make of the figures derived from a similar study of Peruvian remains, because of very small sample sizes (Allison *et al.* 1974). Results from that work showed that a sample combining two groups of coastal sedentary agriculturalists have the highest IM (2.63), while people living at higher altitudes are not so susceptible. These findings seem to contradict those of Alonso's study, cited by Allison *et al.* (1974), in which 70% of individuals from a nomadic Mexican group had lines, while only half the members of a contemporary sedentary group (1200 A.D.) were affected. Therefore, while there is general agreement between the findings of Wells' and Allison's group, there are variations which are contradictory to the general thesis.

McHenry's (1968) study of Harris lines in Californian hunter-gatherers clearly shows the link between the reduction of lines per individual and the widening of the subsistence base of these people through Early (11.32 IM), Middle (7.68) and Late cultural Horizon (5.03). This work pointed to an inverse relationship between lines and infectious disease, morbidity, sedentism and general social and economic intensification, all of which become prominent in the Late Horizon. This, in effect, is the reverse of Wells' findings. McHenry concluded that seasonal reductions in nutrition were probably to blame for the production and pattern of this large number of lines and that adequate nutritional relief brought recovery and, thus, line formation. His evidence, which shows a decrease in the number of lines in the face of a rising incidence of disease and nutrition, point to a nutritionally based aetiology being the most likely cause for Harris lines. A similar aetiology is proposed here for the incidence of lines in Aboriginal groups living on the south coast of Australia. Support for McHenry's findings has come from a survey of the skeletal remains from Dickson Mounds (Goodman *et al.* 1983). These data show the IM for the Late Woodland, Mississippian Acculturated and Middle Mississippian phases actually dropped from 1.3, 1.19 to 1.06, respectively. This seems to contradict the view formed from other palaeopathological evidence, that health got worse as sedentism and agriculture were adopted by the Indians of this region. The possibility that fewer lines are formed in those undergoing constant physiological, metabolic or nutritional stress, however, alters the interpretation of the incidence of Harris lines in populations known to have suffered increases in these factors due to cultural change, population growth and general intensification of their economic and demographic base. Increasingly, therefore, workers support the conclusion that Harris lines cannot be used in mutual support of other skeletal stress markers and probably indicate other characteristics of a society, including, perhaps, the susceptibility of the individual to stressful events (Goodman *et al.* 1983; Hummert and van Gerven 1985).

Summary

Unfortunately Harris lines as skeletal indicators of stress can be intepreted in two very different ways. In an area where there are substantial numbers of lines, it can be concluded that a large number of stress events has occurred. Therefore, it seems obvious that people in that area were stressed. On the other hand, it can be claimed that people are stressed in a second area that has fewer lines than the first because, by invoking the processes of Harris line formation, it can be explained why lines are *not* forming. In other words, many or few lines in a given population may mean the same thing: that the population is stressed. The only difference between these two examples is the first group is recovering from a transitional stress (i.e. seasonal famine) while the other is not, because the nature of its stress is more chronic and/or permanent.

From this study, it seems likely that Harris lines are an indication of acute rather than chronic stress. This makes any association between them and other indicators of chronic stress, such as cribra orbitalia and dental hypoplasia, tenuous. Neither can they be expected to appear in similar patterns to those of other stress markers. This may go some way to explain the differential patterns that arise between these lines and other stress markers not only in this study but in similar studies carried out worldwide. Bearing all this in mind, the need to use other forms of skeletal stress marker in a study of this kind is very clear.

Nevertheless, the conclusion that Harris lines are a sign of stress is almost assuredly correct. Thus, this study is deemed worthwhile in conjunction with other forms of stress marker. Although argument still surrounds the question of what sort of stress is involved, there is strong evidence to suggest that nutritional inadequacy may cause their formation more often and more regularly than disease or trauma, at least among precontact Aboriginal people. Further, I have pointed out above that even if disease does cause growth arrest, a substantial number of the diseases and traumas cited as aetiologies can be discounted in this particular study. Moreover, the distribution pattern of lines points to the regular occurrence of stress, which is better explained by seasonal food shortage than an annual incidence of sickness. It is my contention, therefore, that the majority of lines are a product of nutritional inadequacy of some sort rather than a feature of general pathological processes. Support for this conclusion comes from the fact that those areas with substantial cribra orbitalia and dental hypoplasia are not the same as those which have the highest incidence of Harris lines. The evidence presented here for Australia supports the conclusion reached elsewhere that there is little connection between the incidence of Harris lines and other pathological stress indicators.

The variability of line frequency between the groups discussed in this chapter is surprising, particularly as regards the south and eastern coasts. These show a surprising dichotomy both in the number of lines per individual and the number of individuals having them. I have suggested that seasonal factors of food availability are probably responsible for these differences. A larger sample of individuals from both these areas might, however, provide results quite different from those above.

The Murray, for once, does not show the degree of stress featured in other chapters. Conversely, this may be a false impression caused by the processes of line formation and the high probability that stress among Murray people, as indicated by other markers, is not the same as that on the south coast.

Chapter 5

BONE INFECTION

This chapter is divided into three sections: Non-specific Infection; Treponemal Infection and Tibial Bowing. Non-specific infection refers to three infectious conditions affecting post-cranial remains: periostitis, osteitis and osteomyelitis. The section which discusses treponemal infection presents evidence for the condition as found on the cranium, using descriptive lesion criteria regarded as indicative of the condition. Tibial bowing has been included in this chapter because of its alleged association with treponemal disease or as a symptom of the disease (Hackett 1936a, b; Jaffe 1975; Steinbock 1976).

Non-Specific Infection

Introduction

Osteomyelitis has been recorded in dinosaur remains from the Permian, Jurassic and Cretaceous periods and in extinct groups of mammals from the Pleistocene (Moodie 1923). Various forms of bone infection are known to have occurred in Neanderthal people (Trinkaus and Zimmerman 1982), Neolithic Britons (Brothwell 1963), groups of prehistoric Indians from many parts of North America (Goldstein 1957; Blakely 1980; Shermis 1975; Clabeaux 1976; Goldstein et al . 1976; Walker 1978; Larsen 1981, 1982) and South America (Saul 1972). Therefore, infection of this kind is widespread, has a long history and by itself is not an unusual phenomenon.

Non-specific infection has, however, become a useful indicator of the general health and wellbeing of a skeletal population and used by palaeopathologists in a similar way to cribra orbitalia, dental hypoplasia and Harris lines. The buildup of infectious pathogens is known, for example, to have a positive correlation with socio-economic and demographic intensification among human groups. Any increase in the frequency of infection in skeletal populations can, therefore, provide clues to social change in the past.

Some of the pathogens which produce bone infection include *Staphylococcus aureus*, which accounts for up to 90% of cases, and the streptococcus group which includes *Escherichia coli*, *Salmonella typhi* and *Neiseria gonorrhoeae* (Aegerter and Kirkpatrick 1975:252). These are introduced into the bone either through trauma, following exposure of the bone (e.g. compound fracture) or through wounds or abrasions to the soft tissues. Osteomyelitis can begin as a periosteal infection (periostitis) which eventually causes swelling of the bone cortices (osteitis), or move, via the bone's bloodstream, to the medulla, there forming a site of infection or gumma (osteomyelitis). Over a number of years a suppurative condition arises from sites of infection within the medulla (gumma). Necrotic destruction of the cortex follows, producing holes (cloacae) through which pus from the internal infection drains. Cloacae become, then, a distinctive diagnostic feature of osteomyelitis, together with patches of new, woven bone which often surround them and other small, rarefied foci of cortical destruction.

Periostitis and osteitis can be regarded as primary conditions in the onset of osteomyelitis (Steinbock 1976). Therefore, while it is possible to differentiate between these three conditions in dry bone, they mean, basically, the same thing: that an infectious condition has existed. In this survey these three lesions were recorded following Hackett's definitions (1976:124). Periostitis is inflammation of the periosteum which usually leaves a thin layer of localized sclerotic new bone, gauze-like in appearance (Plate 5-1, Top). Osteitis is produced when an infection of the cortical bone arises, causing it to swell. This process may deform the bone somewhat, adequately distinguishing it from periostitis (Plate 5-1, Middle). Osteomyelitis is defined by infection of the marrow cavity causing a suppurative condition which produces one or more cloacae (Plate 5-1, Bottom). This type of infection can deform and/or destroy the bone and is, therefore, most distinctive in appearance. Chronic and acute forms of osteomyelitis have not been differentiated in this study. Haematogenous pyogenic osteomyelitis is said not to be present in Australian Aboriginal remains and, indeed, none corresponding to the classic symptoms was encountered during this survey (Sandison 1980).

Plate 5-1 TOP:- PERIOSTITIS. Inflamation of the periosteum but lacking swelling of cortical bone beneath.

MIDDLE:- OSTEITIS. Internal and external view of swollen bone cortex which is covered by sub-periosteal new bone formation. Osteitis has altered internal structures, effectively thickening the cortex and imposing an irregularity on the external surface. The lack of gumma within the medulla and cloacae externally, help separate osteitis from osteomyelitis. Note the presence of a Harris line across the medullary cavity.

BOTTOM:- OSTEOMYELITIS. This view shows an abscess or gumma surrounded by perifocal sclerosis opening through cloacae on the external surface. Medullary trabecular bone is becoming sclerotic and beginning to fill the cavity. Below, another tibia displays multiple cloacae from chronic infection. Note coalescence of cloacae causing patches of destruction on the bone's external surface.

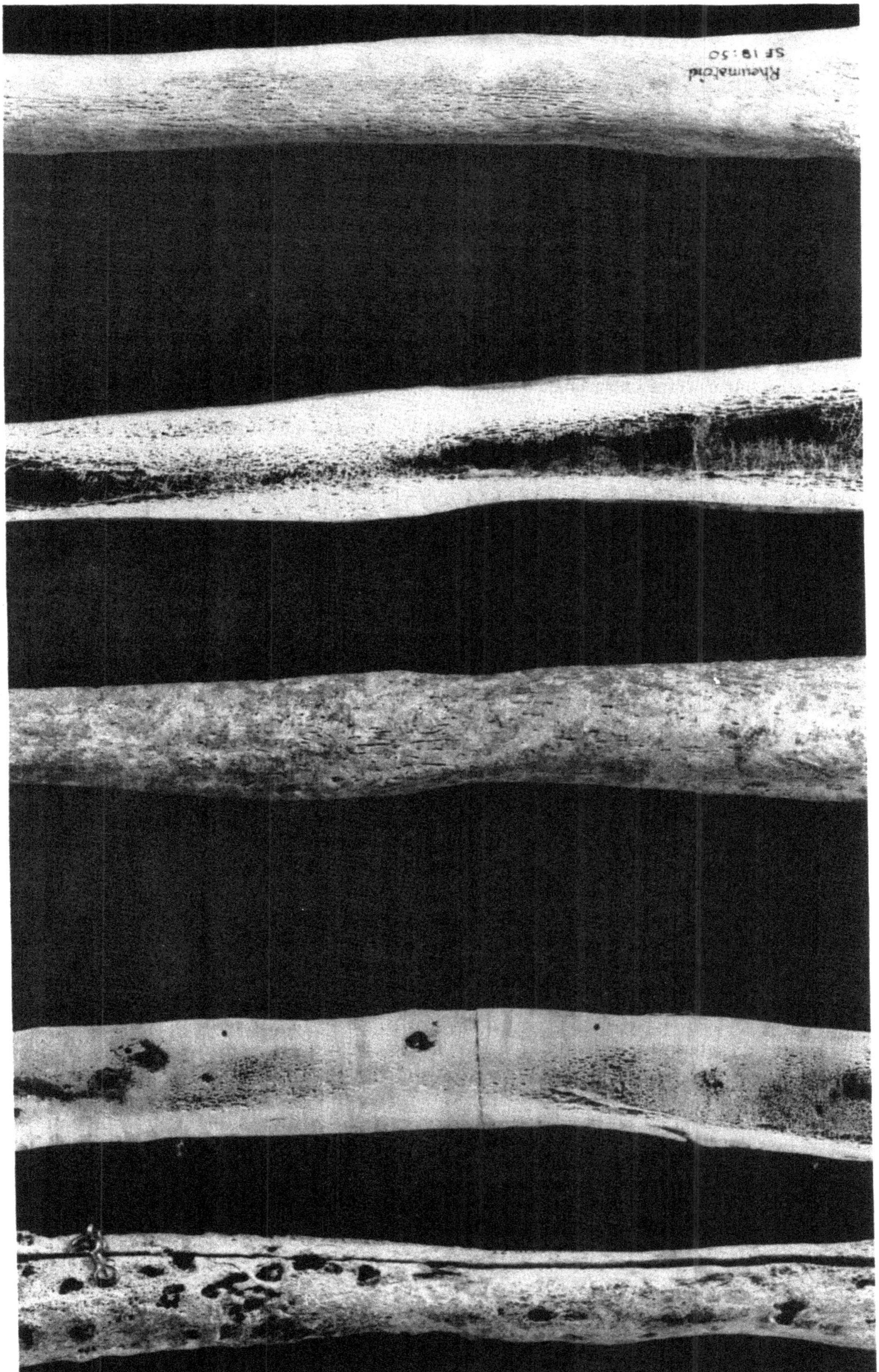
Rheumatoid
SF 19:50

Methodology

Examination of skeletal remains was carried out bearing in mind the pathogenic continuum which an untreated bone infection follows. Sharp differentiation between the three stages defined above cannot, in many cases, be made. It is for this reason that I have combined them to form one lesion termed 'non-specific infection' or 'infection'. Having said this, I am aware that osteomyelitis may not be so 'non-specific'. It can result, for example, from treponemal infection, which is very specific. I am ignoring this particular aetiology, however, because the tertiary stage of treponemal disease that normally produces this would, in most cases, cause distinctive cranial lesions also. The associated cranium is, more often than not, missing from the skeletal remains held in collections and it is because of this that the incidence of treponemal disease will be determined using the cranial evidence only (see below).

The frequency of non-specific bone infection was assessed in the following way. The six major long bones were divided by side and the frequency of infection in both left and right bones determined (Fig.5-1). The highest frequency in either side side of any bone type from one sample (e.g. central Murray) was chosen to represent the maximum number of individuals so infected from that area (Fig.5-2). This method, unfortunately, prevents analysis by 'location' because, for the vast majority of locations, sample sizes of individual bone types are too small. Analysis, therefore, has been restricted to area only. Having said this, some individual locations do have large enough sample sizes to provide some indication of the frequency of these conditions on the smaller scale (Fig. 5-3). Even on the level of area analysis, the Tropics sample is small (n=19) but will be included to give at least some idea of bone infection frequency in the area.

Sub-adult post-cranial remains are few, therefore, they have been included with the adult remains.

Results

Infection in coastal areas is minimal, the highest incidence being 6.1% in right tibiae from the east coast (Table 5-1, Fig.5-1). The frequency rises to 8.6% and 10.3% in tibiae from the New South Wales and Queensland coasts, respectively, if the Broadbeach and southwestern Western Australian samples are removed. When combined the latter three samples have only one individual with this type of infection. The highest frequency on the south coast is 3.3%, in the right radius, while the desert group has 16.7% of its femora affected. Allowances must be made, however, for the latter sample which, while qualifying as a sample (n=24), is small compared to those from other areas.

The difference in the frequency of infectious disease between the central Murray and the Murray Valley as a whole are minimal. The right humerus is the bone most affected in the central Murray (9.5%), while it is the right tibia for the Murray Valley (11.2%) as a whole. Infection rises to a peak of 12.8% in ulnae at Swan Hill and 22.6% in fibulae from the amorphous Murray River group (Fig.5-3). Fig.5-2 is composed of the highest frequencies of bone infection from each of the six areas analysed. From this and Fig.5-1 it can be seen that the right bone is nearly always affected to a greater extent than the left. The reason for this, however, is unknown.

The south coast shows a statistically significantly difference in its frequency from the Murray Valley and desert populations ($p=<0.025$). It is worth noting also that the south coast consists mainly of the Swanport material, which has one of the lowest frequencies of infection (Tables 5-2 and 3, Fig.5-3). Surprisingly, the desert is not statistically different from any other area, which is probably partially due to the small sample size from that region.

Discussion

Although there is no statistical difference between the Murray River and other areas, with the single exception of the south coast, the pattern of frequencies represented in Fig.5-2 follows those for the stress pathologies previously discussed. If the stress marker data was ignored, it could be argued that the differences between the Murray and the coast are more apparent than real. However, the picture we have seen so far, with regards to stress in the Murray, makes these non-specific infection results interesting.

Other research has shown that high incidences of bone infection are an indication of population increase and aggregation and the accompanying bloom in pathological organisms associated with sedentism. Shermis (1975) has pointed out that certain pathological characteristics found in a prehistoric population are closely linked to the human ecology of that group:

> *The relationship is often intimate: osteoarthritis and the economy; infectious diseases and population; oral sepsis and diet; parasites and available animal vectors; nutritional diseases and the level of technology are all examples of the dependency between man and his environment.* (1975:55)

This relationship has been amply demonstrated in a number of North American Indian groups, where increases in bone infection rates have been seen to be commensurate with demographic and economic changes (McHenry 1968; Blakely 1971, 1980; Shermis 1975; Lallo and Rose 1979; Larsen 1981, 1982). These increases have, however, been variable. Thus, at the Dickson Mounds Lallo and Rose (1979) discovered a significant rise in the incidence and severity of non-specific bone lesions between the Mississippian Acculturated Late Woodland group (30.8%), whose hunter-gathering activities were supplemented with some agriculture, and the subsequent Middle Mississippian phase (67.4%), which had an economy based upon maize agriculture. Larsen (1982) found that periosteal infection among Indians living along the Georgia coast increased in all bones between the pre-agricultural and agricultural phases. This increase was highly significant in tibial infection, which rose from 4.5% to 15.0%. Larsen (1982:252) attributes this rise to:*....crowded conditions associated with permanent occupation of large, densely populated sedentary villages....*

It is interesting to note the differences in frequencies between the Dickson maize agriculturalists and corn farmers along the Georgia coast. Shermis (1975:43) has highlighted these differences among other prehistoric Indian groups:

> *Certain aboriginal American groups display low incidences of* [infectious] *disease, such as the Archaic Indian Knoll population of Kentucky or the Pecos Pueblo population of New Mexico (Hooton 1930). In other groups, late or proto-historic or historic populations such as the Fort Ancient population of Ohio and Kentucky or the Arikara of the Dakotas, the disease reached virtual epidemic or pandemic proportions.*

While some groups have higher frequencies of infectious disease than others, Blakely has determined that an average figure for prehistoric Americans lies somewhere '...between ten and fifteen percent' (1980:31). He also points out that bone infections in the Etowah mound burials of northwest Georgia (10.1%) 'compare favourably' with the 11% he found in the Middle Mississippian people from the Dickson Mounds (1980:32). I am unable to explain the difference between Blakely's findings and those of Lallo and Rose, perhaps it lies in lesion definition, but while the former figures seem more realistic this contradiction of results cannot be ignored.

The frequencies obtained in this survey suggest that the difference in the incidence of infectious lesions between the Murray River and the coastal groups is real when we remember similar differences for previously discussed stress markers in these two regions. Also, lesion distribution is different between the Murray and coastal people. In the former there is a more even involvement of all bones from upper and lower limbs, while on the coast upper limb infections are negligible. This pattern suggests either that injuries were more common in the Murray for all limbs, from which more infection might arise, or the chances of infection without injury were high. In the chapter dealing with trauma, I conclude that the Murray does not have high rates of serious post-cranial injury and substantially less than other areas. Therefore, serious forms of trauma do not account for the infection in this area. If the Murray has an 'infectious environment', however, such as that met with in large aggregations of sedentary people, this might make its 'all limb' pattern of infection more understandable.

The tibia seems to be the bone most affected by infection in a number of areas (MUY, R, CTL) and is a close second in all others. A susceptibility noted for elsewhere in both hunter-gatherer and agricultural groups (Clabeaux 1976; Larsen 1982). Continuous blows to and bruising of the lower legs might offer one explanation for this, but it must be noted also that of all the long bones the tibia is the most commonly affected in syphilitic disease (Adams 1976). In the central Murray, however, the humerus is the bone most affected, with the ulna, radius and tibial frequencies being similar to each other. The reason for this may be related to functional stress on the upper limb. Patterns of mechanical stress in the central Murray are reflected in the high incidence of upper limb osteoarthritis, thus pointing to extensive use of the elbow (this is explained in more detail in Chapter 6). Further, the bilateral pattern of arthritis in the Murray suggests an equal use of both arms which is not typical of the distinct handedness often seen in hunter-gatherer populations. Whether this has anything to do with the pattern of upper limb infection cannot be decided although it might be tentatively suggested.

At Rufus River rates of infection are lower than those encountered further up the Murray but greater than in the Swanport population near the mouth of the river. In other words there is less of this type of infection as one moves down the Murray River towards the coast. The eastern coastal group has more infection than people living on the south coast, however, a finding which corresponds to the overall patterns of the stress indicators discussed in Chapters 2 and 3.

The greatest drawback with studying and diagnosing non-specific infection in Australian Aboriginal skeletal remains is the question of what proportion of post-cranial bone infection might be due to endemic or introduced treponemal disease. This will be discussed fully in the Treponemal section below and becomes particularly relevant for an explanation of 'non-specific' infection frequencies in desert people.

Treponemal Disease

Introduction

There is an extensive literature concerning the origin and spread of syphilis throughout the world (see Butler 1928; Hudson 1958a, b, 1961, 1963, 1965; Hackett 1963; Brothwell 1970). The main concern in this study is to establish evidence for treponemal disease in Aboriginal skeletal material and if there is whether it is of an endemic form, an introduced syphilitic variety or a mixture of both. Of equal importance is to gain an understanding of how much 'non-specific' infection can be attributed to treponemal disease.

Non-venereal or endemic treponemal infection has been given a variety of names in many parts of the world. In Scotland it was known as the sibbens (Morton 1967), the radesyge in Norway, saltfluss in Sweden, spirocolon in Greece and Russia (Steinbock 1976) and the frenga in Bosnia (Grin 1953). Other forms of this disease are pinta, which is confined to the Americas and does not manifest itself skeletally, bejel of the Middle East (Hackett 1963) and, in Australia, erkincha (Hackett 1936a) (spelling follows Spencer and Gillen 1899:443). Yaws may be included also as an endemic form of treponemal disease because it is endemic to the tropics and present in a number of areas before the arrival of Europeans. These regions provide a suitable environment in which the spirochaete involved (*Treponema pertenue*) can live and be transmitted (Cilento 1942; Hackett 1947; Manson-Bahr 1961). Similarly *T. pallidum*, which causes venereal syphilis, is believed to be environment-specific, because of its adaptation to a regime of transmission in temperate regions.

It is clear that treponemal disease existed in many prehistoric cultures. In North America evidence for pre-Columbian treponemal disease comes from a number of sites stretching from California to Florida, New Mexico to Illinois (Brothwell 1970; El-Najjar 1979), the oldest of them being Indian Knoll, Kentucky, dated to about 5000 B.P.. Other examples come from the Mississippian culture (1000-1400 A.D.), the Horton and Morris sites, Oklahoma (pre 1400 A.D.), the Changa Mound in South Carolina (approx 770-1120 B.P.), Florida and Indian Knoll (5000 B.P.) (Brothwell 1970). Although there is some disagreement over the evidence, there is, now, a general acceptance of treponemal disease existing among pre-Columbian North American Indians (Cook in press, cited in Klepinger 1983:75). In Mexico, cranial lesions typical of caries sicca have been described in material dated 900-1200 A.D. (Brothwell 1978), and Guatemala 900-1000 A.D. (Goff 1967). It has also been claimed to have existed in the West Indies

(Gejvall and Henschen 1971) although this is doubted by Hackett (pers. comm.). Recently, however, some evidence for treponemal infection among North American bears around 12,000 years ago has been published (Rothschild and Turnbull 1987). It seems likely, then, that the organism has been around for some time, although when it first infected humans still remains uncertain.

In Oceania crania showing Hackett's 'discrete' form of lesion have been found on the Marianas Islands and dated to the ninth century (Stewart and Spoehr 1952), while caries sicca is present in Sarawak in the 16th century (Brothwell 1976). Treponematosis occurs also along the south coast of Papua New Guinea in the 15th century. Further east, Pietrusewsky (1969) has identified treponematosis in prehistoric Tonga and Bowers (1966) has raised the possibility of yaws causing some of the infections he noted among 864 traditional burials from Mokapu, Hawaii.

Hackett's writings over a number of years have concerned the aetiology and epidemiology of treponematoses and bowing in Australian Aboriginal people in the arid areas of the continent (Hackett 1936a, b, c, 1968, 1974, 1975, 1978). His 1974 paper describes a Tasmanian Aboriginal cranium with pathological changes consistent with treponemal infection. A photograph of this cranium appears in an article by Moller-Christensen and Inkster (1965) and shows clearly the lesions Hackett refers to. These changes do correspond to those originating from treponemal disease but Hackett's (1974:442) conclusion that the changes are possibly due to:

> *...a childhood, and hence indigenous, treponemal infection which accompanied the original inhabitants on their migration into Tasmania from Australia about 10,000 years ago*

seems misplaced. Indeed there is no evidence that treponemal disease existed in Australia at the time Tasmania was joined to the mainland. No other lesions of this type have been found in Tasmanian crania (including this survey). Moreover, the Tasmanians were isolated from the mainland for about 7,000 years (during which time endemic treponematosis is thought to have arisen around the world) and Australian crania dated to before this time show no evidence for the disease. These particular lesions, therefore, more persuasively point to an infection of European origin rather than an endemic one.

Primarily, treponematoses in the living are diagnosed not by bone lesion morphology but following serological tests such as the Wasserman, Kahn, Kline, Hinton and VDRL tests (Robbins and Cotran 1979:406). Cutaneous lesions provide an immediate assessment of the stage to which the disease has progressed. Lastly, and by no means least, the patient can provide a history and possible origin of the condition.

Needless to say, none of these are available to the palaeopathologist. Bone infection tells us only that the disease reached the third or tertiary stage, not that it was caused by yaws, erkincha or venereal syphilis. This matter will be taken up again in the Discussion section below.

Methodology

All cranial lesions conforming to Hackett's (1976) 'contiguous' and 'discrete' forms of treponemal infection were noted. These included, in the contiguous series: serpiginous cavitation, nodular cavitation, caries sicca and the presence of stellate cracks within healed scars (Plates 5-2 and 3, see also Hackett 1976, Fig.C, p42). In the discrete series only circumvallate cavitation and radial scars were recorded (Plates 5-4 and 5, see also Hackett 1976, Fig.B, p31). Naso-maxillary alteration or destruction, usually accompanied by periostitis (Plate 5-6, Left), nasal bridge collapse ('saddle nose') (Plate 5-6, Right) and any palatal destruction were noted also. Often, both serpiginous and discrete lesions were observed on the same cranium.

Hutchinson's teeth were looked for to help in the differential diagnosis of congenital syphilis among those individuals having distinctive cranial lesions (Jaffe 1975:922). None were found, however, but some individuals were missing their central incisors and canines which are normally affected by this condition.

Any individual with 'lesions' suspected of being pseudopathological or produced post-mortem was rejected. The outer cranial table can be subjected to various environmental and/or mechanical processes during interment which produce features similar to those derived from treponemal disease. These usually take the form of pitting , and it was for this reason that clustered and confluent pits, forming Hackett's 'initial series', were not included in the survey. In those cases where several types of scarring were present on the one cranium, such as serpiginous cavitation, caries sicca and circumvallate cavitation, the most diagnostic lesion type was recorded. For example, caries sicca would take precedence over serpiginous cavitation.

Results

The central Murray has the highest incidence of treponemal disease for males and females (Fig.5-4 and Table 5-4). This is exemplified in Fig.5-5, which shows the infection at its highest among locations in the centre of the area, peaking at 17.7% in females from the K location

Plate 5-2 Nodular cavitation defined after Hackett's (1976) continguous series of treponemal bone infection.

Plate 5-3 Serpiginous cavitation described by Hackett (1976) as a stage in his continguous series.

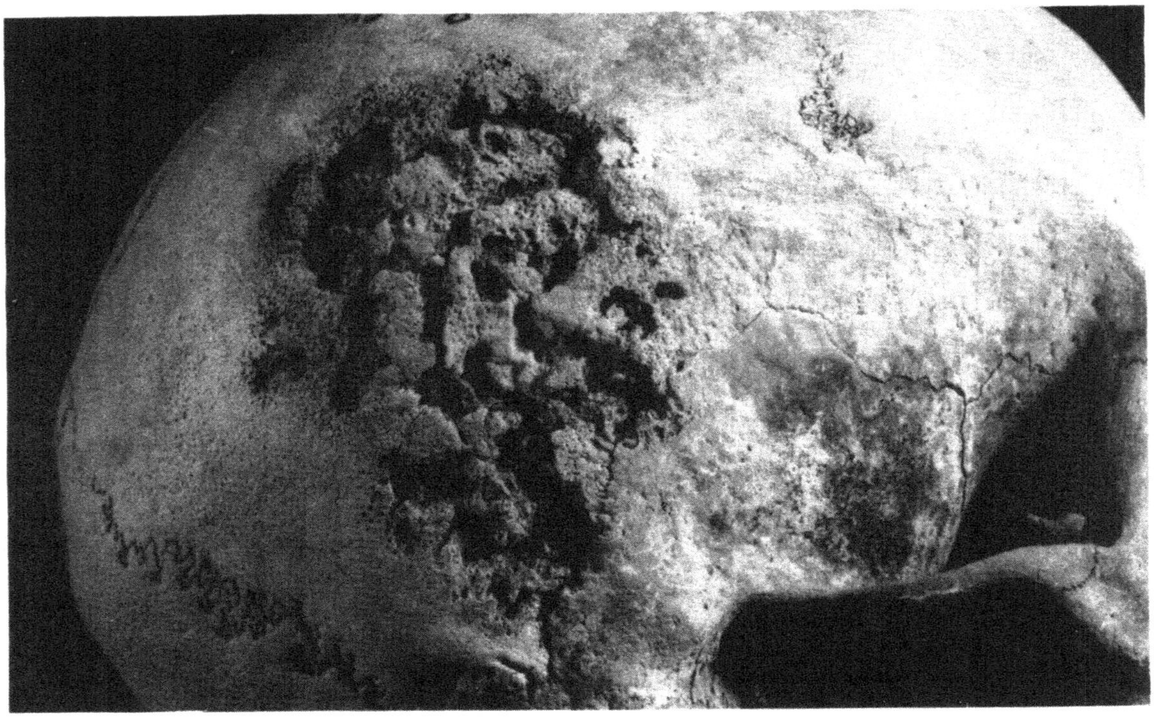

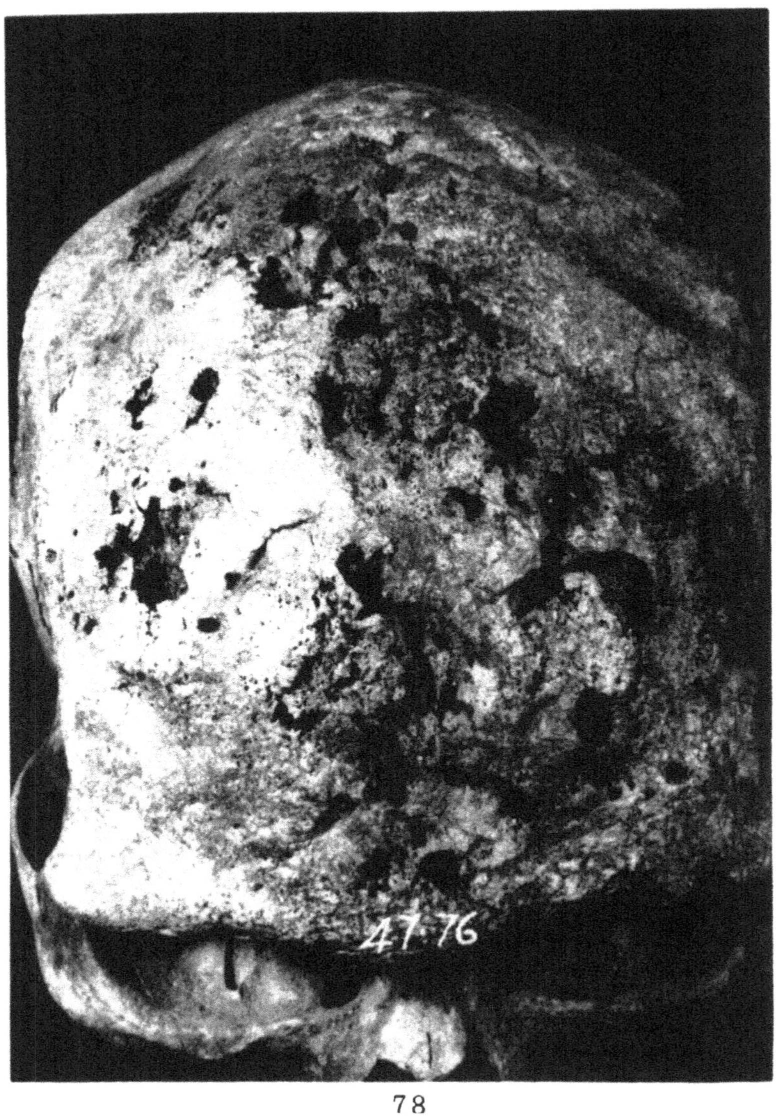
47.76

Plate 5-4 Early circumvallate cavitation from Hacket's (1976) discrete series of bone changes typical in treponemal infection.

Plate 5-5 Late circumvallate cavitation in the discrete series. Radial cracks in the scar tissue have been produced as a by-product of the healing process.

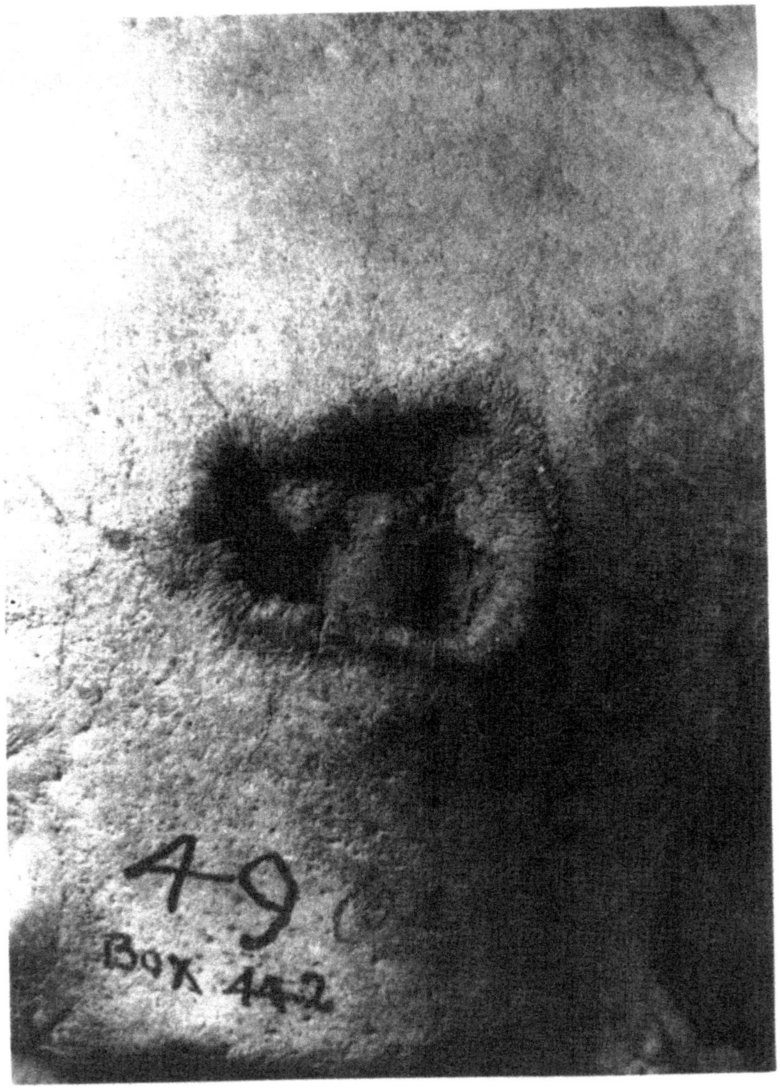
49
Box 442

Plate 5-6 LEFT:- Treponemal infection has destroyed bone in the nasal and anterior alveolar region of this individual, eliminating the definition of the nasal rim and removing internal structures, such as the vomer, conchae, septum and palatine bones. Lesions on the frontal and maxilla (below the zygomatic process) conform to serpiginous cavitation (cf Plate 5-3). A small patch of caries sicca can be seen of the frontal, centre-left.

RIGHT:- Nasal collapse on this individual has resulted in the condition known as 'saddle nose'. Complete resorption of the nasal bones has been followed by further destruction in the lacrimal region.

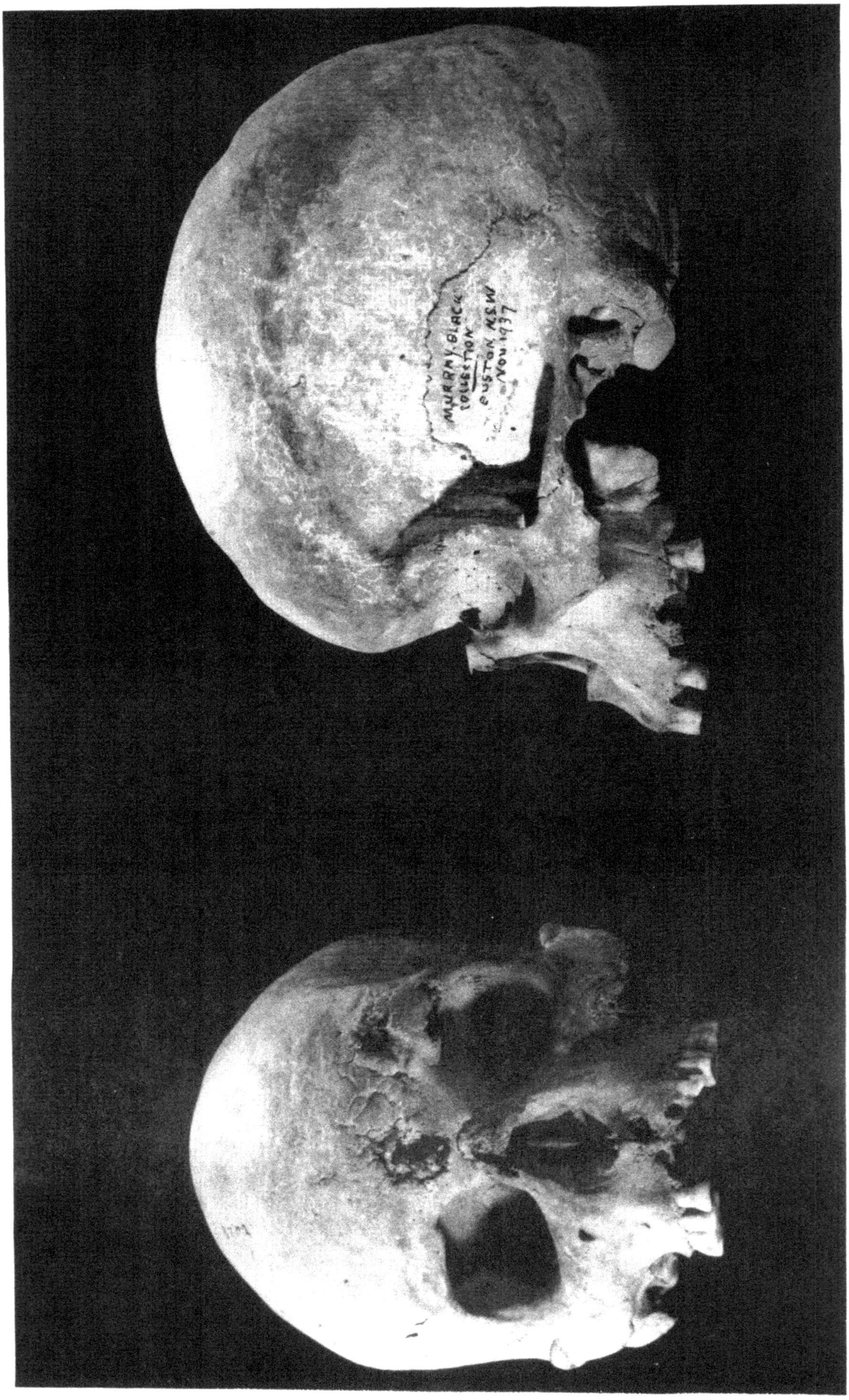
MURRAY BLACK
COLLECTION

and 16.1% in males from Swan Hill (Table 5-5). Elsewhere, there is great variability in the incidence between sexes and locations. This is not so distinctive on an area basis, although females seem to suffer more infection than males, except on the east coast. Only one male exhibits treponemal

infection from each of the T, C, M and W locations and one female from SP, N, and NT (Table 5-5). The overall incidence outside the Murray is, therefore, very small with west central Victoria the single exception. This may be due to the latter's small sample size.

The difference in population frequencies between the central Murray and eastern and southern coastal people is highly significant ($p=<0.005$) but less so between the Murray as a whole and the coasts ($p=<0.01$) (Table 5-6). There are no statistically significant differences between any of the other areas.

Discussion

It is possible, indeed most likely, that some of the non-specific infection observed in Aboriginal post-cranial material has its origin in treponemal disease but how much is difficult to say at present (see previous section). One way to approach this problem is to compare the frequency of cranial treponemal disease with the post-cranial non-specific type. Such a comparison (Table 5-7) shows that less treponemal infection existed than the non-specific form, overall. Two locations within the central Murray area, K and B, show a slight reversal of this trend, however.

Can we say, therefore, that non-specific infection minus treponemal infection equals the *real* incidence of non-specific disease? I think not. It seems more prudent to mearly provide both frequencies without explanation of either than speculate in that way. Crania and post-crania cannot in most cases be matched, so it is not possible to associate their respective lesions. Further, it cannot be assumed, that if 10% of individuals from one location or area have cranial lesions of a treponemal variety, 10% of post-crania from the same place must have lesions originating from the same infection. Further, Steinbock (1976:113) shows that while 42% of syphilis sufferers with osseous involvement have cranial lesions, the next highest bone, the tibia, is affected in only 26% of cases ($n=239$). Using these data, cranial lesions outweigh post-cranial lesions by almost 2:1, which is, in fact, the reverse of the situation in this study. It is, impossible, therefore, to show how much non-specific infection is due to treponemal disease. It can be concluded, however, that not all non-specific disease is the result of treponemal infection.

This brings us to our next problem. Is the treponemal disease in Australian Aboriginal skeletal material of an endemic or venereal (introduced) variety? In 1976 Hackett published his 'Diagnostic Criteria of Syphilis, Yaws and Treponarid (Treponematosis) and of Some Other Diseases in Dry Bones'. After more than 40 years extensive study of the effects of treponemal disease on bone, in many parts of the world, he established criteria by which syphilis might be diagnosed in skeletal remains. I will use an extensive quotation here from his shorter publication of the previous year regarding identification and differential diagnosis of European venereal syphilis and erkincha/yaws. Hackett says:

> *For simple and immediate diagnostic criteria of syphilis the caries sicca sequence of calvarial changes, and nodes/expansions with superficial cavitation in long bones are recommended with confidence. Because the late [tertiary] bone lesions of acquired syphilis cannot be distinguished from those of yaws and treponarid* [endemic treponematosis or erkincha], *diagnostic criteria of syphilis are really evidence of treponematoses. Which of these three related infections is the cause of the bone changes can only be indicated by the provenance of the specimen (humid or arid), its accurate dating, and the way of life (urban or rural) that the population concerned might have lived.* (1975:238).

He then goes on to say:

> *In European bones most calvarial sequestra are labelled syphillis but in a few diagnoses this is accompanied by (pyogenic) osteomyelitis.... The absence of sequestra in Australian calvariae casts doubt on the purely syphilitic origin of these changes labelled syphilis in European calvariae. In Australian long bones sequestra are also absent.*
>
> *Most of the changes in aboriginal* [sic] *bones in Australian national collections are of the caries sicca sequence. A similar occurrence of yaws bone changes and absence of osteomyelitis were found in the 1950s in Western New Guinea by the first medical service doctors.*
>
> *In the Pacific area, skulls from a number of islands have caries sicca changes, e.g. New Hebrides, Solomon Islands, Samoa, and New Guinea, where clinical yaws was present until the recent penicillin mass treatment campaigns.* (1975:238)

These statements raise several questions. Firstly, although caries sicca can be indicative of syphilis, unless it is accompanied by pyogenic osteomyelitis, it is probably not syphilis, merely treponematosis. Secondly, cranial lesions alone cannot be used to indicate what form of treponemal infection caused them. Thirdly, when the distinctive caries sicca lesion does occur, it must be related to the environment in which the individual lived and not assigned immediately to a syphilitic origin. Finally, neither European pyogenic osteomyelitis nor the nodes/expansion tertiary bone changes have been seen in Aboriginal remains in this study or that of others (Sandison 1980). These results point to an absence of overt signs of syphilis in collections of pre-contact Aboriginal skeletal remains.

While based on long term investigation and many comparisons, Hackett's work is not clearcut and demonstrates the problems of diagnosing these diseases in dry bone. Therefore, the avenue of differential diagnosis between bone lesions of various treponemal diseases is not considered possible at present.

Notwithstanding what has already been said, was syphilis present among Aboriginal populations before the arrival of the colonists? While it seems unlikely that this disease was widespread, there may have been a slim chance of its sporadic introduction into northern Australia, through contact with Asian and Indonesian fisher-traders. But syphilis is not usually present in a population in which yaws is endemic, such as Indonesia (Hackett pers. comm.). In this case its spread from Macassans to Aborigines in the north of the continent, where yaws is encountered also, was probably even more unlikely. Even if it was introduced from time to time, it most likely had little effect on the continent as a whole or even regionally and was confined to one or more groups in northern coastal areas (Cleland 1928). Except for this slight possibility, it is more reasonable to accept that there is little chance that venereal syphilis existed in Australia before first white settlement and that any disease causing similar alterations to the bone is more likely to have been tropical yaws, endemic in countries to the north of Australia. Under this scenario erkincha was either introduced or, perhaps, stemmed from yaws in northern Australia.

In the absence of any evidence to the contrary, I feel confident in concluding that there was no venereal syphilis among pre-contact Aboriginal people. Any pre-contact bone or cranium having lesions of a treponemal origin must, therefore, indicate an endemic form of the disease, whether it be yaws or erkincha. With yaws confined to tropical areas (i.e. northern Australia), any specimens from central or southern Australia must display endemic erkincha. Of course, the best evidence for such disease in the south would come from material firmly dated to before 1788. Prokopec (1979:12) has claimed that lesions in juvenile long bones and cranial lesions in an adult from the prehistoric Roonka population, were caused by yaws. If true, because of the location of Roonka near the Murray River, these lesions must surely indicate erkincha, not yaws. These particular specimens were not available during my inspection of the Roonka collection.

In reviewing some of the earliest observations of treponemal disease in this country, Hackett (1936c) was persuaded that most accounts were probably wrong in their diagnoses that European syphilis was prevalent among Aborigines. He believed a large number of the symptoms described were more reminiscent of yaws which, as many observations were recorded in southern areas of Australia, must have been an endemic form of syphilis rather than tropical yaws. Hackett pointed out also that reports of 'leprosy' and 'cancer', believed by early settlers and explorers to be causes of facial destruction, were almost certainly cases of gangosa-like lesions which often accompany the tertiary stage of yaws and erkincha. He cites Mackillop's observation as an example of this. Mackillop says: *Cases of cancer are frequent....* and *There are cases also of something very like leprosy....* along the Daly River, Northern Territory (1893:258-259). Bone cancer affects between one and three people per 100,000. I have found only one case of multiple myeloma and possibly two cases of metastasic carcinoma among the Australian skeletal collections. It seems, therefore, extremely unlikely that Mackillop is seeing 'frequent' cases of cancer: more likely they were lesions from an infectious treponemal condition.

Similarly, Cadell's account of women 'without noses' near the Blyth River, Northern Territory, was probably gangosa, resulting from yaws, rather than the syphilis he attributed it to or the leprosy which a medical aquaintance of Cadell's thought it could have been (1868:21). The presence of yaws among Aborigines living in the Northern Territory at the turn of the last century is supported by medical officers residing in the area (Breinl 1912; Holmes 1913; Breinl and Holmes 1915) and photographs taken by Baldwin Spencer (Vanderwal 1982) and others (Basedow 1932).

The tradition of the erkincha, as related to Spencer and Gillen (1899:443) by the Aranda, is a myth associated with the transmission of an endemic form of disease resembling treponemal infection. Part of the myth is concerned with native cat or Achilpa men who were the first to be seen with erkincha. According to Spencer and Gillen, the Aranda employ a restriction on the eating of native cat (*Dasyurus geoffroyi*), not just by the members of that totem but everyone. Only small amounts are allowed to be eaten by the old people. One reason for this is that: *...it is supposed that any one, save an old man or woman, eating Achilpa would be afflicted with a special disease called Erkincha* (1899:168). The way this myth is related and the food taboo imposed suggests knowledge concerning the epidemiology of erkincha. This is implicit in the fact that young people only are thought by the Aranda to be susceptible to it. Eyleman (1908) confirms that Aborigines said they could only contract the disease once and that was during childhood. Endemic treponematosis, of course, is contracted only by children and young adults.

Apart from the tradition of erkincha reported by Spencer and Gillen (1899) and Eylmann (1908, cited in Hackett 1936a:738), a number of reports from medically trained individuals has confirmed the existence of gangosa in Central Australia (Basedow 1932; Black and Cleland 1938), Coopers Creek and Port George IV, north Western Australia (Basedow 1932), the Musgrave Ranges (Hackett 1936a) and Cockatoo Creek (Gray and Cleland 1933). Cilento (cited in Cleland 1928:143) states that syphilis was rare around Townsville and the treponemal disease of Aborigines there was due to yaws. Erkincha may also have been the disease the explorer Stuart encountered at Attack Creek (Northern Territory) in 1862, when he saw one of two elderly men who was *...very much diseased and lame.* (1865:440).

It has already been pointed out that this type of early observation concerning Aboriginal disease must be carefully assessed. However, the presence of a myth to explain a disease present before contact with Europeans, traditional cures for endemic yaws in the Northern Territory (Foelsche 1882; Webb 1933) and syphilitic-like disease present at the time of first contact (Sturt 1833) suggest that the existence of endemic treponemal disease cannot be ignored, at least in the arid and northern parts of Australia.

I will return to this again below. For now, the high rates of infectious disease in the desert group may be either an indication of the presence of endemic treponemal disease, especially with the minimal likelihood of staphylococcal infection in the hot, dry desert environment, or an artifact of the small sample size. The conclusion, however, that it did exist there, traditionally, seems unavoidable.

In the southern parts of the continent early observations of treponemal disease are conflicting. Some indicate endemicity, while syphilitic infection of Aboriginal people by whites is fully recognised. Aborigines around Melbourne denied having this disease before Europeans arrived (Eyre 1845) and Eyre states that within three years, from 1841 to 1844, Aborigines around Murundi had become badly infected by their association with the townspeople. Moorehouse (1840, cited in Basedow 1932:11) says the Aborigines themselves believed the disease was known to them before Europeans mingled with them. Westgarth, writing a little later, seems to come to a similar conclusion:

> *But by far the most terrible of these accompaniments of colonization has been the spread of syphilitic disease. It has indeed been doubted that the introduction of this malady is due to Europeans, as it has been*

found to exist among tribes which had never previously communicated with the colonists, and who, when questioned upon the subject, could afford no satisfactory information as to how or when it appeared amongst them. But it is thought to be of a milder character and of less frequent occurrence among those tribes that have not mixed with Europeans. (1848:81-82)

In his well known description of Aborigines inhabiting the Murray, Lachlan and lower Darling rivers, Beveridge says that:

> *...in general, it is supposed that the venereal disease amongst the aborigines* [sic] *is entirely due to the Europeans, but a greater error than this never had promulgation, for long before the advent of the white man it was one of the vilest scourges this..... people had to bear. The probabilities are that the trepang-hunting Malays and Chinese first introduced it on the northern coast, centuries ago, whence it spread from one tribe to another, until at last the foul disease became a national calamity.* (1883:22).

While of course, it can be understood why these early settlers might not want to accept that syphilis was introduced by Europeans, it is interesting that the disease was reported to be of allegedly a milder character and that Aborigines could not recall when it was first introduced. Collins writing at the time of the First Fleet says:

> *It was by no means ascertained whether the lues venerea* [venereal syphilis] *had been among* [the Aborigines] *before they knew us, or whether our people had to answer for having introduced that devouring plague. Thus far is certain, however, that they gave it a name, Goo-bah-rong; a circumstance that seems rather to imply a pre-knowledge of its dreadful effects.* (1798:496).

Interestingly, Collins tells us also that the Aborigines around Sydney gave smallpox a name (Gal-gal-la): ample evidence that an Aboriginal name for a disease does not necessarily prove its endemicity. Nearly 40 years later Cunningham made an interesting observation:

> *I have often observed the men, too, labouring under the eruptions of the skin resembling syphilis, and open tumours also in their groins apparently of the same nature; but time with them cures all disorders.* (1827:45).

A number of comments can be made with regard to these observations. The observers do not seem to be sure of what they are seeing, otherwise they would, one might expect, have made positive identifications of at least venereal syphilis. Or, perhaps Aborigines were suffering two diseases, syphilis and an endemic form of treponematosis. Some individuals do not have, perhaps, the cross-immunity believed to take place between syphilis and yaws/erkincha (Cannefax *et al.* 1967; Garner *et al.* 1972; Jacobs 1978). From this situation, there could have stemmed a plethora of symptoms which confused description by Europeans.

Is the cure of which Cunningham speaks suggestive of endemic treponematosis, rather than the more fatal syphilitic type? This, together with the fact that some Aborigines believed they had the disease before the arrival of colonists, surely points to the presence of an endemic form prior to the arrival of the colonists. It could be argued that Aboriginal recovery, in the majority of cases, implies a familiarity with it. Further, if the disease has a milder effect than the imported syphilis, this too might suggest a long standing association. The last point to make is that any newly introduced infection, particularly one that can severely maim, would surely still be fresh in the minds of older Aborigines if it was recent and not forgotten as implied in some accounts.

These arguments cannot be taken further. They are imponderables to which there are no suitable answers at this time. One thing is certain, however; the confusion concerning this disease encountered in almost all early observations prevents a clear picture being formed from that evidence alone. Nevertheless, it is worth recollecting one more early written account. One of the earliest and most famous accounts of a treponemal-like disease, among what were then remote Aborigines, is that of Sturt during his exploration of the Murray River between Euston and Rufus River in January 1830:

> *The most loathsome diseases prevailed among* [the Aborigines]. *Several were disabled by leprosy, or some similar disorder, and two or three had entirely lost their sight.* (1833:94).

A little further west he remarks again:

> *The most loathsome diseases prevailed throughout the tribes, nor were the youngest infants exempt from them. Indeed, so young were some, whose condition was truly disgusting, that I cannot but suppose they must have been born in a state of disease; but I am uncertain whether it is fatal or not in its results, though, most probably it hurries many to a premature grave. How these diseases originated it is impossible to say. Certainly not from the colony, since the*

> *midland tribes alone were infected. Syphilis raged among them with the most fearful violence; many had lost their noses, and all the glandular parts were considerably affected.* (1833:124-125).

First, the fact that Sturt sees a disease, which certainly resembles a form of treponematosis, in people separated from the colony by uninfected groups seems to suggest that it was not spreading from the settlements in the east or seafarers at Encounter Bay to the southwest. Secondly, even if the odd convict escapee carrying syphilis fled into the interior and survived, it seems highly unlikely that such a lot of tertiary stage infection could have originated from a single individual at this early stage in the history of the colony. Thirdly, the symptom described by Sturt as 'leprosy' suggests severe destruction both of soft tissue and bone similar to those already noted from elsewhere on the continent (see above). I suggest that Sturt is describing the syptoms of an endemic form of chronic treponemal disease rather than leprosy. To date no clinical or palaeopathological evidence for leprosy occurring among Aboriginal people has come from the southern half of Australia.

Regarding Sturt's observation that ...*the midland tribes alone were infected,* Black's (1975) data concerning the prevalence of endemic treponematosis in Amazonian Indians has shown that treponemal infection can be confined to a group of tribes, while those living around them remain unaffected. Similar results have been obtained in Papua New Guinea, where great variability in sero-reactivity has been observed between different villages and language groups (Garner and Hornabrook 1973).

It seems safe to say that some of the normal confusion surrounding early observation of treponemal disease might be avoided if we emphasise lesions of distinctive destruction. These constitute inordinately stronger evidence of tertiary treponematosis, than interpretations based on ulcerations, pustules and other skin lesions in groins, necks and arms. Also, it is clear that because obvious signs of this disease were not reported by the majority of early explorers, this does not mean that it was not present in a subclinical sense. All cutaneous lesions and swellings that would be prominent enough for description by early explorers and settlers are of little value in differentiating between endemic and introduced treponemal infection as these can occur in both conditions (Manson-Bahr 1961; Robbins and Cotran 1979). *Granuloma pudenda,* swellings in the groins and under the arms, gangosa or facial destruction, rashes, curved 'boomerang' tibia (see below) seem to occur in or be associated with most of the treponematoses. Further, *granuloma pudenda* (*granuloma inguinale*), which is often associated with treponemal disease in Aborigines (Basedow 1932), has been reappraised:

> *Because it is so often associated with other venereal infections and because the usual site of involvement is the genital region, it has been called a venereal disease. However, this designation might well be inappropriate since the evidence that the disease is spread by sexual contact is somewhat equivocal. Concurrent infections in husband and wife are uncommon.* (Robbins and Cotran 1979:420).

So, as a basic statement, early observations and descriptions of Aboriginal disease cannot be considered firm evidence for discriminating between the presence of veneral syphilis or other forms of treponemal disease. This, of course, must be tempered by considering the area in which the observations are made, such as the Northern Territory, where encounters with yaws were and still are likely, and the time at which observations were made. At present one can only point out that treponemal disease of one kind or another was more prevalent on the Murray and in the arid regions than elsewhere. Therefore, it is only with future archaeological work and the examination of precontact skeletal material that the problem of whether indigenous treponemal disease existed in southern Australia before 1788 might be resolved.

One point of interest that needs highlightling is the 4.4% frequency in the Northern Territory people where yaws is endemic. This low frequency might, in itself, suggest that higher frequencies from the south do, in fact, reflect the contagion of a newly introduced disease like venereal syphilis. I have, however, observed 36% of individuals from coastal Papua New Guinea dated to the fifteenth century with tertiary treponematosis. This tends to suggest that treponemal disease can display a high frequency of tertiary stage symptoms even if it is endemic.

Finally, with the strong likelihood of treponemal disease being present in North America and elsewhere from an early period, it is pertinent to ask how valid is the 'non-specific infection' category which is often used there? Surely one cannot separate 'non-specific' periostitis and osteitis from that caused by treponematosis if the latter is present in the population?

Obviously some bone infection may come from open wounds and bruising, but the proportion of this can never be known so long as a more specific origin for that infection is possible. Of course, if the post-cranial bones have it while associated crania do not, a non-specific aetiology may apply. If, however, post-cranial bones bearing lesions are analysed in isolation from their associated crania, treponemal disease cannot be ruled out until the latter has been inspected also. I emphasise that here I am talking of skeletal remains having no other obvious cause of infection, such as pyogenic arthritis or traumatic infection. Now we must ask how valid non-specific infection is as an indicator of socio-economic and demographic change such as sedentism, population growth and overcrowding. In fact, whatever the origin of the lesion, whether it be treponemal or staphylococcal, any increase in its frequency over time could be closely associated with factors of population growth and crowding. Similarly, when two populations are compared, that with the higher frequency of infection of whatever variety exhibits greater susceptibility, which, in terms of precontact human ecology, normally means population aggregation and larger numbers of people. Therefore the implications and interpretations of infectious bone disease are valid, though questions concerning accurate palaoepidemiology remain.

Tibial Bowing

Introduction

Palaeopathological and modern medical data indicate that bowing of the leg has occurred in a number of prehistoric groups (Stewart and Spoehr 1952; St Hoyme 1969; Goldstein *et al.* 1976) and modern populations from all over the world (Bahr 1914; Chesterman 1927; Wilson and Mathis 1930; Hackett 1936a, 1951). One of the first reports concerned with sabre-like or bowed tibia is that of Topinard (1894:299). He says: *The first time it was observed was in the tibias of the family buried at Cro-Magnon,...'* (1894:299-300), but as far as I can ascertain, however, this has never been corroborated. He mentions also that 14% of 200 tibiae from Parisian cemeteries dated from the fourth to the tenth centuries were bowed and makes an association that needs some explanation with regard to this study. His words are: *The character which the tibia sometimes presents, and which bears the name platycnemia, or sabre-like,...* (1894:299).

It is not clear whether platycnemia, or the lateral flattening of the tibia, appeared to Topinard sabre-like or whether there was a real bowing of the shaft. Platycnemia often accompanies bowing but in this study only a distinct curvature of the tibia is accepted as diagnostic. Wood (1920) has pointed out that more than 34% of Aboriginal tibiae show platycnemia to some degree and treats this trait as an anatomical variation rather than a pathology. Moreover, he does not mention bowing as an accompanying feature and while I too have seen many specimens with platycnemia, these were not classifiable as bowed.

Stirling (1894) referred to Topinard's findings when writing about the tibial curvature he noticed in Central Australian Aborigines. Spencer and Gillen (1899) also observed the condition in the Aranda, who attributed it to a bend that the legs of a child sustained as they grew and hung ever farther out over the edge of wooden food containers in which they lay. Being soft, the leg allegedly would take on the shape of a boomerang (Basedow 1932:21-22).

On the Upper Darling, Newland (1926) knew an Aboriginal man who had antero-posterior bowing of the shins, while Spencer says that: *...an anterior curvature of the shin bone ... gives rise to what the white settlers have described by the apt term 'boomerang leg.* (1928:194).

Other references to tibial bowing have come from a number of medically trained individuals who saw it in Central Australia (Basedow 1932; Hackett 1936a; Black and Cleland 1938), northern Australia (Breinl and Holmes 1915; Breinl and Priestly 1915; Binns 1945) and Torres Strait (Elkington 1912); in the latter two areas, it was found in association with yaws.

It is believed that the bowing observed in late congenital syphilis is caused by longitudinal growth, stimulated by irritation of the epiphysis. Because the fibula remains unaffected by this process the diseased tibia bows antero-posteriorly between the two fixed points represented by the ends of the fibula (Jaffe 1975:921). Subperiosteal and endosteal deposition of new bone (hyperplastic osteoperiostitis) of late congenital and acquired syphilis are other processes which cause tibiae to bow in this way (Steinbock 1976). Both these tend to thicken the tibial cortex, usually in the middle third of the diaphysis, blunting the anterior margin and giving the bone a bent look (Plate 5-7).

Methodology

All tibia were surveyed for pronounced antero-posterior bowing. Because there is a range of natural bow in the tibia, only those having a posterior displacement of greater than nine degrees from the perpendicular were deemed bowed. To eliminate those specimens that had retroverted heads, only those displaying a bowed diaphysis were included. All bowed tibiae were included whether displaying infectious lesions or not.

Results and Discussion

Firstly, no area has a high frequency of this condition (Table 5-8). Secondly, the only statistically significant difference between areas is that between the Murray River and the south coast ($p=<0.01$). Lastly, there is a statistically significant difference between bowed tibiae with and without infectious lesions in the central Murray, along the Murray River as a whole and in the sample from the east coast.

Only 17 of the 65 bowed tibia have infectious lesions and 13 of those have lesions (osteitis and osteomyelitis) which cause 'pseudobowing' rather than a real bending of the diaphysis (King and Catterall 1959; Jaffe 1975:937). The remaining 38 tibiae are bowed in the classical fashion (Plate 5-8) and there is a highly significant statistical

Plate 5-7 Remodelling of internal and external structures of this tibia by infection has produced the appearance of bowing.

Plate 5-8 These bowed tibia are without any trace of infection. Sub-periosteal bone apposition has occurred internally at the middle one third of the diaphysis. The bone in the centre shows that this process has almost filled the medullary cavity.

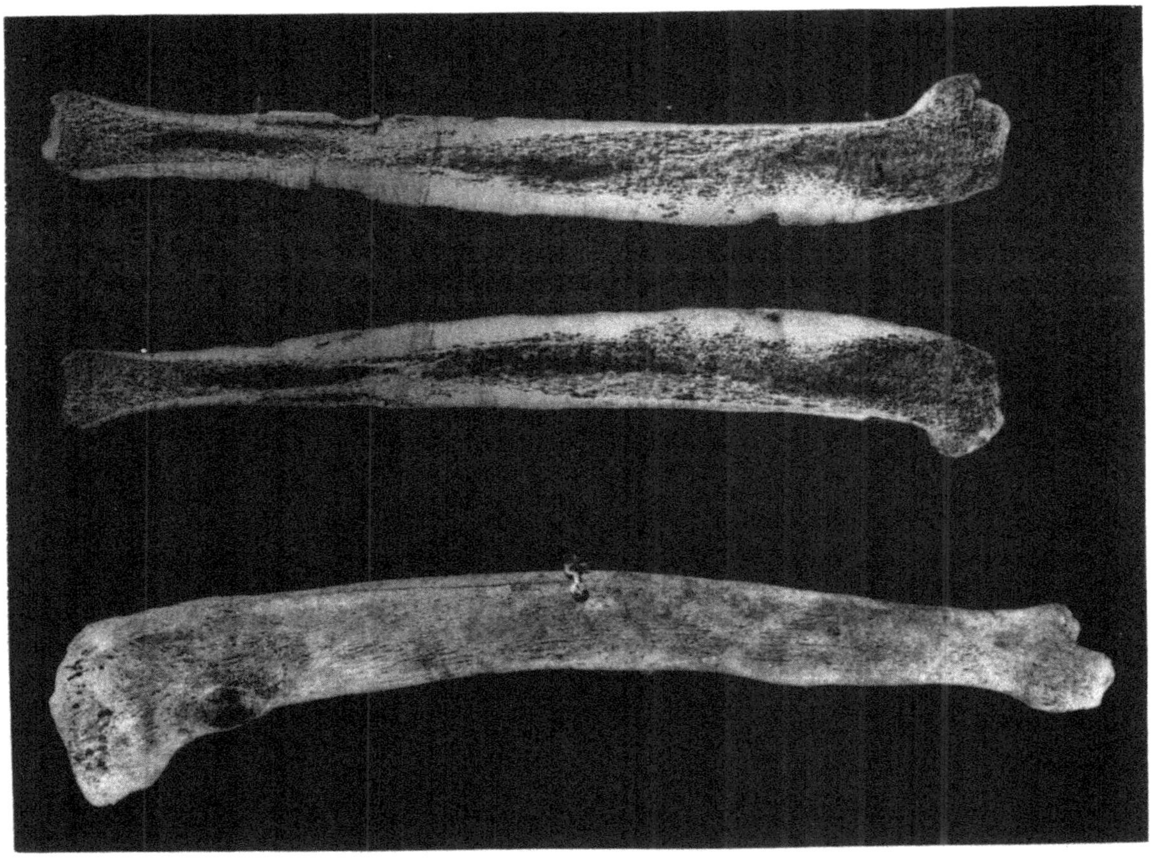

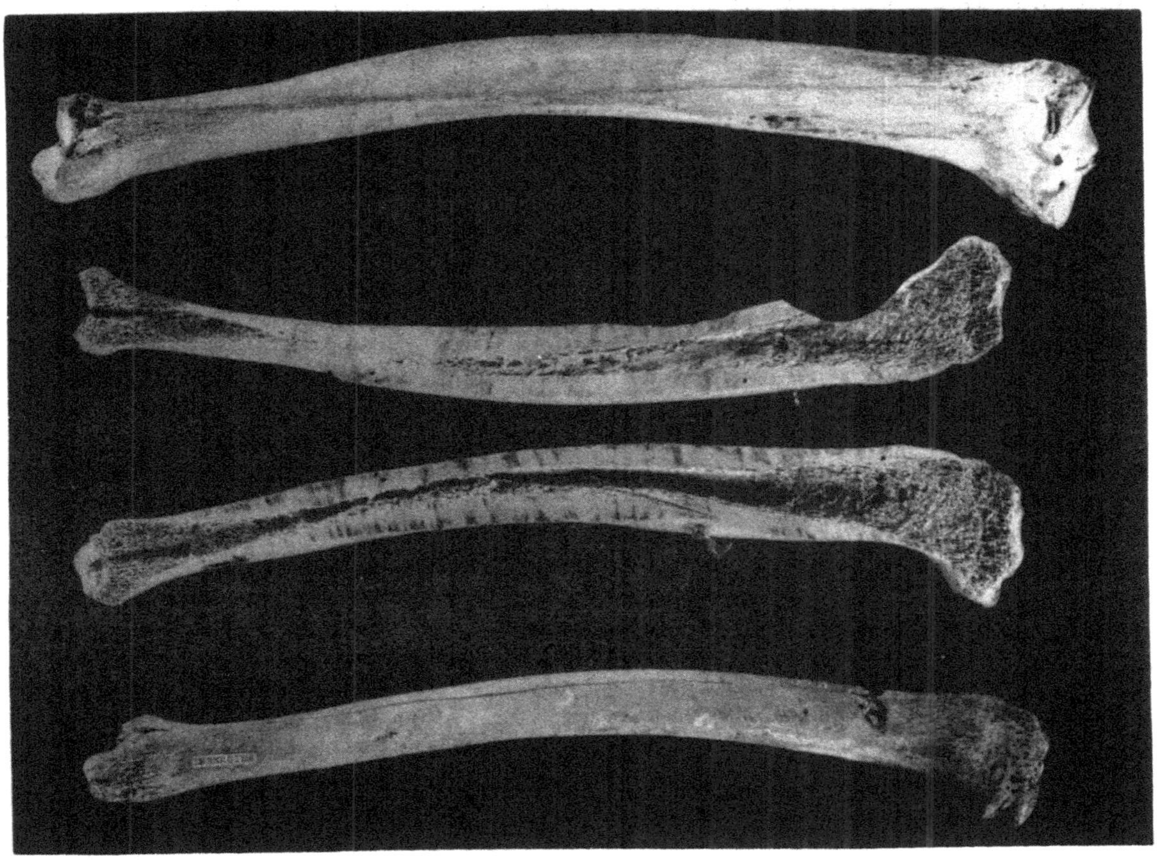

difference between the two groups ($p=<0.001$) (Table 5-8). It is interesting to note that these results support those of other workers who have noted the condition to the exclusion of other outward signs of disease, particularly treponemal disease. Such deformation may be due, of course, o subclinical pathological processes in those having treponematosis but who, to the observer, look completely healthy. Serological testing (TPI) in the Northern Territory has shown that 23% of 1542 Aborigines were positive for treponemal infection and in the vast majority of cases they were suffering endemic erkincha and yaws (Garner *et al.* (1972). Only *...a few cases of marked anterior bowing of the tibia were seen....* but *There were many cases of slight bowing of the tibia....* (Garner *et al.* 1972:287). Unfortunately the link between those with bowing and those who were TPI positive was not explained. Hackett's data from the Northern Territory show that 49% of 322 Aborigines had boomerang leg and associated scarring from tertiary treponematosis and only 9% had boomerang leg alone (see Hackett's Table VIII, 1936a:22). Of interest here is the extraordinary high frequency of bowing. The palaeopathological evidence gives little indication of this, although the samples representing areas of endemic erkincha and yaws are, compared with others, small, especially the sample from the tropics. On the other hand, this condition may be more obvious in the living through the use of different diagnostic criteria than those used in this survey.

Earlier research proposed a definite association between bowing and treponematosis. In a paleopathological sense, however, those bones having no sign of infection cannot be categorized one way or another. On the other hand, they may have belonged to an individual with clinical or sub-clinical treponemal disease, something skeletal assessment cannot verify. Alternatively, this deformity may have nutritional or metabolic causes, either stemming from or indirectly related to treponemal disease. One avenue of analysis might be to see whether those individuals having treponemally induced cranial lesions have bowed tibiae also. In the majority of cases, past processes of collection and curation, which include cranial and post-cranial separation and different subsequent numbering of remains, have made this type of association impossible to verify. Cross checks on the eastern and southern remains have shown, however, that there is not one single case where an association between bowing and cranial lesions can be proven. Obviously the aetiology of bowed tibia demands further investigation. Recent advances in bone chemistry diagnostics and general techniques of osteological and related analysis may well provide some answers to its origin in future.

The evidence presented by Topinard (1894) concerning bowed tibia from burials predating the supposed introduction of treponemal disease into Europe underscores this necessity. Spencer raised the question of whether bowing (and camptocnemia) was pathological or not:

> *To what extent either, or both, of these conditions are pathological is difficult to say. They do not seem to be associated with any indication of debility in the man or woman in whom they are present.* (1928:194).

Similarly, those describing Bedouin remains in Israel, have said: *Whether the pronounced bowing in the long bones can be considered pathological, is not certain* (Goldstein *et al.* 1976:628). This opinion is supported here. It seems, then, that only three things can be said regarding the palaeopathological evidence for tibial bowing among Australian Aboriginal populations:

1) the low frequencies obtained suggest pronounced bowing affected people in a minor way;

2) it is evident that bowing was present in almost all parts of Australia but greatest along the Murray and east coast; and

3) in the majority of cases it seems unconnected with other signs of infectious disease, either on the tibia itself or associated crania.

There is one final comment that should be made. Cecil Hackett (1936a) was the first to draw attention to reports of boomerang leg in more southerly parts of the Australian continent and he referred to a number of bowed specimens from many regions in the south that were housed in museum collections. Some of these specimens are included in this survey but none have been dated. Therefore, it is possible that if bowing was caused by treponemal disease, in the south this could be due to European congenital or late acquired syphilis. There seems, however, no evidence that all bowing is caused by treponemal disease. Also, it has been argued in the Treponemal section above that satisfactory evidence for pre-European treponematosis along the Murray and coasts will come only from dated archaeological material.

Recently, a bowed tibia was recovered in the Robinvale Aboriginal burial ground situated on the Murray River (my location E) (Howard and White in Bowdler 1983, Appendix 1). The curvature is believed to have been caused by warping processes during interment. I have not personally inspected the specimen, as all human skeletal material was reburied at the request of the local Aboriginal community. It seems, however, unlikely that a single robust tibia would have been warped by natural processes, leaving all the others completely normal. If, indeed, the tibia was bowed in the same sense as those described in this section, I have no doubt it would be the first solid evidence for this condition in precontact Aboriginal society.

Summary

1) Although non-specific bone infection is found almost everywhere in Australia, the highest rates occur along the Murray.

2) The highest incidence of treponemal disease is found along the Murray River.

3) The extent and prevalence of treponemal disease in pre-contact Australia cannot be determined from skeletal material held in collections around Australia until some of it is dated.

4) Although such evidence is equivocal and confusing, a few early descriptions of bodily disfigurment in Aborigines suggest that an endemic form of treponemal disease was indeed present in the southern parts of the continent.

5) While tibial bowing can be seen in all regions, it is in low frequencies.

6) No palaeopathological connection between tibial bowing and treponemal disease has been established in Australian Aboriginal skeletal material.

7) The combined evidence presented above suggests that both treponemal disease and bowing were present among Aboriginal people prior to European contact.

Chapter 6

OSTEOARTHRITIS

General Introduction

The first indications of arthritic disease appear in aquatic Mesozoic reptiles, *Platycarpus* (100 million years), an armoured reptile from Britain, *Polycanthus foxi* (140 million years), which is thought to have had ankylosing spondylitis, and *Diplodocus longus* (100 million years) which has vertebral osteophytic spicules (Moody 1923: Karsh and MacCarthy 1960; Zorab 1961; Hollander 1962; Steinbock 1976). Similar conditions occurred in Miocene crocodilians and Pleistocene cave bears, sabre tooth tigers, mammoths, cattle (*Bos primagenius*), hyenas and the large flightless birds, Moa and Aepyornis (Wells 1964, 1973). Evidence for osteoarthritic disease in humans has been observed in Neanderthals, inflicting lesions severe enough to cause initial misinterpretation of the form of their posture and carriage (Boule 1911-13, 1923; Smith 1924 cited in Straus and Cave 1957). Neanderthal remains from La Chapelle-aux-Saint, La Quina and La Ferrassie show arthritic changes in the temporomandibular joint, spine and hip, and vertebral osteophytic exostoses appear in Late Palaeolithic remains from China (Weidenreich 1939; Straus and Cave 1957).

Studies of recent populations show many forms of arthritic disease occuring in varying frequencies and severities in people from all over the world. Some of the skeletal populations studied include Egyptians, both ancient and modern (Ruffer and Rietti 1912; Smith and Dawson 1924; Wells 1964; Bourke 1971), Byzantines (Angel 1979), Yugoslavs (Edynak 1976), Peruvians (Hooton 1930), Eskimos (Blumberg 1961; Ortner 1968; Jurmain 1977a, b) and Anglo-Saxons (Wells 1965). Other studies have reported various types of arthritis among modern hunter-gatherers and these have been used to give some assessment of biomechanical stress derived from that lifestyle (Chapman 1962, 1968; Angel 1966; Ortner 1968; Jurmain 1977a, b; Hartney 1978; Kelly 1979; Martin *et al.* 1979; Pickering 1979; Larsen 1982).

Palaeopathological analysis of skeletal populations from various parts of the world have discovered a variety of arthritic conditions. These include pyogenic arthritis (Steinbock 1976), ankylosing spondylitis (Aufdermaur 1957; Zorab 1961; Hudson *et al.* 1975; Hochberg *et al.* 1978;

Beute 1979), vertebral osteophytosis (Stewart 1966; Pietrusewsky 1969; Clark and Delmond 1979), rheumatoid arthritis (Blumberg *et al.* 1961; Short 1974; Klepinger 1979; Leisen and Duncan 1979; Riddle 1979), gout and pseudo-gout (Kuzell and Gaudin 1956; Bluhm 1979), osteoarthritis (Duncan 1979) and arthritis of the temporomandibular joint (Pales 1930; Blackwood 1963; Alexandersen 1967; Seward 1976; Alpagut 1979; Richards and Brown 1981). For general overviews of the palaeopathology of arthritis and its aetiology see Karsh and McCarthy 1960; Bourke 1967; Steinbock 1976; Jurmain 1977a, b; Beute 1979; Cassidy 1979; Cockburn *et al.* 1979; Reyman 1979 and Ortner and Putschar 1981.

This chapter discusses in order: **Post-Cranial Osteoarthritis; Pleistocene/Early-Holocene Osteoarthritis** and **Osteoarthritis of the Temporomandibular Joint.**

Post-Cranial Osteoarthritis

Methodology

This section deals with osteoarthritis or degenerative joint disease in the knee and elbow joints only. To do this I have used the proximal ulna, radius and distal humerus as prepresenting the elbow and the proximal tibia and distal femor as the knee. Three basic changes to bone architecture have been used as indicators of osteoarthritis in these joints (henceforth referred to as arthritis). These are marginal osteophytosis (Plate 6-1A), eburnation, (polishing of the articular surface), and erosion, (or pitting of articular surfaces) (Plate 6-2).

Opinions differ as to whether osteophytosis indicates the presence of arthritis (see Bourke 1967 and Jaffe 1975). Nevertheless, most researchers in this field accept osteophytes as indicative of arthritic changes and use them when assessing the frequency of this disease in both modern clinical and skeletal populations (Copeman 1955; Chapman 1968; Jaffe 1975; Adams 1976; Steinbock 1976; Jurmain 1977a, b; Angel 1979; Clark and Delmond 1979; Cassidy 1979; Pickering 1979; Martin *et al.* 1979; Ortner and Putschar 1981). Jaffe's view, for example, is that osteophytes *...appear so commonly in primary osteoarthritis that they constitute a characteristic feature of its pathology.* (1975:751).

Osteophytes are exostotic bone outgrowths located at the joint periphery, following the outline of the joint capsule insertion. Grading them by size is a subjective task because of their almost infinite morphological variability and frequency between similar and different bones and on any one bone. Grading systems have, however, been employed to provide some uniformity in recording and comparability with

Plate 6-1A Osteophytic lipping (osteophytosis) around the trochlear notch of an ulna.

Plate 6-1B 'Scroll' lipping on the edge of the medial condyle of a femur. This exostotic proliferation usually follows the course of the medial and lateral edges of the articular joint surface and can measure up to 12 mm at its widest point. A subchondral cystic defect has appeared on the medial condyle.

Plate 6-2 Eburnation with erosive pitting on the medial articular condyle of a right tibia. Some eburnation appears on the lateral articular surface and remnant osteophytosis occurs around the edge.

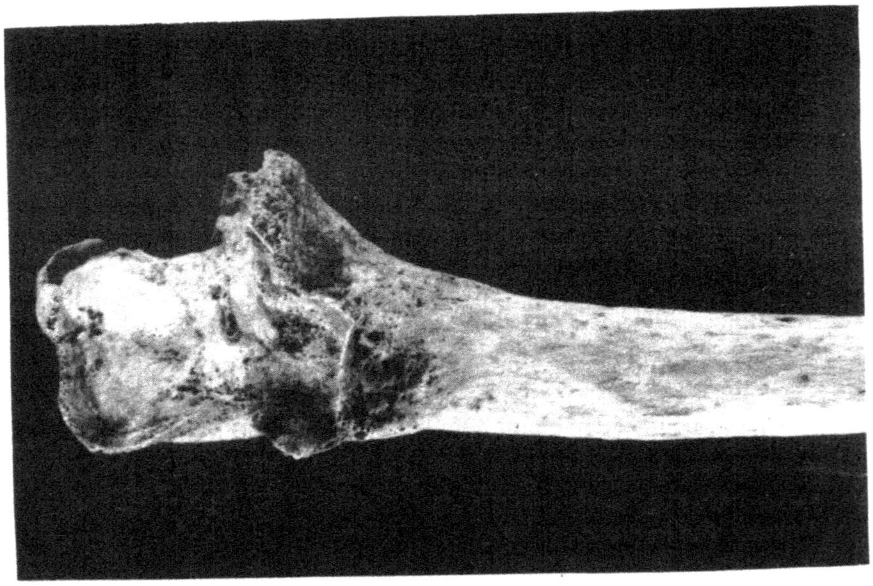

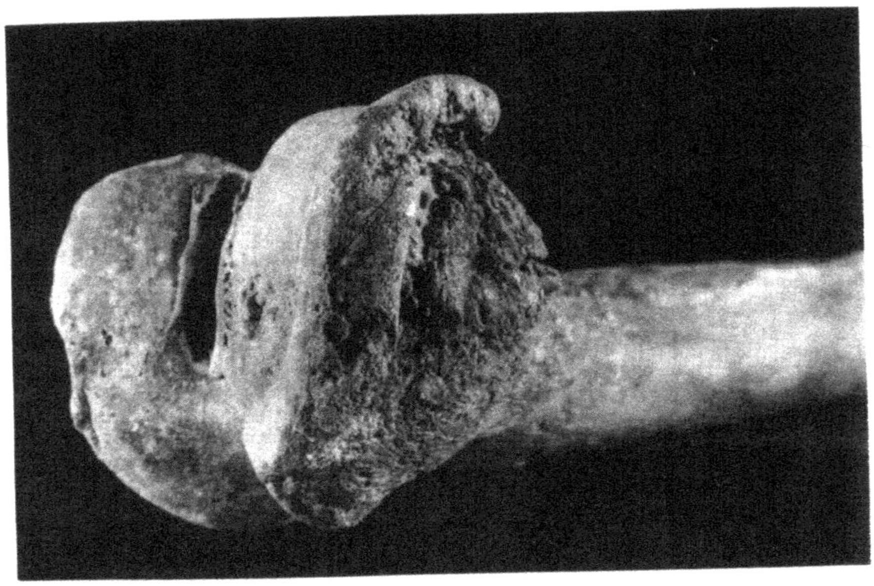

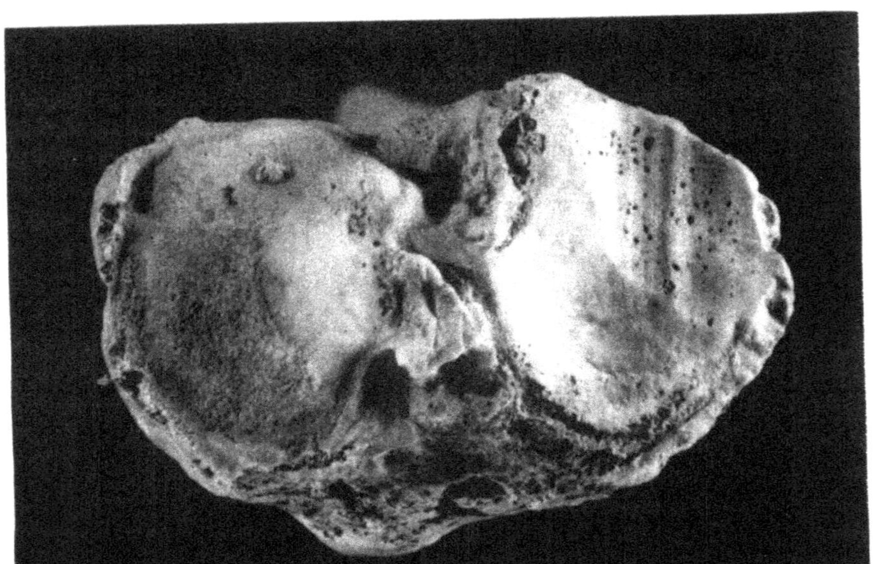

future studies (Angel 1979; Martin *et al.* 1979). The grading used in this survey is outlined in Appendix 1. An extreme form of osteophytosis was encountered which I have termed 'scroll' (Plate 6-1B) and is similar to that described by Wells (1965) in Anglo-Saxons. These boney outgrowths were seen to project by as much as 8mm on the distal humerus, 10mm on the proximal tibia and 12mm on the distal femur during this survey.

Any eburnation or erosion were recorded separately, together with their position on the joint (medial or lateral) and the approximate size of the affected area (see Appendix 1). It must be emphasised that these areas were not accurately measured so much as roughly estimated. Because of time constraints and data manageability, however, depth of analysis was somewhat reduced to allow a comprehensive but more general picture of arthritis to be presented. This reduction has eliminated analysis of arthritis by its position on the articular surface and the size of the affected area. It was felt that rather than give a summary presentation of these aspects, which may provide a more intimate interpretation of behaviour they should be reserved for publication at a later date. Some compacting of recording parameters has taken place also. Two grades of osteophytosis have been merged to form 'present' or 'absent'. Likewise, all grades of eburnation and erosion have been ignored and only presence or absence noted. While this compaction makes the results of the survey rather coarse, some refinement is provided by using osteophytosis as an indication of the presence of arthritis, and eburnation and/or erosion as evidence of the chronicity of the condition.

Post-cranial sample sizes for the desert group are small, particularly for females. The latter has been included, however, in order to give some indication of arthritis in desert women. In 42 combined left and right, male and female and unsexed bones of all types from the tropics only three (7.1%) had any sign of arthritis. No arthritis was evident in 40 male/female right bones. The very small sample size from this area prevents further subdivision into bone type and sex and for this reason the whole area has been excluded from the analysis below. Similarly, the Murray locations of T and K have minimal numbers of lower and upper limb bones, respectively. They too have been eliminated from the location analysis for these respective limbs. There are not enough sexed post-cranial bones from Euston (E) to make analysis by sex meaningful. Here males and females have been merged to form a sample adequate for inclusion on a population basis.

Results

Overall, arthritis affects males more than females, elbows more than knees and Murray people more than anyone else (Fig. 6-1). The basic pattern of arthritis shows that the left knee is affected more than the right and the right elbow more than the left. The latter pattern can be seen on both an area and a location basis and is probably to be expected with the world wide predominance of right handedness. The only exception to male susceptibility is the east coast where females have more arthritis in both the elbows and knees and both sexes have higher frequencies of left arm involvement than right. Not surprisingly, chronic arthritis, denoted by eburnation and/or erosion, reflects the pattern of osteophytosis. The latter affects east coast females in the right elbow and the knees (Fig. 6-2, Table 6-1) and Figs 6-3 and 4 show clearly that chronic arthritis of the knee is confined to women from coastal Queensland. The only other area where females have predominant arthritis of the knee is along the Murray, but it is not of the order seen in coastal Queensland women. Instead, Murray Valley females have much higher frequencies of severe (chronic) elbow degeneration. With the exception of coastal Queensland, this phenomenon is confined to the central Murray and is statistically higher than in the Rufus River and south coast groups ($p=<0.05$) (Table 6-2).

The highest frequencies of arthritis in men are found in the Murray and Desert groups. Arthritis of the knee appears to affect all Aboriginal men to a varying but similar degree (Figs 6-1 and 2) and there are no statistically significant differences between areas (Table 6-2). Coastal Queensland males have one of the highest frequencies of severe arthritis of the knee anywhere (Fig. 6-3, Table 6-3). These men, together with those from coastal New South Wales and central Australia, have this condition to a somewhat greater extent than Murray men, with the one exception of the Kow Swamp/Loddon/Kerang (K) location in the central Murray.

Extensive degeneration of the elbow joint occurs much more frequently in Murray men than in those from other areas (Figs 6-1 and 2). This is statistically significant for both the left ($p=<0.05$) and right (<0.02) arms (Table 6-2). Desert men, while having less arthritis of the elbow than Murray men, are not significantly lower. This may be due, however, to the large differential in sample size between these areas. There is a significant statistical difference between the central Murray and the south coast for both arms ($p=<0.02$) and a highly significant difference between men from the central Murray and those from the east coast ($p=<0.001$). The locations of Tocumwal (T), Baratta (B) and Swan Hill (S) emphasise the elbow involvement in the Murray by having the three highest frequencies of chronic arthritis for both arms seen anywhere in Australia.

On a population basis, elbow degeneration is most common in Murray people (Fig. 6-4, Table 6-4). For the knee, however, there is proportionally less arthritis than elsewhere (Fig. 6-3). Lastly, Fig. 6-5 and Table 6-5 give some idea of the different osteophytic frequencies found on the various bones which make up the left and right elbow and knee joints in men and women. These patterns emphasise the differences between inland and coastal groups, particularly in males. Handedness is highlighted in the Rufus River and coastal samples, while in both the desert and central Murray groups elbow degeneration has an almost bilateral pattern. Central Murray females use both elbows almost as much as the men and certainly more than for women from other areas.

Only five of 4777 assorted post-cranial bones from the Murray (0.1%) and two of 2770 from all other areas (0.1%) showed obvious signs of traumatic arthritis. These consisted of two male, one female and two of undetermined sex in the Murray River group and two females in the other. The only cases of infectious arthritis come from the Murray, where eight male, three female and eight of undetermined sex (0.4%) were observed.

Discussion

In modern populations arthritic changes appear as early as the second decade of life and by 40 years of age 90% of all persons will have them, however minor, in their weight bearing joints (Sokoloff 1974). The incidence of arthritic disease rises with age, from about 4% at 18-24 years to around 85% between 75-79 years (of which 23% have moderate to severe changes). An American survey erevealed that nearly all cases under 45 years were mild. In England 85% of individuals between 55 and 64 years were found to have arthritis in at least one joint (Moskowitz 1974). Men under 45 years were affected more than women; a situation reversed after reaching 55 years. After 55 male/female patterns were found to be similar. Equal numbers of blacks and whites had arthritis, although the knee joint of Black people were affected more often. Contemporary North American Indians have a higher prevalence than either white or black people. Rather than a racial susceptibility towards the disease, Indians and blacks are thought to reflect the fact that they hold proportionally more manual occupations than whites.

Although biological ageing and mechanical stress play important roles in the onset of arthritic disease, environmental factors are thought to contribute also. Studies in Northern Europe and America have suggested that the frequency falls with increasing latitude. Evidence for this has emerged from studies in Finland and Holland which have shown that Finns have far less arthritis than Dutch people (Moskowitz 1974). A study of Jamaicans and Britains revealed, on the other hand, an equal incidence. Contrary to popular belief, there is no proof that cold wet environments predispose individuals to arthritic disease; rather these climates stimulate muscular pain mechanisms which tend to simulate 'rheumatic' pain. Moskowitz tells us also that obesity may be an important cause of joint degeneration, particularly in the knee, since 61% of obese people were affected, compared with 26% of non-obese individuals. Whatever the complicating effects of other factors, overall it is probably the amount of wear and tear on a joint that determines whether it becomes affected by mild or severe degeneration at an early or late age. A number of studies have established clear links between occupation and arthritis (Tempelaar and van Breeman 1932; Lawrence 1955; Kellgren and Lawrence 1958; Brodelius 1961; Anderson *et al.* 1962; Solonen 1966; Kouba 1967). The functional stress hypothesis seems, therefore, the most sensible way to approach the analysis of arthritis in this study.

In any palaeopathological analysis certain variables must be assumed to be constant. In this survey, for example, I can only assume that age predisposes Aboriginal people to arthritic degeneration, as it does elsewhere and that this degeneration is compounded by activity differentials between males and females and groups living in similar and different environments. Therefore, it is expected that the range of arthritic variation will tend to reflect the type and amount of stress applied to joints in everyday life and the overall mechanical stress experienced by different groups.

It is surprising to see the small frequencies of arthritis of the knee joint, bearing in mind the almost constant activity of the hunter-gatherer. The vast difference in arthritic disease between women from coastal Queensland (around 15%) and those from the Murray (between 3-5% from Baratta) must be a reflection of their differing lifestyles, activity patterns, age or a combination of all these. The basic round of female activities, with the gathering of vegetable foods and shellfish and some minor hunting, were probably similar for almost all groups of Aboriginal women, wherever they lived. What is different, however, is the variation between groups of females with respect to early death. Figure A1 shows that the death rate among young adults from coastal Queensland and Baratta are about the same. Therefore, an explanation for the difference in terms of age associated arthritis is not possible. Moreover, Swanport females who are living longer than women in both groups do not show any incidence of arthritis of the knee. This suggests that the frequencies derived reflect a mechanical stress on this joint that for coastal Queensland seems rather high compared to other populations.

More arthritis might be expected in males than females for two reasons. One, activity levels are expected to be higher and, two, in all locations men live longer than women. With almost all locations showing a less than 10% involvement of the knee, I suggest that this can be accepted as a normal percentage of knee joint arthritis for males in any population, tribal or urban.

In modern populations, arthritis: *Almost never involves the first metacarpophalangeal joints, wrists, elbows, or shoulders, except after trauma* (Moskowitz 1974:1043). Adams (1976:249) confirms this view: *Osteoarthritis seldom occurs in an elbow that was previously normal.* It seems, therefore, any incidence of degeneration in this joint would be a reflection of unusual and/or excessive use, particularly with respect to occupation and repetitive movement. Palaeopathological work has generally supported this view but has had to rely on explanations derived from known activities of the individual groups studied when arthritis in the elbow joint has been found to be prevalent. Jurmain (1977b) has shown that arthritis of the elbow is not necessarily age related, has a slightly earlier onset than in other joints and is found more often in right arm of males than females, although some variation in this pattern has been observed between groups leading different lifestyles. Eskimos, traditionally experiencing a great deal of functional stress, have higher frequencies of arthritis in the elbow joint than other groups, together with severe involvement of the right arm at an early age in males (Jurmain 1977b). This is true also for females, though to a lesser extent, and reflects their equal participation with men in the arduous tasks connected with living in the Arctic.

Changes in the frequency of arthritis over time in both the knee and elbow joint have been discovered in people undergoing socio-economic change. In a sample of pre- and post-agricultural individuals from the Georgian coast, a reduction from 13.7% to 6.1% for men and 9.6% to zero percent for women has been recorded (Larsen 1982). This reduction in joint wear and tear corresponds to reduced activity levels. Agriculture is probably no less strenuous than hunting and gathering, but the duration of arduous tasks is often shorter and this, together with periods of minimal activity during growing seasons, tends to be reflected in skeletal remains. Moreover, the degree of stress upon joints would not be the same as that experienced by a hunter. Larsen's data tie in with those of Jurmain (1977a,b) which show that Pecos Pueblo Indian farmers had substantially less arthritis for all sections and ages of the their population than for the Eskimo.

The elbow is probably the best indicator of the amount of work stress in a population because the knee is subject to a lifetime's weight bearing and daily repetition stress from walking, running, squatting and standing and, thus, is susceptible to a certain amount of wear and tear in all people. It is interesting to see, then, the overwhelming involvement of the elbow for both sexes in the Murray with respect to other groups around the continent. The prevalence of the right elbow is something mirrored often in human groups, but the stark inter-regional variation in the amount of use to which the upper limb is put is interesting, especially in the light of the commonly held belief that Aboriginal people were living basically similar lifestyles in most parts of the continent with a homogenious activity regime. Both the bilaterality and the extent of elbow use observed in men and women from the central Murray is probably the most sigificant result to come out of this particular survey.

The frequency of radial head involvement among desert men is quite different from that seen in the central Murray. This suggests a variation in the use of the upper limb between these two areas. Repetitive arm rotation, in using a spear thrower for instance, could be one reason for the difference (Webb 1989). Moreover, the obvious right hand bias of the condition tends to support the belief that this activity may be responsible for the pattern. Whatever the reason, male activity in the central Murray seems to be different from that in the desert. The substantial difference between the left and right distal humeri adds further support to this conclusion. In the central Murray, for example, the overall wear pattern of the elbow emphasises the use of the arm as a hinge joint, with the associated articular surfaces of the ulna and humerus showing the higher frequencies of arthritis. This counters the emphasis on rotation seen among desert dwellers. Moreover, the similarity of arthritic involvement in the left and right ulnae and radii from the central Murray suggests equal emphasis of use for both arms. Locations within the central Murray confirm that there is a far greater use of the arm, by both sexes, in this area than in locations outside. Although handedness is apparent, a comparison of elbows emphasises the bilateral nature of upper limb stress in this region that is particularly obvious in females.

Even though we have seen that age has little to do with arthritis of the elbow joint in a modern population, some association would more likely exist in those that work hard. Age, however, would not seem to be the reason for higher frequencies in the Murray people, although this argument could be used to explain the situation in the desert, at least for men.

The wear patterns of the desert and the central Murray are not seen elsewhere. On the east coast, stress at the elbow is almost non-existant. This indicates that people here were not using the joint in the same way or to the same extent as Murray and desert groups. I am going to include the people from the south coast in this statement, ignoring for now the large frequency of arthritis in the proximal ulna, because all other bone involvement in this group is similar to that of the east coast, which is minimal. What is most extraordinary is that, with the sole exception of the right elbow in males there is no indication of this condition among these people. An immediate explanation of the differential pattering between coastal and Murray River people must lie in different subsistence strategies and/or the amount of work involved in procuring food.

Pleistocene/Early-Holocene Osteoarthritis

WLH 3

WLH 3 is a male individual from Lake Mungo, western New South Wales and dated to between 30-32,000BP. Besides arthritic lipping on the thoracic and lumbar vertebrae, which has been described elsewhere (Bowler and Thorne 1976), there is an extreme form of arthritic degeneration and ankylosing of the right elbow. This condition, however, has been described in detail elsewhere (Webb 1989).

Kow Swamp and Coobool Crossing

A full anatomical description of the Kow Swamp remains, including minor arthritic changes, has been made (Thorne 1975) and little can be added to this. It is interesting to note, however, that humeral capitular erosion and other changes similar to those of WLH 3 can be seen in specimens of the Kow Swamp.

The Pleistocene sample, consisting of Kow Swamp and Coobool Creek material, shows 48.0% of males (n=25) and 46.7% of females (n=15) with arthritis of the glenoid fossa which extends to the mandibular condyle in 55.5% of males (n=9). Four of the six young males show category 1 (66.7%), none has category 2, eight of 19 old males (42.1%) have category 1 and three show advanced degeneration (Table 6-7). Two factors of immediate interest are: one, more young than old males affected, and two, the mandibular frequency is greater than that of the glenoid fossa (Figs 6-6 and 7). Both these features are completely opposite to patterns seen in the more modern groups (see below), though this may be due to the small Pleistocene sample.

Murrabit

This individual is undated, but on morphological grounds is believed to be contemporary with the Coobool Creek and Kow Swamp remains.

Category 2 osteophytosis occurs around the distal joint of the right humerus but there is no sign of accompanying erosion or eburnation. No arthritic changes are visible on the right ulna, however, and the associated radius is missing. The right humerus is more robust than the left, which implies greater use of this arm and, therefore, a greater susceptibility to arthritic conditions. The left humerus has slight osteophytosis at the distal joint (category 1) and on the distal left radius. This bone has, however, been fractured (see Chapter 7) and subsequent misalignment has probably caused irregularity at the wrist joint, thus predisposing the bone to arthritic processes (traumatic arthritis).

There is vertebral osteophytosis (category 2) on the inferior margins of L4 and L5, but this may be caused by the combined factors of age and an active lifestyle.

Keilor

There is severe arthritic degeneration of the left glenoid fossa on this individual believed to be 13000 years old (Macintosh and Larnach 1976). The wear corresponds to category 2 or advanced erosion. The right fossa is missing. The degeneration has obliterated the fossa, filling it with proliferative bone which has formed a 'plate' upon which the mandibular condyle would have pivoted. Moderate osteophytosis is evident around the periphery of the plate. Unfortunately the mandible missing so the extent to which this process may have affected the associated mandibular condyle cannot be assessed.

Museum of Victoria No. 31837

Although this individual is not dated, the heavy mineralization and calcium carbonate cover is not unlike that found on the Coobool Creek and Kow Swamp remains. Thus, a late Pleistocene/early Holocene date is possible for the individual. The only description of this extremely dolichocephalic male from the Parish of Nyang in the Riverina is that of Fenner (1938). He concluded that the complete fusion of all cranial sutures, plus the elongated, narrow and high vault, were consistent with scaphocephaly (glabella-opisthocranon 212mm, bi-parietal 121mm, basion-bregma 147mm). He noted also that the *...glenoid fossae show advanced arthritic wear* (1938:153). The wear is an extreme form of category 2, morphologically similar to that already noted on the Keilor cranium, with both glenoid fossae filled with new bone deposition and displaying marginal osteophytosis. The mandibular condyles show the typical mushrooming, flattening and eroded articular surfaces often observed in extreme arthritic degeneration.

Osteoarthritis of the Temporomandibular Joint

Introduction

Temporomandibular arthritis occurs in Neanderthal remains from Krapina, La Quina, La Ferrassie and La Chapelle-aux-Saints (Pales 1930; Alexandersen 1967), in neolithic groups from Turkey (Alpagut 1979), North American Indians (Leigh 1925; Rabkin 1942; Goldstein 1957) and a variety of other prehistoric peoples (see Alexandersen 1967). Population frequencies are not commonly found in the literature, but the following examples derived from various skeletal populations allow some comparisons to be made: Negroes 4.5%, New Hebrides 5.6%, New Caledonia 24%, Loyalty Islands 26.4% (Pales 1930:166-167), and 43% and 33% in male and female Aborigines, respectively (Richards and Brown 1981). In modern Europeans, frequencies of 16% for males and 33% for females have been found (Oberg *et al.* 1971:380). The incidence is, therefore, very variable but standardized categories enabling direct comparisons are few.

Other than these studies the frequency and aetiology of arthritic disease of the temporomandibular joint in skeletal populations has attracted minimal attention from the palaeopathologist, particularly as regards its aetiology and interpretation of its frequency (Ortner and Putschar 1981).

Methodology

Arthritis of the temporomandibular joint was surveyed using the right and left glenoid fossae and mandibular condyles. Degenerative attrition of the glenoid fossae was noted whether it occurred in the fossa itself, on the articular tubercle or both of these areas were involved. Arthritic changes were divided into two categories, 1 = minor arthritic wear and/or erosion and 2 = major arthritic wear and/or erosion. It should be noted here that 'wear' is used synonymously with flattening.

Assessment criteria basically follow those used by Richards and Brown (1981) with some minor changes. For the glenoid fossa these are:

Category 1 - Localized erosion or wear, little or no arthritic lipping (Plate 6-3A). This corresponds to Richard's and Brown's Grades 1 and 2.

Category 2 - Major erosion or wear. This includes any erosion, eburnation, grooving and osteophytic proliferation, as well as those individuals with substantial alteration to the joint area or obliteration of the fossa(e) (Plate 6-3B). This category is equivalent to Grades 3 and 4 in Richard's and Brown's survey.

For the mandibular condyle, category 1 is the same as Richard's and Brown's Grade 1 but Category 2 is represented by their Grades 2, 3 and 4.

Category 1 - Localized erosion and/or minor wear (Plate 6-4A).

Category 2 - Lipping plus substantial wear, erosion and/or eburnation (Plate 6-4B).

TMJ will be used as an abbreviation denoting the frequency of arthritis in the glenoid fossa and MAC (MAndibular Condyle) will refer to arthritis on the mandibular condyle. It should be noted also that MAC sample sizes are almost always smaller than those of the TMJ because there are always less mandibles than crania surveyed. (A mandible was not included in this survey if its associated cranium was missing). These already small MAC samples are reduced even further when they are divided into males and females, then into young and old adults. It is for this reason that analysis of MAC frequencies will be confined to 'area' except for a comparison of population totals between upper (TMJ) and lower (MAC) jaw frequencies by location (Fig. 6-6).

Results

In the locational comparison of TMJ and MAC frequencies a strong pattern emerges showing the susceptibility of the glenoid fossae to arthritis over the mandibular condyle (Fig. 6-6). With the exception of the Cape York/Cairns group, this trend occurs throughout Australia. Nevertheless, the overall trend of MAC arthritis follows that of the TMJ, with central Murray frequencies prominent (Table 6-6). TMJ arthritis also features strongly in groups from Adelaide and coastal Victoria. The rather high frequency among late Pleistocene remains from the Murray River is comparable to later populations in that area (Baratta (B)). The central Murray groups have a minimum frequency of 52.5% whereas Tasmania, the next highest frequency outside the Murray, has 38.3% (Table 6-7), a statistically significant ($p=<0.05$) drop.

Plate 6-3A Category 1 osteoarthritis of the temporomandibular joint (TMJ). Minor localised erosion appears on the articular tubercle and in the glenoid fossa.

Plate 6-3B Category 2 osteoarthritis of the temporomandibular joint (TMJ). Severe erosion, pitting and some eburnation has destroyed the articular emminence and bone proliferation has filled and almost obliterated the glenoid fossa.

Plate 6-4A LEFT:- Category 1 osteoarthritis of the mandibular condyle (MAC). There is some localised erosion anteriorly with minor flattening of the condyle.

Plate 6-4B RIGHT:- Category 2 osteoarthritis of the mandibular condyle (MAC). Advanced erosion with lipping around the edge of the condyle.

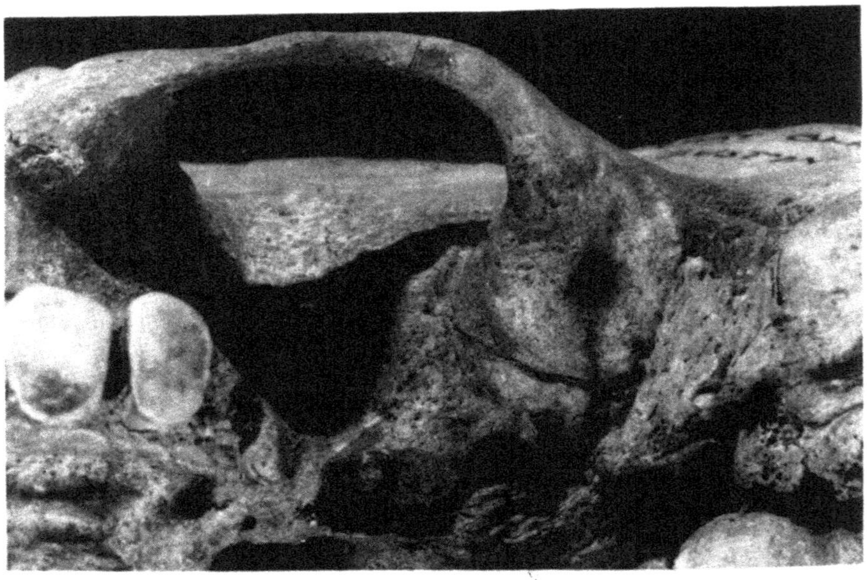

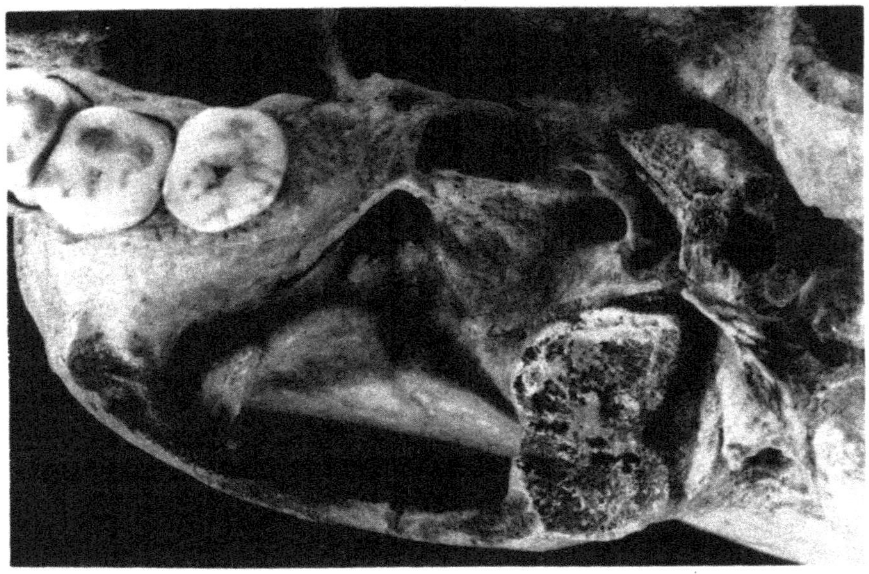

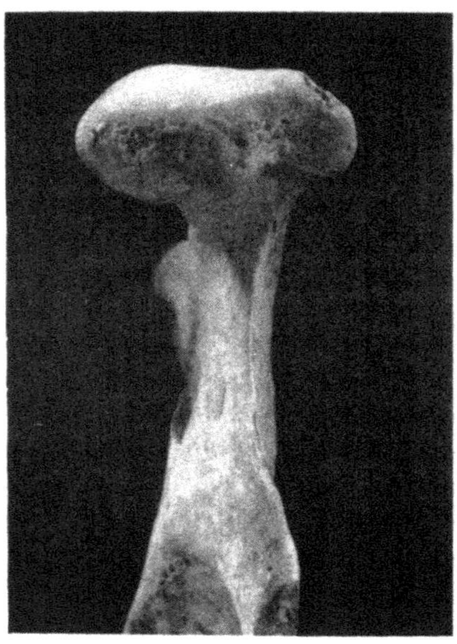

For reasons of sample size, male/female comparisons will be confined to TMJ frequencies dealt with on a location basis, unless otherwise stated. Figure 6-7 shows few significant differences between male and female frequencies for TMJ arthritis although, more often than not, males have a somewhat at higher incidence. With the exception of Swan Hill, the central Murray has a more even distribution of the condition between men and women than noted elsewhere.

The proportion of the frequencies noted in Fig. 6-7 that can be attributed to the advanced form of arthritic degeneration (category 2) can be seen in Fig. 6-8. The largest incidence, by far, occurs in Murray River people and females from Cape York/Cairns region. Figs 6-9A and B show the distribution of category 2 arthritis between young and old adults. More young adults from the Murray River have the condition than those in all other locations, although young men from Tasmania and coastal Victoria and young females from coastal New South Wales are prominent (Fig. 6-9A). These frequencies are compounded with age, particularly in the Murray where frequencies among old adults are inordinately greater than in the same group living in other areas (Fig. 6-9B). Of course, when adult samples are subdivided in this way, sample size is effectively reduced causing many to fall well below the minimum size considered desirable for interpretations to be made. This includes some of those locations mentioned above (see Table 6-7).

Analysis by area is, therefore, a more suitable way to highlight trends when these sorts of sub-divion are called for because of the increase in sample size (Fig. 6-10A and B). Fig. 6-10A clearly shows the sharp dichotomy between the Murray-south coast areas and the rest of Australia, which is highly significant for males ($p=<0.001$) and very significant for females ($p=<0.02$) (Table 6-8A and B). There is also an underpinning of the trends suggested in the age and category breakdown among locations. The Murray, particularly the central Murray, and the south coast stand out as areas with most category 2 arthritis in both males and females (Fig. 6-10B) as well as places where it develops at an early age in both sexes (Fig. 6-10C). This ultimately leads to very large amounts of category 2 arthritis in old adults in the Murray, which stands out in comparison with the rest of Australia (Fig. 6-10D). It is worth noting that besides women from the Murray River, young women from the east coast form the only other female group which has more advanced arthritis in young than old women (Fig. 6-10C). The opposite trend can be seen in women from the tropics, where no severe arthritis occurs before their mid-thirties. The large frequency in old women from this area is found mainly in females from Cape York, not those from the Northern Territory.

Patterns of arthritis on the mandibular condyle (MAC) are less clear, but Fig. 6-11A shows that the trend for males is similar to that already seen for TMJ, although there are fewer statistically significant differences (Table 6-8C and D). The most significant of these is that between the central Murray and the Desert ($p=<0.025$) and the central Murray and the east coast ($p=<0.02$). The female pattern is more random than that of the males, except for the tropics where the largest frequency is seen again in old women. (compare Fig. 6-10C and D and Fig. 6-11C and D). There is a commensurate increase in severe arthritis with age (Fig. 6-11D), which follows the pattern for the upper jaw and is, therefore, to be expected.

Past research has shown that tooth attrition is common in many groups of hunter-gatherers (Leigh 1925, 1929; Rabkin 1941, 1943; Roydhouse and Simonsen 1975; Scott and DeWalt 1980), including Aborigines (Campbell 1925, 1937, 1939; Campbell and Barrett 1953; Beyron 1964; Barrett 1969, 1977; Seward 1976). No survey of dental attrition was undertaken during the present study, but it has already been linked to temporomandibular joint arthritis in Aborigines (Seward 1976; Richards and Brown 1981). Tooth loss was recorded, excluding those individuals with avulsed incisors, and the association between individuals having one or more missing teeth and temporomandibular joint arthritis is shown Table 6-9.

The highest numbers of individuals having both conditions live along the Murray River (CM, MUY, R) as well as men living in the desert. Over 70% of individuals which have one form of either condition also have the other (Fig. 6-12). When those with missing teeth only are plotted (Fig. 6-13), the Murray/Rufus River area is the least effected. In the central Murray males and females have frequencies of 14.2% and 12.0% respectively, as opposed to a combined average of 28.7% for males and 27.2% for females in the areas outside the Murray (SC, DES, TRO and CTL). The difference between Murray people and the rest is highly significant for both sexes ($p=<0.001$). When we look at young and old adults, tooth loss with increasing age tooth loss is less of a problem in the Murray than in areas outside (Fig. 6-14A and B). Bearing in mind that the Murray River has the largest frequencies of TMJ arthritis, Fig. 6-15A and B shows how this affects those suffering tooth loss. Almost all young adults from the central Murray who have lost teeth have arthritis of the temporomandibular joint. Young men from the desert and tropics and young men and women from the east coast are entirely uneffected in this way. Young men from the desert have not lost any teeth (Fig. 6-14A), so they will not register. It is interesting to note, however, that while tooth loss occurs far more often in young women from the tropics than in those from the central Murray, only 25% have arthritis as opposed to 80% of the central Murray women (Tables 6-10 and 6-11). Fig. 6-16A and B underlines the prevalence of TMJ arthritis in the Murray even in those with a full complement of teeth, as opposed to groups outside (see also Table 6-12).

Discussion

It has been suggested in recent years that there is a possible genetic component to arthritic disease (McDevitt and Engleman 1976). Whether this component extends to arthritis in the temporomandibular joint is unknown, but for the purposes of this work such an aetiology will be ignored. It is well known, however, that age is an important factor in the onset of arthritic disease (Jaffe 1975; Adams 1976). This is no less true for arthritic degeneration of the jaw mechanism where, in Europe, it has been found to be rare before the fifth decade of life (Blackwood 1963). With more than 40 or 50 years of use, tooth loss and dental attrition, breakdown of the synovial joint can be expected (Blackwood 1963). This age association is clearly demonstrated among pre-contact Aboriginal people.

Although not completely understood, the breakdown in the dental integrity of the masticatory apparatus, which includes tooth loss, dental attrition and facial trauma, and mechanical overuse in the form of repetitive overloading of the joint complex are two other causes of temporomandibular joint degeneration, although these are not necessarily mutually exclusive. Excess in any of these areas can initiate damage to the joint, followed by arthritic degeneration at an earlier age than would be expected normally. The result of this process is the ultimate destruction of facial structures, malfunction of the jaws and loss of the beneficial use of occlusal integrity: in other words, a maladapted chewing mechanism. The use to which the jaws and teeth are put, therefore, is an important consideration in the analysis of arthritis in any population and may give a better understanding of why it appears in the form and frequency observed. Rates of attrition and tooth loss depend on a number of factors. These include the amount of fibrous food consumed, how much grit or other abrasives are contained in the food, the type of abrasive, how food is prepared, the environment in which people live, the type of food, eating habits (whether bones are consumed, for instance) and to what extent the mouth is used as a third hand or involved in the manufacture of material culture (Molnar 1972; Barrett 1977).

The main finding of this study is that most temporomandibular arthritis occurs along the Murray and south coast. The main difference between these two areas is that tooth loss is far greater on the latter. Age at death is similar for young males and females from coastal Victoria and the Murray but people living at Swanport and around Adelaide, who make up the bulk of the south coast sample, were living longer (Fig. A1). The expectation is that with more old people these populations would have more severe arthritis of the temporomandibular joint. The problem is that although the Adelaide group live longer than those at Swanport, it does not have as much category 2 arthritis as Swanport. It is obvious that the neat area grouping called

SC (south coast), examined profitably in other chapters, is inadequate for the purposes of this study. Adelaide does not have the stress on this joint that Swanport does even though it has a similar proportion of old people. Clearly, coastal Victoria is more like Adelaide than Swanport with respect to severe arthritic disease. The pattern at Swanport, therefore, seems closer to that found in the upper parts of the Murray River. I will return to this matter later.

Desert people do not have the frequency of jaw arthritis that might be expected considering their use of grindstones, fibrous foods and amount of hard, gritty inclusions that enter their food considering the dusty environment in which they live. Indeed, wear and tear of this joint is, as one might expect in any group, related more to age. Even though tooth loss is the highest for young and old females anywhere in Australia, this is not markedly associated with arthritis of the temporomandibular joint. The exception to this is old males, in whom 75% of those with tooth loss have associated arthritis, but again this is to be expected in the old. It seems, then, that tooth loss in this group does not mean an almost automatic susceptibility to arthritic disease of the temporomandibular joint as it seems to along the Murray River.

Severe arthritis of the jaw affects only the old in the tropical group, which is to be expected. Moreover, age onset may explain, in part, the rather high incidence of arthritis in females from Cape York/Cairns who share one of the lowest young adult female death rates in Australia. The association of missing teeth and arthritis is low, with no young males and only one quarter of young females with tooth loss having arthritis also. Again, with the onset of age, both sexes become affected, losing teeth and suffering joint degeneration but not necessarily together.

It is unusual that severe arthritis occurs in more young than old females on the east coast. There seems to be no explanation for this, unless it is just one of the vagaries of dealing with an archaeological population of limited numbers. There is minimal tooth loss in young adults but, as with all other areas outside the Murray, this increases dramatically in older people, for example, half of the old females are getting arthritis. There is no associated temporomandibular joint arthritis in young adults who have lost teeth, although there is a minor association between these two factors in old men. As with other groups this can be regarded as a 'normal' condition in old people.

I want now to return to the Murray Valley populations. Why are Murray people more prone to temporomandibular joint arthritis than those living elsewhere? The obvious answer to this lies is the amount of mechanical stress to which the temporomandibular joint is subjected. It has already been established that in the Murray old age is not necessarily a causative factor in the aetiology of jaw arthritis. Therefore, joint degeneration must be a product of joint use, the frequency of use and overall joint morphology. The latter will, of course, change in relation to the amount of tooth attrition and the teeth involved (Richards 1984). In short, the stress on this joint in Murray River people is greater than elsewhere and any tooth loss amidst such high stress rates will, automatically, render those individuals more susceptible to joint degeneration. Other groups are not immune from these processes but is seems as though they are not operating as strongly in them. Further, the Pleistocene evidence suggests that these stresses may have been operating in the Murray for a long time. This begs the question: would not such stress on the orofacial complex trigger selection towards reinforcement of that complex?

Osteoarthritis: A Selective Stress in Aboriginal Cranial Morphology ?

Selection towards more rugged cranio-facial features was the essence of Wright's (1976) argument to explain why late Pleistocene human remains recovered at Kow Swamp (9-12000 B.P. approx.) were more robust than earlier gracile remains from Mungo (26000 B.P. approx.). With the discovery of robust human remains from possibly a much earlier time period than Kow Swamp, this situation may no longer need explaining (Caddie *et* al. 1987; Webb 1989; Thorne in prep.). Rather than the *selection towards* argument of Wright's to explain the distinctive orofacial morphology of the Kow Swamp and, presumably, the Coobool Creek remains at such a comparative late stage is Australian prehistory, a *persistence by selection* approach might provide a better explanation of the continuation of rugged facial features in this areas of southeastern Australia.

In his metrical analysis of these remains, Brown (1982:308) basically supports Thorne's long held belief that there are a number of features that are ...*morphologically archaic, and indicating a morphological continuum with the Indonesian Homo erectus material* (Thorne 1975, 1977; Thorne and Wolpoff 1981). Brown (1982:300) believes also that over the last 10,000 years there has been ...*an overall reduction in the size of the orofacial skeleton and dentition*. Although there have been no studies to date on the variability of facial morphology throughout Australia, those that have been undertaken on the dentition alone have shown that while there is great continental variation in tooth size, the largest teeth are to be found in the central Murray (Townsend and Brown 1979; Brace 1980; Freedman and Lofgren 1981; Smith *et al.* 1981).

In recent central Murray people extensive and advanced tooth attrition is common, a condition which is believed to have acted as a selection process favouring *...large, slow-wearing teeth,....* (Brown 1982:304). Brown goes on to say: *...it difficult to believe that there has been any relaxation in the selection pressure for large teeth in* [the central Murray] *over the last 10,000 years.* (1982:305).

The persistence of selection for large, hard wearing teeth has, I believe, similarly acted for the retention of teeth also. This seems to be firmly established by the stark contrast in tooth loss between Murray Valley groups and those from outside, and the susceptibility of Murray people who have lost teeth to have arthritis of the temporomandibular joint also. The ability to retain teeth is seen on a smaller scale in Fig. 6-17 which shows the division of the south coast area into its component groups, as already mentioned above. The strong affinity of the Swanport group with the upper Murray with regard to tooth retention is amply demonstrated, while the Adelaide and coastal Victoria groups take on the tooth loss characteristics of other groups around Australia.

It seems highly likely that early Australians were heirs to an ancestral genetic compliment that showed a particular robusticity of facial and dental characteristics and this was particularly adaptive for Murray River conditions. (It is worth noting here that it was the general rugged cranial morphology of the Murray River people that prompted Brown (1981a) to develop a multivariate sexing technique for crania from this area which dealt with this ruggosity. This differed significantly from the discrete trait technique developed and used to sex crania from all over Australia for the previous twenty years (Larnach and Freedman 1963)). The persistence of some of the robust traits was maintained by selection processes acting particularly upon the orofacial complex. The later reduction in the robusticity of some of the dental and facial features, rather than indicating a relief in the processes of selection, might be more easily explained by an increasing frequency of contact with groups having a less rugged facial morphology outside the Murray and gene exchange with them. The somewhat higher tooth loss in old people from Swanport over that generally seen in the upper Murray groups may well be an indication of the size reduction of the adaptive dentition of lower Murray people by gene flow comming up from coastal areas. The difference between the adaptiveness of the dentition of Murray Valley people and outside groups, typified in coastal and tropical peoples, is exemplified by greater tooth loss, smaller teeth and, overall, less rugged orofacial morphologies in the latter. This may be due to genetic 'swamping' of small pre-existing populations by later influxes of genes into Australia representing lighter oro-facial complexes, who kept largely to the coasts (Bowdler 1977), and/or intrinsic population growth in the middle and late Holocene (Lampert and Hughes 1974; Leubbers 1978; Ross 1981; Lourandos 1977, 1983; Webb 1987b).

Wright's (1976) proposition that environmental pressure could have been the trigger for selection towards a robust morphology in the central Murray remains unproven, but the idea that selection pressures could have been that strong and maintained a rugged morphology seem sound. It is likely that the late Pleistocene/early Holocene inhabitants of the Murray Valley already had a genetic endowment from their ancestors that allowed them to cope with such stress. This endowment provided the facial buttressing and dental morphology best able to cope with high stress rates acting upon the orofacial complex and which are exemplified in the frequency of degenerative arthritis in the area. It is possible, indeed likely, that the familiar robust 'Murrayian' facial morphology, which included large teeth, was maintained because of its functional importance in combating the destructive potential of stress met with in the area and this morphology has continued, although in a slightly less robust form and subject to natural proportional reductions which have occurred in all human populations, until quite recently.

I realise I have proposed the above hypothesis while ignoring other factors that I have duly recognized play a part in the cause of temporomandibular arthritis. I feel, therefore, that surveys of the amount and extent of attritional wear and adaptive jaw and joint morphology should be carried out in an effort to understand more about the type of stress acting on this joint in the Murray.

I am, at this point, unable to shed light on the ultimate stress(es) which caused the high frequencies of arthritis in Murray people. It is worth considering, however, the net industry of these river people and the extensive stresses imposed on the jaws and teeth of these people during net manufacture. Nets for hunting and fishing were made from the chewed roots of the bulrush (*Typha* sp.) (Krefft 1862:361-362; Beveridge 1882:42; Gott 1982). Their size varied depending upon their use. Emu nets were 150 yards long with six inch mesh and those for catching ducks were *...usually 100 yards long, by 2 yards deep..* with a mesh of four inches (Beveridge 1882:45). Fish nets were the same length but had a three inch mesh, while those used for crayfish were 10 feet by two yards with a quarter inch mesh. There were variations on these patterns, together with the production of a range of smaller baskets and carrying bags which employed the same processes of manufacture.

Part of the preparation of the bulrush for twine involved *...everybody chewing the roots most vigorously....* (Krefft 1862:361). Over a number of years, the enormous amount of chewing on this fibrous material must have produced mechanical stress on all facial mechanisms, not the least teeth and the temporomandibular joint, in both men and women. I need only quote Krefft further:

> *If we take into consideration the large nets for catching water-fowl in use, it is indeed astonishing how great the perseverance of these people (and how sound their teeth) must have been, and it is not to be wondered at that the possession of one of these nets has always been considered to be a sort of fortune to its owner.* (1862:361-362).

We have no idea how long nets have been used by those living on the Murray River. It seems likely, however, that the findings of this survey with regard to male/female prevalence of temporomandibular arthritis and its early onset may have fairly solid foundations in the stress imposed by this sort of cultural activity. The persistance of a robust oro-facial morphology typical of Murray people which, I propose, includes large teeth with broad grinding surfaces, firmly embedded in deep and well constructed sockets, is an adaptation to the stress encountered in the everyday lifestyle of the population which has persisted for a very long period of time.

Chapter 7

TRAUMA

Introduction

This chapter deals with fractures of the cranial and post-cranial skeleton. In the original survey trephination was included, since then this section has been published separately (Webb 1988).

Fractures of the limbs are almost a natural accompaniment to life. From Permian dinosaurs (Moodie 1923), to apes (Schultz 1939), Australopithecines (Wells 1964), *Homo erectus* (Weidenreich 1939) and more modern human populations (Steinbock 1976; Ortner and Putschar 1981), all have shown evidence of trauma. As Wells has remarked: *Life without injury can hardly be imagined* (1964:45). It is interesting that the first palaeopathological description of any kind, made in 1774 by Esper, was a misdiagnosed fracture of the limb of a cave bear (Jannsens 1970:25).

My primary concern here is with fractures among pre-contact Aboriginal populations, what bones were affected and what interpretations can be made concerning the origin of the injuries. It is hoped that this type of study will reveal also something about the behavioural and social characteristics of prehistoric Aboriginal groups which would not be available from other methods employed in the reconstruction of prehistory.

Methodology

As mentioned in previous chapters, male and female post-cranial samples from the tropics and the female sample from the desert are too small to use to form any firm interpretations. The former samples are omitted from this chapter, the latter has been included only so that a balance is provided for the male sample from the desert. In many locations, sample sizes for individual bones types are below the 20-25 minimum normally used in this study, thus precluding them for analysis by location. Discussion, therefore, proceeds by area only.

Various forms of trauma were recorded on the humerus, ulna, radius, femur, tibia and fibula. X-ray examination is normally considered necessary for positive identification of the type of fracture but this was not possible given limitations on the time available for this study. Many of the original classifications contemplated for recording, therefore, were abandoned. Instead, the distinctive 'parrying fracture' of the left ulna will be the only classified fracture discussed, with all others being lumped together. Moreover, the overall number of fractures was so small that analysis of individual types would have been meaningless. Some examples of post-cranial fractures can be seen in Plate 7-1.

By far the most common cranial injury is the depressed fracture. This usually takes the form of an oval or circular depression and was divided, initially, into two types: <2cm and >2cm at the greatest diameter (Plates 7-2 and 3). These were, however, combined later to form one type, together with crushed or dented zygomatic arches (orginating from a blow to the side of the head) and any linear (split) fractures (Plates 7-4 and 5). Technically, split, linear and circular cranial fractures can all be classified as depressed fractures, so combining all cranial trauma in this way was deemed acceptable. Cranial trauma was recorded by its position on the vault, although in the case of fractures that crossed sutures the it was allocated to the bone most affected it. Four of the six flat bones of the calvarium were used in this way, which included the frontal bone, the left and right parietals and occipital. Recording in this way brought a more intimate undestanding of the cause of injury and the direction from which it came. The recording sheet had eight separate boxes, four each for the left or right bone (or left or right side of the frontal or occipital bones - see Chapter 1, Fig. 2-1). In this way multiple injury could be accommodated.

The 'T1' analysis refers to trauma recorded in the first box, regardless of side, 'T2' the second two boxes and so on. A T1 frequency gives the number of crania from an area or location that have at least one incidence of cranial trauma. The 'T2' frequency represents those crania having a second bone affected (be it left or right) and so on. Only rarely were all four boxes used (T4) denoting trauma on all four cranial bones of the one individual. These separate

Plate 7-1 LEFT:- A well united fracture of the left tibia accompanied by a malaligned fibula fracture. The generally good alignment of the tibia and minimal callus formation suggests splinting and aftercare of the patient together with an opportunity for substantial resting of the limb.

MIDDLE:- A complete fracture of this bone has resulted in its shortening. Little or no reduction of the break has resulted in a shortened arm. Ill-defined muscular markings and diaphyseal wasting point to atrophy, probably derived from minimal use of the arm following the trauma.

RIGHT:- This bone shows a well healed fracture that occurred between the middle and distal thirds of the shaft. This position seems to be the most common site for tibial fracture among traditionally living Aboriginal people.

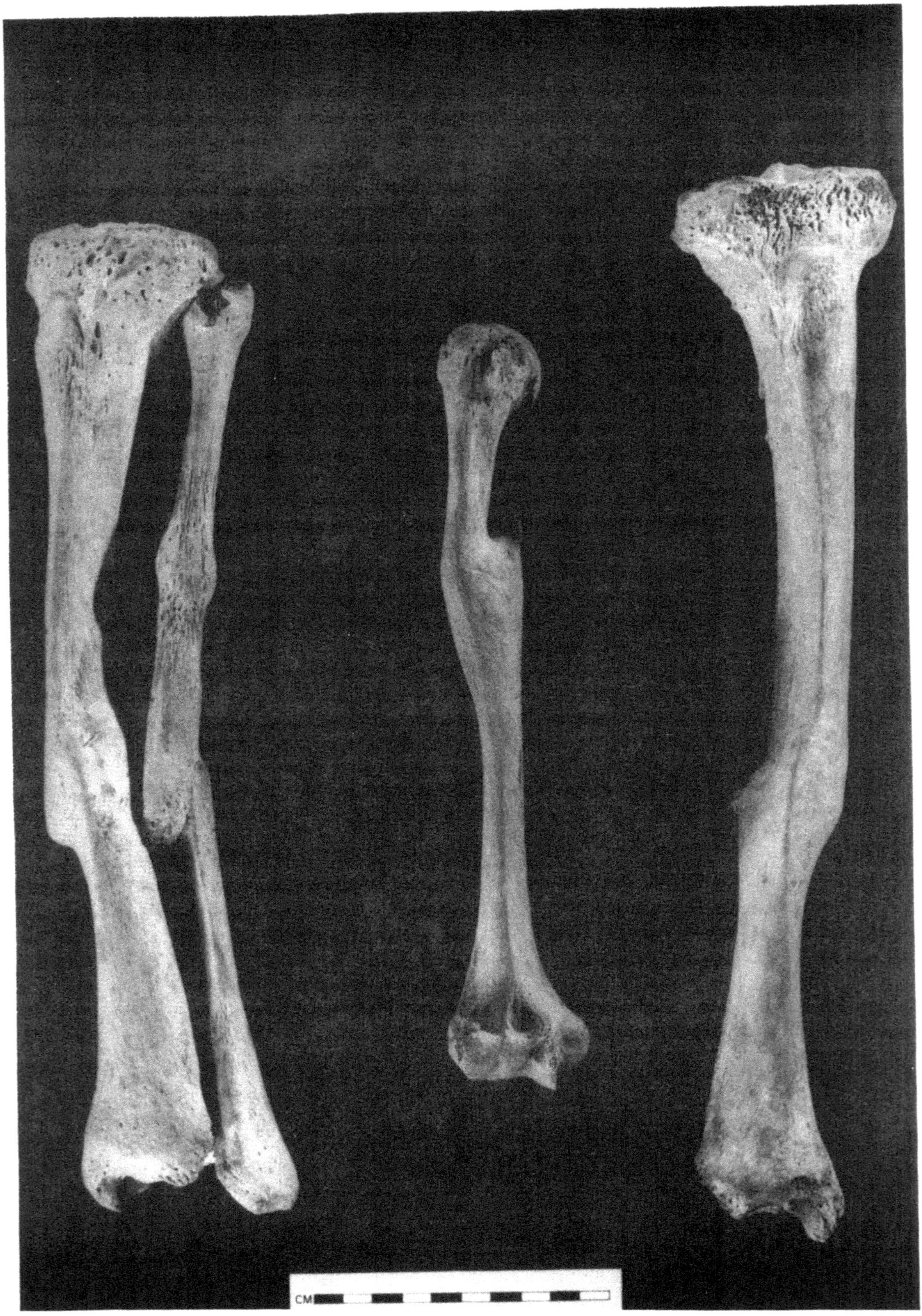
CM

Plate 7-2A — A depressed fracture on the right parietal with a localised periostitis around the wound suggests minor infection possibly associated with damage to the scalp.

Plate 7-3 — The depressed fracture of the frontal bone of this male individual measures 10 mm in depth. No internal or external fracturing of the cranial tables can be detected.

Plate 7-4 — This buckled zygomatic process was probably caused by a blow recieved from a right handed person during fighting. Complete fracturing might have been prevented by the cushening effect of the underlying temporalis muscle.

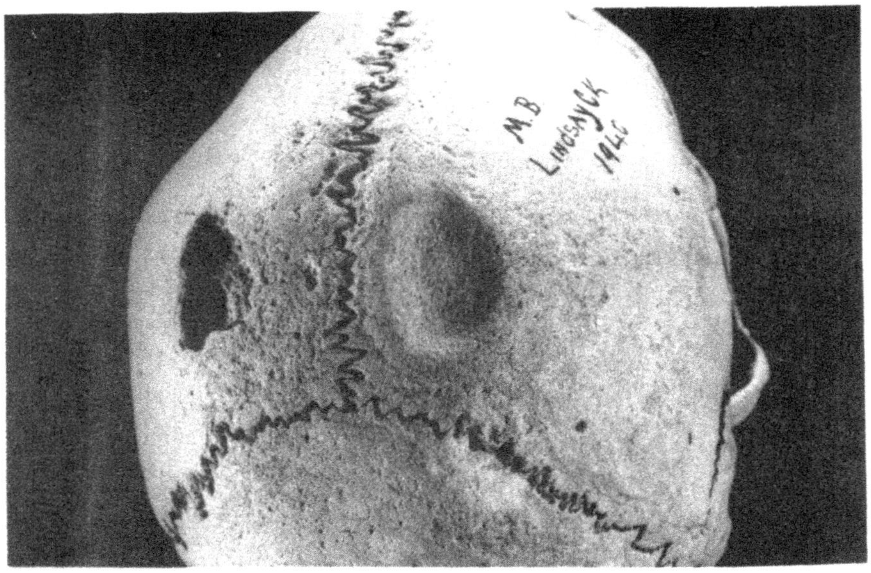
M.B
LINDSAY CK

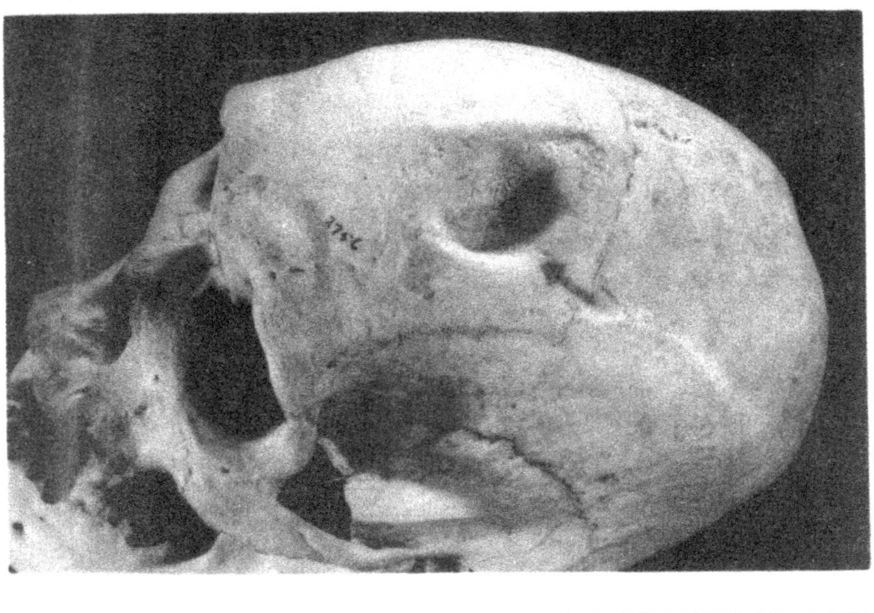

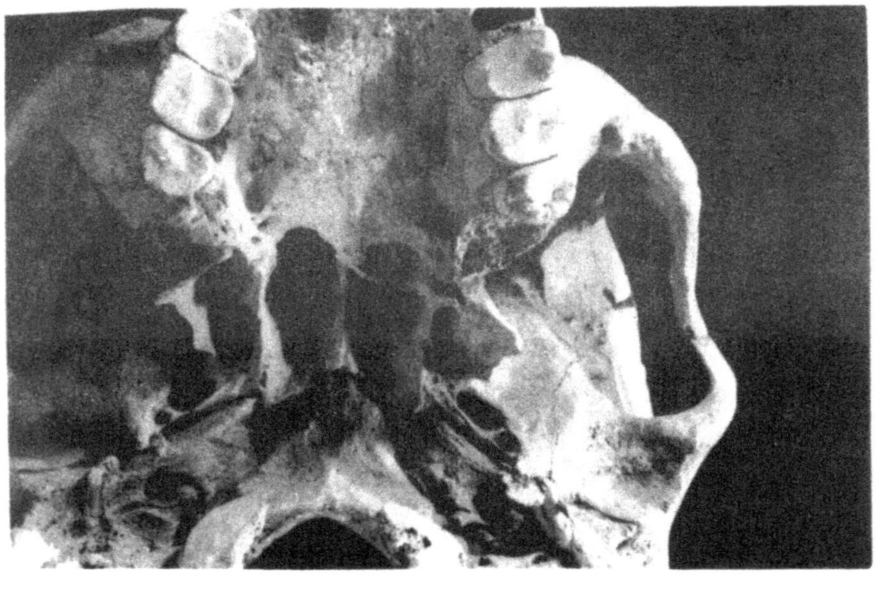

Plate 7-5 The V-shaped split fracture covers a large portion of the left side of the head. Healing has formed a bridge at the lower and anterior borders of the break and testifies to continued life following the trauma.

Plate 7-6 An oblique and lateral view of a pseudoarthrosis of a distal ulna from the late Pleistocene Kow Swamp population. This is likely to have resulted from a parrying fracture that did not heal. Continued use of the arm, without any form of splinting, probably allowed movement of the broken ends, preventing healing.

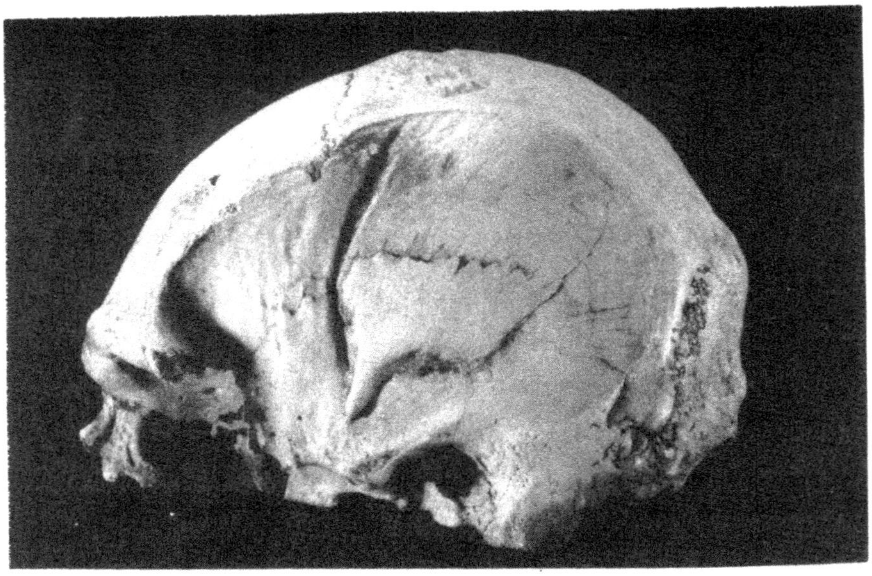

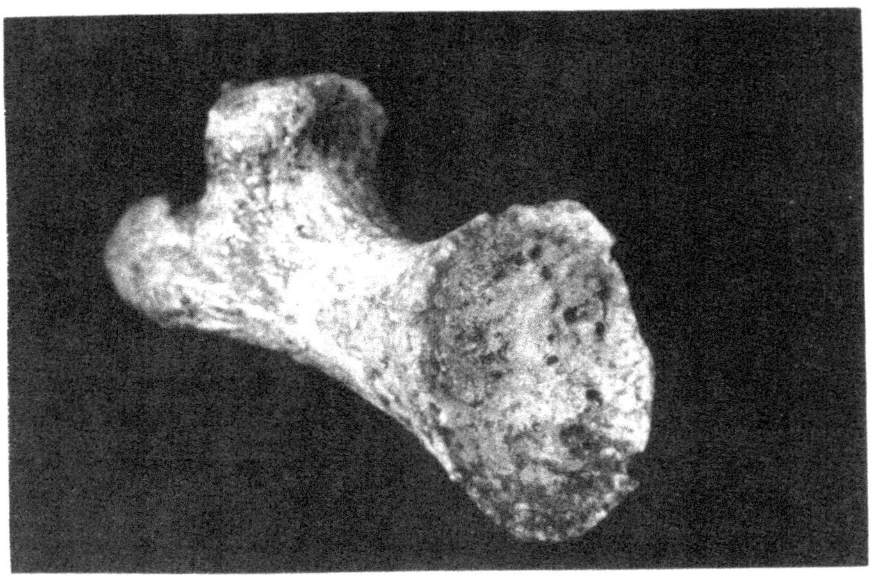

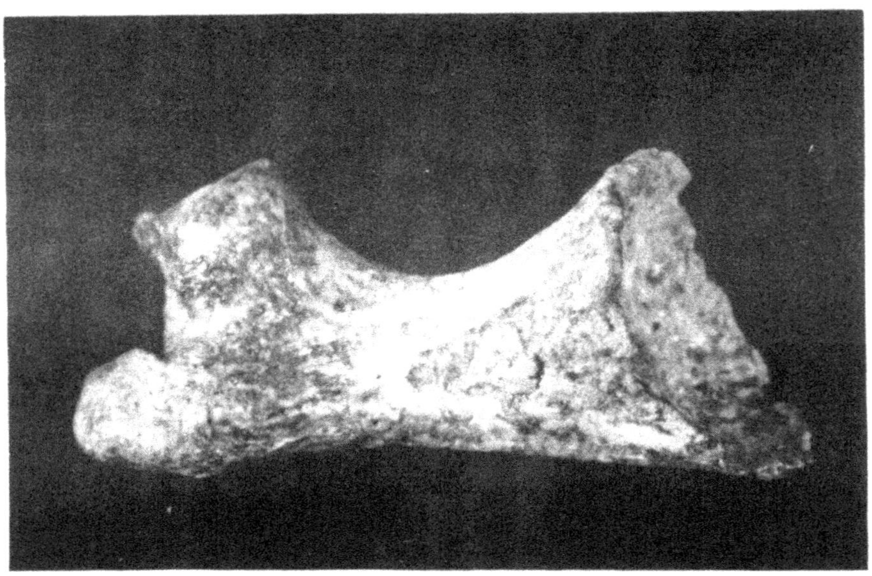

frequencies can be added together to form an idea of which geographical area or location has the most cranial trauma per individual and who was receiving it, rather than merely highlighting how many individuals in a given population were so affected.

Post-cranial Trauma

Results

No classical Colles's or Pott's fractures were detected, which corresponds to the findings of others (MacKay 1938; Wells 1964). Because of the unknown personal age of post-cranial material, there will be no discussion of fractures associated with age, or those resulting from pathological or metabolic causes (Aegerter and Kirkpatrick 1975).

Comparatively speaking, post-cranial fractures are few (Table 7-1). All areas display more fractures of the upper than lower limb and more left than right, which is statistically significant ($p=<0.005$) (Figs 7-1 and Tables 7-1 and 7-2A and B). Upper limb fractures consist largely of parrying fractures. Even so, their frequency is not extremely high in any area and with the exception of desert males and females from the east coast, who have 10.5% and 19.2% respectively, there are minimum differences in thier frequency between areas and sexes. The difference between right and left arms in females is negligible except, again, for those living on the east coast. Because there is a large number of unsexed ulnae from the Murray (MUY), I have included an additional frequency for these in Figure 7-2.

Left arms seem to have mended better (70.7%) than rights (57.7%), which is probably a reflection of the predominant and necessarily continued use of the right arm (Table 7-2B). A similar pattern in the legs is not as easily understood, however. The bones of the lower arm mend with greater alignment than other fractured long bones. The femur and humerus nearly always show a greater propensity for malunification, but these bones are notorious for misalignment because of the massive muscle spasms which take place in the vicinity of the fracture during and after injury and the need for sophisticated nursing (see below). The need for prolonged immobility and some form of traction to avoid malunification is paramount, particularly for the femur.

Discussion

Pleistocene/Early Holocene Material

The Murrabit skeleton has fractures of both lower arms and right leg and the left radius is broken nine centimetres above the distal end. Delayed union has caused a diarthrodial pseudo-arthrosis to form, similar to that noted in one of the Kow Swamp individuals (Plates 7-6A and B). These are the only cases of pseudo-arthrosis observed during the study. The olecranon process of the left ulna was missing pre-mortem leaving, some exostoses medially on the proximal end. There is a healed parrying fracture about 3.5cm above the distal end of the ulna which has left the bone slightly misaligned with a small callus. The right ulna is broken at approximately the same position as the left radius and is surrounded by a large callus of bone. The difference in the size of the callus on these bones could be a product of the continued use of the right arm over the left, which indirectly provides us with a good indicator of the handedness of this man and a similarity of behavioural pattern with modern findings (see above). Some overlap of the broken ends of the right ulna lends further support for this idea. A well healed fracture of the right tibia and fibula has caused ossification of the interosseous membrane, resulting in ankylosis between the two. This process, however, probably aided reasonable unification of the fracture.

The bilateral nature of this man's injuries, together with the overlap of bones in the right ulna and the broken olecranon process suggests extreme violence. These injuries were sustained, perhaps, from a bad fall or a severe beating while he was unable to defend himself. The fact that he lived some considerable time after suffering these extensive injuries suggests that some care must have taken in providing for him both nutritionally and in a nursing sense. After all, a man with two broken arms and a broken leg is not up to hunting for himself.

A description of the trauma in the Mossgeil and Green Gully skeletons has already been published (Macintosh 1967). A more detailed analysis of Mossgeil, including a histological examination of an unusual cortical thickening in the left femur, has also been published (Freedman 1985).

The cranium of Bourkes Bridge 1 (?17,000 B.P.) has a piercing wound just below the boss of the right parietal (Plate 7-7). The resultant wound has produced a 2mm hole through the bone, causing stellate fracturing of the outer cranial table and localized chipping of the bone on the periphery of the wound. The hole itself lies near the groove for the posterior branch of the meningeal artery. Any

Plate 7-7 This head wound on the right parietal of the late Pleistocene individual from Bourkes Bridge was probably caused by a spear point. Note the stellate fracturing radiating from the wound. Small bone fragments around the hole suggest that the trauma occurred pre-mortem while there was soft tissue to hold them in place.

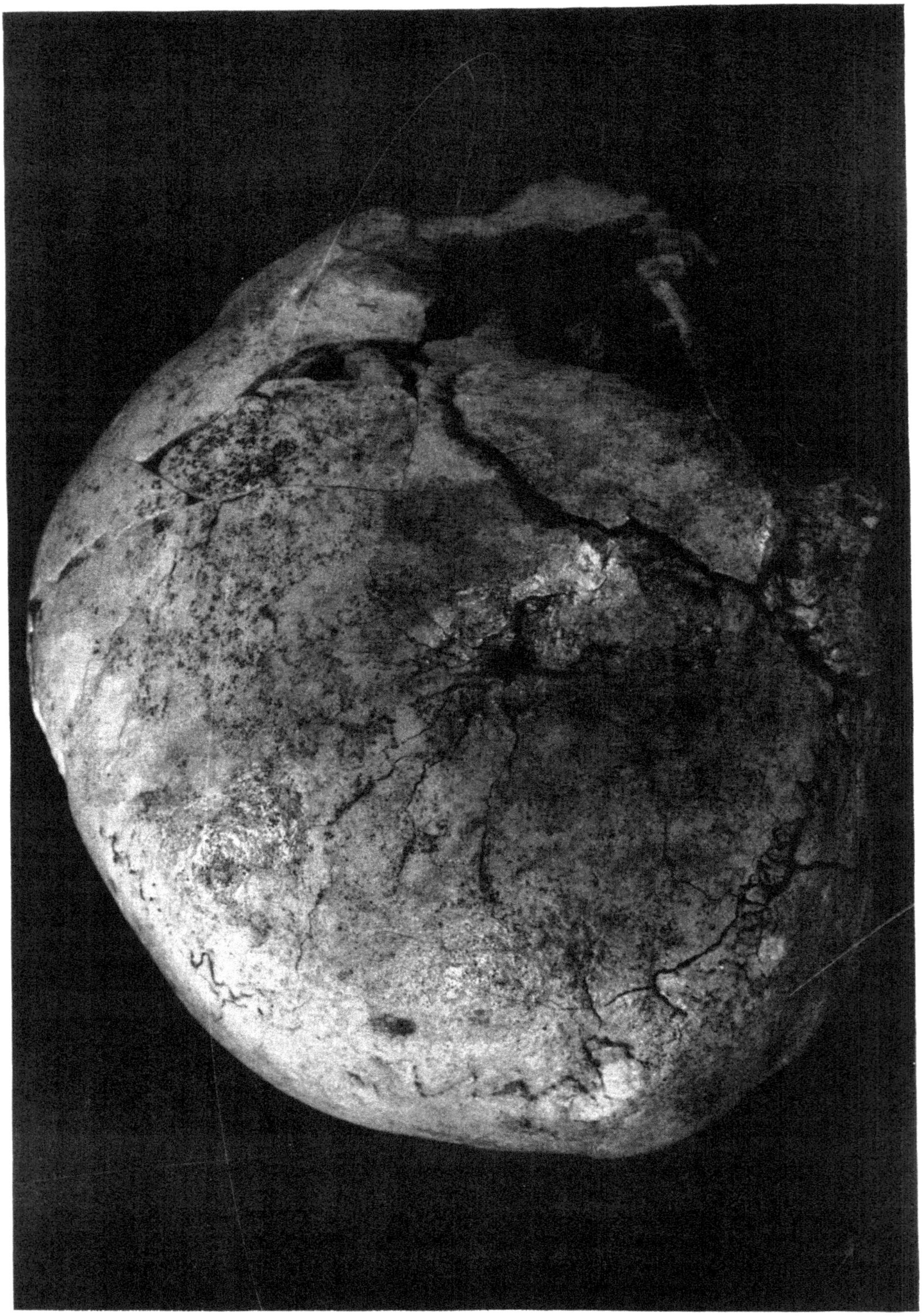

perforation or severing of this vessel would have resulted in cerebral haemorrhage and death. It is uncertain whether this was, in fact, the case, but no signs of healing are visible, indicating that death occurred immediately or soon after injury.

Late Holocene and Recent Material

No overt sign of traumatic injury was found on any juvenile material surveyed.

Although fractures in Aboriginal skeletal material have been noted by a number of researchers (MacKay 1938; Dinning 1949; Macintosh 1967; Wood 1968, 1976; Sandison 1973a, b, 1980; Jurisich and Davies 1976; Prokopec 1979; Freedman 1985), severe malformation following fracturing seems to be restricted to a few examples. Overall, the pattern of injury is biased towards the arms, with 75.9% of all breaks occurring in the upper limb (see Table 7-2B). This is somewhat higher than the 69% revealed in MacKay's (1938:3) study but similar to Dinning's figure of 77.6%, obtained using a sample of 2295 from the South Australian Museum (1949:712). The frequency of upper limb trauma in prehistoric and tribal societies is nearly always greater than for the lower limb (Wells 1964; Clabeaux 1976; Lovejoy and Heiple 1981). The ratio of upper to lower limb fractures in this survey varies with area, which probably reflects a variation in lifestyle and behaviour. The desert group has a similar incidence of upper and lower limb fracturing between upper and lower limbs. Here, 45.5% of all fractures take place in the legs, which is similar to that for North American hunter-gatherers (40.3%) (Lovejoy and Heiple 1981). Fractures are, of course, more likely to occur in people like the desert nomad who roams over large distances in search of food, water and suitable camp sites. Moreover, the considerable hunting activities of desert dwellers, particularly over rocky terrain, would tend to cause more leg fractures than would be incurred by more sedentary people.

Lower limb fracturing is prevalent on the east coast, but five of the six females involved come from coastal Queensland. This is interesting with regard to what has already been shown in the previous chapter concerning the high incidence of knee arthritis in these women. Female activity for this stretch of coast seems to cause large amounts of stress on legs. It is reasonable to assume that 10-15% of any population will sustain leg fractures merely by the vagaries of living. The low frequency for the south coast and particularly the complete absence of such injury in the central Murray does, however, contrast strongly with the incidence at Rufus River, in the desert and on the eastern seaboard. The contrast of frequencies between the southern and eastern coast communities leads to the conclusion that the former led a less strenuous life, particularly with regard to activities that put the lower limb at risk. Perhaps this was the result of a greater degree of sedentism. If such is a valid interpretation, the central Murray has an even greater propensity towards sedentism.

A variety of researchers has argued that trauma patterns are linked to the behavioural, social, economic and demographic disposition of a society. Steinbock (1976:23) has shown that the overall frequency of fractures falls with a move to sedentism and the adoption of agriculture. However, just the opposite occurs at the the hunting/agricultural Dickson Mounds (Goodman et al. 1938). Here adult fracture frequencies rise from 20.5% during the hunter-gatherer phase to 32.4% when these inhabitants became full agriculturalists, although for sub-adults there is an inverse relationship. In western society the incidence of post-cranial fractures is twice that of people living in the ancient eastern Mediterranean (Angel 1974:17).

There is, then, no simple rule for interpreting what fracture patterns represent in terms of behaviour. An essential element in this type of inquiry, therefore, is some prior knowledge of the kinds of activity people were likely to engage in.

For reasons of topography alone it is clear that Aborigines led different lifestyles in different parts of the country and these fundamental behavioural differences put some people at greater risk of, say, leg fractures than others. I have already talked about the strenuous activity of men hunting in the desert and Tindale describes this:

> *Sighting a kangaroo* [a young man] *would chase after it and relying on the curved course taken by such animals the second youth would take up the heavy running in the heat and over the heavy ground.* (1981:1878).

This provides us with reason to suppose that a high frequency of leg fractures in the desert area is a reflection of energetic, nomadic hunting-gathering behaviour. Conversely, the sedentary lifestyle I have reconstructed from other palaeopathological evidence for the central Murray is probably reflected in the zero incidence of leg fractures observed there. Clearly, the result for the central Murray does not mean that broken legs never occurred there, only that they were much less frequent than elsewhere in Australia. This, I believe, is a reasonable conclusion, based as it is on the examination of 400 left and 377 right leg bones from the area. The only people that reflect a high incidence of parrying fractures are east coast females and desert males. This leads to a conclusion that women on the east coast were subject to violence, either from males or from each other (see Cranial Trauma below) and that fighting between desert men might be a more likely origin for this than in groups from elsewhere. Males in all areas do have more fractures of the left than right ulna but in most cases this difference is minimal.

Parrying fractures in males were the commonest fractures found by Angel (1974). Wells (1964:55) has said, however, that:

> *In many peoples known from their history to have been warlike and truculent these* [parrying] *fractures are surprisingly uncommon.* (1964:55).

It would be usual if a man, no matter to what society he belongs, has had a severe disagreement with another at some time during his life even if it did result in a broken ulna. It is difficult to know, therefore, whether a small frequency of parrying fractures is indicative of the degree of aggressive behaviour within a given group or society. Nevertheless, a high percentage must suggest that it is.

Generally, the unification and alignment of broken bones is good. Humeri, as mentioned above, are difficult to set correctly, but exact end-to-end apposition is not believed to be essential for the resumption of natural upper arm function (Adams 1978:132). Only two of the 18 fractured humeri indicate reduction to a near-normal position. On the other hand, while recognising a bias against good unification taking place in the the right arm, 70% of lower arm fractures show good unification.

> *In fractures of the forearm bones accurate reduction is important because even slight displacement may disturb the relationship between radius and ulna, with consequent impairment of the rotation....* (Adams 1978:155).

The high percentage of good results observed in this survey is, I believe, ample proof of the concern Aboriginal people had for fracture repair and their ability to effect and maintain treatment of it, particularly of the arm.

Only two fairly well unified femoral fractures out of six testify to the difficulty encountered in the adequate treatment of this injury (see example of tibial unification on the extreme right of Plate 7-1). A similar situation exists for tibial fractures. If this bone remains unbroken while the fibula is fractured, it can provide a natural splint for the fibula. The reverse does not apply because of the much larger tibial mass.

> *In fresh fractures of both bones of the leg attention should be concentrated solely on the fracture of the tibia. The fibular fracture may be disregarded because it always unites readily, and in any case the bone is of such secondary importance that the position of the fragments is immaterial.* (Adams 1978:249).

Seven of the nine broken fibulae (77.8%) were well aligned and 69.9% of fractured lower arm bones displayed good unification without undue distortion. Stirling (1911:12), when referring to healed fractures noted at Swanport, said that they '...*would have done credit to a skilled surgeon*', while Dinning (1949:712) commented that many of the fractures he examined showed...*anatomical restoration that would be considered good even by modern orthopaedic standards.* Dinning was surprised at these findings because of Love's comment that Aborigines in north western Western Australia were ignorant of splinting (1936:202). Basedow (1904) also observed badly knitted fractures in north western South Australia which had obviously not been treated by splinting of any kind. But both Basedow and Love's experiences appear to be exceptional. The practice of setting broken limbs has been noted in tribal groups from North America, Africa and parts of Oceania, including New Zealand (Ackerknecht 1967). Aborigines were fully cognisant of the techniques of reducing, setting and splinting limb fractures. They performed this in many parts of the continent, although it may not have been universal. Bark bound with vegetable string or sinew was used in the Gulf country (Winterbotham 1951), north eastern New South Wales (Macpherson 1902) and the Gascoyne district of Western Australia (Bates n.d.). Sticks tied tightly with possum or kangaroo skins formed a splint in central Queensland (Roth 1897), Victoria (Anon. 1861) and the Murray district of Western Australia (Bates n.d.). Grass, leaves and clay were applied in the form of a plaster case in north central New South Wales (Parker 1905), central Australia (Giles 1882), on the Murray River (Hoets 1949) and in South Australia (Bartels 1893). In the Northern Territory Aborigines used bamboo splints (Foelsche 1882), while in northwestern Western Australia human leg and arm bones were used on fractures of the respective limb (MacKenzie cited in Winterbotham 1951).

Worsnop illustrates an arm sling used by Aborigines in south eastern South Australia to support a broken wrist (Worsnop 1897, Plate 71). There are other reports verifying Aboriginal knowledge of bone setting and testifying to the adequate results obtained (Anon 1823; Eyre 1845 Vol II:361; Black and Cleland 1938).

Cranial Trauma

Results

The results of this section can be considered in three ways:

(1) how much cranial trauma is occurring in any one area or location;
(2) which locations have the most cranial trauma per individual; and
(3) which part of the cranium is most commonly involved.

Generally women suffer more head injuries than men on both a locational (Fig. 7-3) and area basis (Fig. 7-4). This is the reverse of Angel's findings in samples ranging from the Neolithic to modern American whites. He says: '*Females have fewer fractures throughout, especially in the head and face....*' (1974;17). Some locations show an almost even distribution of cranial trauma between men and women for single bone involvement (T1), while others (SP, A, N, CA, NT, CR and CQ) have disproportionately higher female frequencies (Fig. 7-3). Males rarely have higher frequencies than females and when they do, they are only slightly higher. It is interesting also that in a number of locations the amount of trauma per individual is high (T2 and T3 in Fig. 7-3), particularly for females. Because only four locations have trauma involving a fourth bone (T4), these were not plotted but their frequencies can be seen in Table 7-3.

By area, men and women from the central Murray suffer less cranial trauma than any other area, except males from the tropics (Fig. 7-4), while the highest frequencies are to be seen in males at Rufus River and females from the desert and east coast.

Almost all traumatized crania have some parietal involvement (Table 7-4). Generally, this type of injury occurs more on the right in females and on the left in males (Fig. 7-5, Table 7-4). The exceptions to this are males from the tropics, who have no parietal damage, and females from the central Murray, who follow the male pattern. The second most affected bone is the frontal, which accounts for 20-25% of all head injuries (Fig. 7-6). With regard to occipital trauma, there are large differences between areas, most noticeably between Murray river groups and those from other parts of the continent (Fig. 7-7).

Discussion

Cranial injury was not uncommon in many prehistoric and tribal societies (Moodie 1923; Wells 1964; Courville 1967; Jannsens 1970; Steinbock 1976; Ortner and Putschar 1981) and has been noted in Aboriginal skeletal material (Morton 1845; Wood 1968; Prokopec 1979). There is, however, very little reference in the general literature, and none in the Australian, to possible patterns of head injury and, if they exist, how they differ between groups and between males and females (Angel 1974). There are,on the other hand, descriptions of the weaponry used by various societies which could cause head injury (Courville 1942, 1951, 1952). The fact that clubs, maces, cudgels, axes, sling shots, boomerangs, digging sticks and other weapons can inflict trauma on the head is not surprising and, indeed, hardly needs illustrating.

Rather than provide mere description of cranial trauma and the offending weapon, it seems more worthwhile to try and give some interpretation of what the trauma might mean in terms of tensions within and between groups. It is my contention that the variance in the orientation of cranial trauma may be able to tell us something of its origin. This type of analysis has, as far as I am aware, not been explored to any extent and, because of this, my conclusions are tentative.

Head injuries on central Murray women indicate that frontal assault is more common than that from behind. The higher frequency of injury to the left parietal is evidence of a blow from a right handed person when the recipient is face to face with an assailant. In other regions the situation seems to be reversed. They, without exception, show more right parietal damage, that attacks from behind were more common, although females from the east coast approach an almost even distribution of parietal trauma. In addition attacks from behind increas the chances of occipital trauma and Fig. 7-7 clearly shows this to be the case.

The highest incidence of female cranial trauma occurs in the desert. This mainly involves the right parietal and occipital, with few injuries to the frontal. The most obvious interpretation is that this pattern is the result of attack from behind, probably originating in domestic squabbles. The pattern is different in the central Murray, where females have more frontal and less occipital injury than those living in the desert. This is further support of previous evidence for frontal attack being common amongst these women and could indicate that female head trauma in the central Murray originates from arguments among women. Equally, this could be said to apply to the east coast, but here the larger frequency of female trauma overall, together with the even distribution of parietal damage and substantial frontal and occipital involvement, suggests head battering comes from all directions. The tropics show a similar trend to this, although there is less occipital damage. The greater degree of violent behaviour among east coast communities could indicate a generally higher degree of friction there than in the central Murray. Coastal Queensland has more individuals affected than coastal New South Wales, but the extent of damage per individual is greater in the latter location. There is, then, some balancing out between these two coastal areas.

Certainly, it is females living along the east coast who have, by far, the highest frequency of parrying fractures anywhere in Australia, including males although this difference is not statistically significant (Fig. 7-2). Moreover, with the exception of the central Murray and Rufus River, the highest death rates among young adult females occur along the east coast (Fig. A2). Female cranial injury on the south coast is among the highest there is, but most of this takes place in the Swanport and Adelaide groups. Again, domestic argument, denoted by right parietal and occipital injury, seems to be the main type of aggression, although frontal bone trauma indicates assault from this direction also. This pattern, however, do not reflect aggression to quite the same extent as the patterns obtained from the east coast sample.

The fracture patterns at Rufus River are almost identical for men and women in all areas of the cranium. Men do not have trauma of the occipital but follow the general trend for males around Australia in having more injury to the frontal bone and left parietal. Violence in this area, therefore, seems to be distributed rather evenly between the sexes and extensively per individual affected (see Fig. 7-3). Attacks from behind being rare, frontal confrontation seems most likely, perhaps in the manner of warfare with both sexes involved. The predominance of left parrying fractures in the men lends some support to this. The combination among south coast men of high left parietal and frontal injury witht a small overall amount of head injury points to belligerent acts restricted to minimal bouts of hand-to-hand fighting. The small amount of injury to the occipital and right side of the cranium adds further support to this conclusion. Parrying fractures in these men point to the same pattern, as they do in Rufus River men. In desert men the even pattern of right and left parietal damage and the added high frequency of occipital involvement suggest all round attack, from the front and the back, but the degree of head trauma is, overall, lower than for Rufus River and the east coast. The frequency of frontal assaults in desert men is emphasized by them having the second highest frequency of frontal bone trauma and the highest frequency of parrying fractures.

I cannot offer any explanation for the low frequency of cranial trauma in males from the tropics. With one frontal and five occipital injuries, all that can be said is that aggression between men, in the form of close hand-to-hand combat, must have been minimal.

Discounting pathologically induced fractures, broken bones are a direct indication of accident or violence. In the case of post-cranial bones, only the parrying fracture can be directly attributed to aggressive behaviour. A circular injury to the cranium severe enough to cause a lasting depression in the skull can usually be regarded as the result of deliberate aggression, especially when it is positioned superiorly on the calvarium. Of course, tripping backwards over a log can cause occipital injury and banging the side of the head on the back wall of a rock shelter can cause parietal injury, while demonstration of grief by striking the head with stones and other objects, as practised by some Aboriginal groups when in mourning, can cause injury to frontal areas (Krefft 1862:363; Meggitt 1962:125; Berndt and Berndt 1972:455; Tonkinson 1978:84; Elkin 1979:338-339).

These alternative causes of head injury are not ignored here. Rather, they are accepted as being less likely to have caused the depressed fracturing and denting of the zygomatic arch observed in this survey. The reasons for this belief are that almost all depressed fractures were small and round (between one and three centimeters in diameter), typical of that made by a blow from an instrument with a small striking surface. Moreover, although certain types of accidents and falls can produce serious cranial fracturing, these are normally associated with industrial societies (Miller and Jennett 1968). The average accidental 'bang on the head' derived from everyday living leaves very little, if any, trace on the bone, although the actual incidence of this type of injury is hard to assess because if no other clinical signs transpire there will be no formal report or record made. At the same time, the expertise and agility of hunter-gatherers, within an environment known in detail by them reduce the chance of accidental cranial injury, such as those suffered in modern society.

Chapter 8

DISCUSSION

The six preceding data chapters have each dealt with a skeletal pathology or pathological indicator and its frequency among Aboriginal groups living in different parts of Australia. I want now to reverse this approach and discuss the cumulative pathology in each area. By drawing the pathologies together, a clearer picture of the ecological, biological, social and demographic circumstances of each regional population will become more apparent.

The Desert

There are no major surprises arising from the skeletal sample I have termed the 'desert' group. The lowest incidence of anaemia in Australia for both children (30%) and adults (8.3%) demonstrates that the causal factors of this condition (population aggregation and parasitism) posited for other groups are minimal. Although young children are affected to some degree, there is a rapid reduction in the frequency as adulthood is reached. The dental hypoplasia survey confirms the overall reduction in stress as children grow. While under six's from all parts of Australia suffer stress to some degree, both boys and girls in the desert have less than those those living elsewhere. Relief from the stresses of infancy arrives early for this group also, as indicated by the very small frequencies of dental hypoplasia of the third molar. The ever decreasing number of Harris lines with increasing age lends further support for the general alleviation of stress in these people, although the overall adult index is in the middle of the range. The explanation of the apparent contradiction posed by the evidence of Harris lines lies in their aetiology, which is probably associated more with acute stress as opposed to cribra and hypoplasia, which are indicators of a more long-term or chronic stress.

These results show that life in an arid environment is hazardous only for the very young but no worse overall than for infants in other areas: it is probably better in fact. There is a notable reduction of stress from early to late childhood. Periodic episodes of acute stress, as indicated by the Harris lines, emphasise times of famine often encountered in the desert (Peterson 1978; Tindale 1981; Long cited in Nathan and Jabanangka 1983:73). To offset this precarious situation, desert people have learned to utilize a wide range of food resources, though shortages still occur (Sweeney 1947; Lawrence 1969; Cane 1984).

Water is the crucial element in desert life. Too much rain may flood large areas, isolating both fauna and people and drowning vegetable foods. Too little water confines people to those places where permanent water is available in the form of springs and/or wells. During these times hunting away from permanent water becomes non-viable, game dies and the necessity to stay near a water source causes the hunting-out of an area. Moreover, high temperatures force people into inactivity in an effort to conserve body fluids and energy because of the low hunting return for energy expended. As Tindale (1981:1878) reminds us, even the first rain may cause stress: *The first rains bring starvation as birds and mammals disperse.* But he goes on to say: *Soon... insects appear and birds begin to breed providing eggs and tiny morsels of chick as food.*

It is precisely a short-lived stress of this type followed by a period of recovery that is believed to initiate the formation of Harris lines. The fact that over 70% of individuals surveyed have these lines indicates that the population as a whole is affected and that general living conditions are the cause.

The desert group shows the highest frequency of non-specific infection of post-cranial bones but, as I have discussed above, this is probably due to an endemic form of treponematosis (Nathan and Jabanangka 1983). The epidemiological factors necessary for this bone infection to be genuinely 'non-specific', such as population aggregation and sedentism, are not part of the desert lifestyle. Moreover, if these factors existed they would have shown up in higher frequencies of anaemia (cribra) in adults and children and a persistence of stress during childhood in the form of higher frequencies of third molar hypoplasia.

Osteoarthritis is frequent in the knees and elbows of men (the female post-cranial sample is too small to analyse). A nomadic existence which involves extensive hunting activity may have more to do with this than anything else. The equal incidence of fractures to the upper and lower limbs may also reflect the activities of a nomadic hunter. It is worth noting, however, that more men achieve old adult status in the desert than they do in any other group and this may be a factor in the observed incidence of arthritis in the major joints.

The pattern of cranial trauma in both sexes and the frequency of parrying fractures in men highlight aggression, to the extent that female cranial trauma here is worse than for any other group. Such behaviour, however, cannot be said to affect mortality rates among young women, because these are the second lowest among all samples. The cause of early death in desert women is probably different from that in young women from other areas. From this survey there seems to be less metabolic and physiological stress among desert women than those from from the central Murray. If this is a true reflection of their status, then we have little by which to judge the desert situation and to explain early death in young women from this area. Moreover, to make a direct extrapolation of the results of female cranial trauma to the early death of young females is probably misleading. The evidence for violence in the desert group does, however, suggest that the socio-economic balances needed in a generally harsh environment could create social stresses within and between groups leading to the trauma observed.

There must have been some variation in the degree of nomadism in various groups. The nature of the desert sample, however, is not such as to reveal any of the inter-regional differences that probably existed across this extremely large area which has, for the sake of expediency and sample size, been treated as one large population. Nevertheless, the palaeopathological findings do fit a pattern of high mobility and low population concentration.

There seems to be a paucity of data world wide concerning the palaeopathology of desert-dwelling hunter-gatherers. Therefore, these results cannot be directly compared with those of a similar nature from elsewhere. Because of this I must assume that the data presented here reflect this type of lifestyle. It is my contention that we have now a basic palaeopathological template for desert-living hunter-gatherers and a measure of the stress which effects these people. Therefore, in the light of the view often held that the desert living Aborigine is the archetype of all Aborigines and, for that matter, all hunter-gatherers, I will use these results as a kind of yardstick against which the results from other areas of the continent may be compared. I am aware of the differences in sample size between the various areas to be compared but feel that the smaller samples are still generally representative of the areas from which they come.

The Central Murray

In dealing with the central Murray I am moving to what is clearly the other end of the palaeopathological spectrum. The central Murray consists of samples from Euston and Balranald in the west to Echuca and Kow Swamp in the east, a stretch of river 350km long.

Even on a world scale the people living in this region have extraordinary high rates of anaemia. The children have only marginally less (59.3%) than those from Rufus River (62.5%), a difference which, at this level of frequency, is inconsequential. Hypoplasia of the third molar emphasises the persistence of stress throughout childhood in both males and females and this picture is supported by the low, if steady, amelioration of chronic anaemia as children grow.

As noted above, the stress causing hypoplasia, while not necessarily correlated with that precipitating anaemia, is, nevertheless, a measure of the general stressful conditions existing within the society. It is, therefore, valid to use both these indicators as mutually supportive on that general level. The limits on recovery from anaemia are adequately demonstrated by the persistence of the condition into adulthood. It remains high, 33.7%, over four times the frequency in desert dwellers.

The frequencies of cribra and hypoplasia might suggest the likelihood of a very high incidence of Harris lines also. Instead, the incidence of these lines is not very different from that found in some other areas. While at 1.64 the index of morbidity is the second highest, it is far below the 3.18 of the south coast. One explanation for this could be that stressful conditions exist more or less constantly in the central Murray and the formation of lines was limited because the recovery phase needed to produce them was irregular, infrequent or non-existent for extended periods. We know that there was seasonal variation in food resources and weather patterns along the Murray River and these must have produced times of general hardship and, possibly, nutritional insufficiency for many (Allen 1972). Although a number of individuals received their lines at weaning, some later nutritional or metabolic insult affected an equal number of individuals and for the population as a whole constancy of stress probably acted to prevent the continual formation of these lines. Thus while superficially the frequency of Harris lines is similar to that for the desert, the reason why they were produced, or, more correctly, why they were not produced, is completely different.

In view of such results, I introduced earlier the concept of an 'infectious environment' existing in the central Murray. It follows, therefore, that it should be no surprise to encounter non-specific bone infection there. With the exception of the desert, the highest percentage of people so affected do live in the central Murray, although some instances of non-specific bone infection may in fact be due to treponemal disease (see Chapter 5).

The fact that men and, in particular, women die at an earlier age in the central Murray than they do elsewhere suggests also that age is not necessarily a factor in the high incidence of arthritis in both men and women. The wear and tear on joints, therefore, indicates a pattern of hard and constant work, with the upper limbs bearing the brunt of physical activity. Supporting this hypothesis for high stress loads on the upper limbs is the complete absence of leg fractures. This contrasts with the equal distribution of upper and lower limb trauma seen in the desert group.

The osteological evidence for aggression indicates that social tensions must have been of a low order. Patterns of head trauma show that women probably fought among themselves, but domestic violence is comparatively less obvious than it is elsewhere. Fighting between men is also minimal, judging by the patterns of head injury and the low incidence of parrying fractures. High death rates among young adults from this part of the Murray might, then, have less to do with violence than might be the case, for example, among desert people. A more likely explanation is that the human ecology of the central Murray has created an infectious environment which over time causes contributes to mortality in the population. The strength of these adverse conditions is displayed in the high frequencies of cribra orbitalia and dental hypoplasia and inversely in the case of Harris lines. This situation could only have been compounded by the work regime already mentioned.

Major factors contributing to the build up of an infectious environment are usually population increase and aggregation, overcrowding, demographic confinement and sedentism. These results are considered not to have resulted from episodic nutritional deprivation, seasonal hardship or the effects of group/family isolation on small portions of high ground during flooding, although these, no doubt, contributed. Rather they are believed to be a clear indication of the close proximity and aggregation of people living in the central Murray prior to 1788 which was a direct result of steady population increase during the last two thousand years (Webb 1988b). While the palaeopathological data cannot, by itself, fully support these conclusions, they are hard to explain in other terms. If anything, they emphasise the need for a complete reappraisal of the prehistoric lifestyle and palaeodemography of the central Murray and a full archaeological study which I alluded to in the introductory paragraphs to this chapter.

Archaeological work is only now coming to grips with the prehistory of the central Murray. Preliminary research has revealed that oven mounds were built in large concentrations; at least 95 occur in one particular 2 square kilometre area (A Berryman in prep.). The mounds range in size from eight to 48m in diameter, with 85% averaging around 18m, and rise from 0.5m to 1.5m above the ground. Spatial analysis has revealed that they were built with a deliberate regularity, as though personal space was conciously sought. They are believed to be communal cooking ovens owned by a group of families, where up to one ton of *Typha* sp. was cooked at one time (also see Curr 1886). Observations by early settlers recount that a fishtrap located near a mound would be owned by the group who occupied the mound. Such mounds are found over an area of nearly 20,000 square km, but they are only in dense concentrations, as far as is known, in the central Murray. None has as yet been dated earlier than 4000 B.P. and the dates that are available suggest that those farther away from the central Murray area are younger.

This work is obviously of great importance for the future understanding of the human ecology and demography of this part of the Murray and other large river systems such as the Lachlan and Murrumbidgee. While there are difficulties in the use of archaeological evidence to imply population size (Hassan 1981), the biological evidence presented here would support the interpretation of the archaeological indications in terms not only of a larger population than was previously considered but also high levels of sedentism and cultural organisation.

Finally, it is possible that a diet high in carbohydrates, which can be inferred from extensive use of the root of the *Typha* bullrush and the existence of the ovens used to cook it, would increase the chances of nutritional stress. It was noted also during this survey that there was more dental calculus on the teeth of people from the central Murray than elsewhere. Calculus is usually formed by the consumption of large quantities of sticky foodstuffs such as carbohydrates, without adequate removal of residues from the teeth. Future microscopic and biochemical analysis of this calculus may enlighten us further about what food plants were eaten, by whom and in what proportions. If high carbohydrate consumption did occur among these people, it could partly explain some of the palaeopathological findings presented here. Could such a diet be a further indication of high population densities? Obviously, an answer to this must wait for further palaeobiological analysis.

The South and East Coasts

The incidence of anaemia (cribra orbitalia) is lower in both these areas than for the Murray. The highest frequencies are found along the New South Wales (31.5% in adults) and Victorian coasts (28% in adults), but there is greater relief from childhood stress on the south coast than the east. In the latter area the incidence of dental hypoplasia in the older group of sub-adults points to a persistence of stress. While this is not of the same magnitude as in the central Murray, it could be interpreted in a similar way - due to high populations. The major difference between this coast and the Murray River, however, is that while the range of subsistence resources is similar (Bickford 1966; Lawrence 1969), a coastal larder remains relatively plentiful and varied all year round and is not subject to drastic seasonal alteration, droughts and floods.

Poiner (1976:191) argues that this 'plentiful and variable' larder would still have had a low point of species availability during winter, when shellfish, fish and plant categories were reduced and that while:

> *...this change in availability of all food resources both in quality and periodicity is not marked it is probably sufficient to induce an attendant change in food procurement strategy.*

Some examples of this species reduction are: shellfish, from nine in February to five from April to May inclusive; fish, from 14 in February to seven from June to September inclusive; and plants, from 20 in November to 10 in July (Poiner 1976:191-193). Even with resources halved, this situation does not force people to rely on one or two carbohydrate staples in the same way that central Murray probably had to.

Poiner (1976) depends heavily on early observations along the south coast of New South Wales for evidence of seasonal variability in resources and the corresponding movement of people to procure them. In the light of possible population disruption and depletion along the New South Wales coast following the 1789 smallpox outbreak at Sydney Cove (Collins 1798 Vol 1:53-54,496; Butlin 1983), these references may not be as useful as first thought. Seasonal movement inland and along the coastal fringe has not been archaeologically demonstrated (Poiner 1976; Sullivan 1982). While Poiner believed mobility to have been one response to the seasonal variability of resources, this must have differed between the north and south coasts of New South Wales. Sedentism is thought to have arisen from a littoral-based economy on the north coast, at least as far as Moreton Bay in Queensland (Hall 1982; Sullivan 1982). Historical evidence for large, solidly built, permanent dwellings centred around extensive fish traps along the north coast would support this idea (Coleman 1982).

On the south coast, Poiner has proposed that life for people revolved around 'semi-nomadism' in spring, summer and autumn and 'full nomadism' in winter, undertaken by smaller groups (1976:201). It has been contrarily argued, however, that people would be inclined to remain closer to a littoral economy, for which also they were specialized, and perhaps take advantage of the variety of micro-environments located near coastal home bases (Lampert 1971:130). Presumably this would mean a more sedentary lifestyle than that proposed by Poiner. Evidence for a sedentary lifestyle among south coast people similar to that observed along the north coast, however, has yet to be presented.

The overall uncertainty concerning human response to 'seasonal stress' on the south coast of New South Wales might be enlightenend by the biological data presented here. The low incidence of Harris lines among these people suggests either that seasonal stress did not occur to any extent, or if it did Aboriginal strategies for coping with it were excellent. This means that whatever the seasonal strategy was, there is no skeletal evidence for seasonal shortage. There is, however, evidence for some considerable degree of sedentism. Cribra (anaemia) should not be as prevalent as it is, with the high levels of iron freely available in birds, fish and particularly shellfish (a frequent component of the diet) all regular components of the diet along this coastline. It is possible that while the evidence for shellfish consumption is overwhelming, dense populations might reduce the amount consumed per person and, in addition, predispose people to greater chances of parasitism and anaemia.

Therefore, while coastal New South Wales and the central Murray share some aspects of the palaeopathological spectrum, the differences seem to lie in the comparative all-year-round abundance of the New South Wales coastal environment. This would have tended to ameliorate a stress status somewhat similar to that of the central Murray, which for both these populations may have had a common origin in their respective demographies.

The picture for the south coast (SC) shows that persistent stress in the form of cribra orbitalia and dental hypoplasia is of a low order for the Adelaide/Swanport section but the Victorian component presents a similar picture to that of coastal New South Wales. The evidence for seasonal stress, however, quite possibly leading to malnutrition, is more prominent for the whole of this area than for the east coast or, for that matter, anywhere else in Australia. A comparison of the 0.8 Harris lines per person found on the east coast and the nearly 3.2 per person in this area quickly identifies the major difference between these two coastlines. In addition, if stress is seasonal, it should affect most of the population and this is what we see here, with 80% of individuals having Harris lines, whereas only 43% of people on the east coast have them. Most of the latter are boys under six years and it is worth remembering that at least one line might be caused by weaning, which has already been noted as a physiologically traumatic time for toddlers anywhere and particularly for males. The presence of repetitive seasonal stress on the south coast is further emphasised by nearly 70% of men and women having more than four lines, whereas no individual in the coastal New South Wales sample has as many as four. What, then, is the reason for such a major difference between adjacent coastlines?

The southwestern districts of Victoria have, in recent years, been the centre of a number of archaeological studies and subsequent hypotheses concerning Aboriginal economic intensification, sedentism, and social complexity (Coutts *et al.* 1976, 1978, 1979; Lourandos 1976, 1977, 1980, 1983; McBryde and Harrison 1981; Williams 1985, 1988). While it has been postulated that seasonal stress was experienced in this region (Coutts *et al.* 1978; Lourandos 1980), it has also been postulated that *...there is no record of extreme scarcity in any season* (Lourandos 1980:249). The evidence from Harris lines now renders this view open to question. Lourandos accepts that *...unpredictable variations (e.g. droughts) would have had a serious effect on resources* (1980:249), but such sporadic events would not account for the regular periodicity of Harris lines observed among people living in this region. As Coutts (1978:37) and others have commented concerning the estimation of population size, we *...require access to biological data that do not exist at the present time.* Hayden (1975:18-19) also emphasises that the osteological biology of archaeological populations is a necessary factor in establishing resource over-exploitation and, thereby, indirectly telling us something about population size. It is clear, now we have begun to look at such evidence, that populations from this area seem to have been under a persistent stress, but of an order that does not obscure the formation of Harris lines which have been formed by regular seasonal shortages.

My interpretation of these biological data is that the notion of sedentism claimed from the archaeological record is supported by the biological results seen here. Moreover, the fairly large population implicit in the intensification argument is also likely to have existed given the frequency and combination of pathology noted . Further, it seems reasonable to propose, as others have done, that the economic intensification inherent in the construction of extensive water channels and fish and eel traps was a strategy born out of the necessity to exploit resources to the full by extending energy extraction potential. This in itself suggests that the people in the area were living close to the ceiling of existing food procurement strategies and that, periodically, they may have needed to go beyond it. Regular seasonal stress in the form of food shortages must also have added impetus for economic intensification.

Luebbers (1978) has suggested that patterns of land, campsite and resource use along the eastern Coorong (South Australia) point to sedentism and clustering of people in the late Holocene. This is similar to the proposed changes taking place in western Victoria at about the same time. With such demographic changes, he argues, would have come an increasing use of marginal resources, which would, in turn, have worked to *'...intensify exploitation of all available resources in the swamps and the coast* (1978:307). He goes on to say:

> *With downturns occurring in food supply and Aboriginal populations being forced to exploit a comparatively limited set of resources, the need to equalize energy flow throughout the year became stronger. This is not only because of the seasonal flux in supply but because access to foods in times of stress were [sic] being denied by neighbours and friends alike.* (1978:307)

Although Head (1983) has criticised Luebbers' interpretation of the archaeological record, the biological evidence presented here, particularly the Harris lines, supports it. If it is possible to use the palaeopathological data as a kind of measure of the magnitude of population size, however coarse this might be, it could be suggested that the population of the Victorian coast was probably less than that of New South Wales and certainly less than the central Murray, but larger than those in the centre and north of the continent.

The pattern of osteoarthritis among Queensland females shows that they are working much harder than women or men from southern coastal groups. In fact, there is no arthritis in females from the New South Wales and south coast groups, although the latter is a small sample. Biomechanical stress is minimal in all male samples, but those from Adelaide are particularly small. The complete bias towards male right arm involvement might support the belief that spearfishing off rocks was at least as important as other forms of hunting, particularly on the New South Wales coast (cf. Lampert 1966; Lawrence 1969; Bowdler 1975). The pattern of arthritis observed here is in complete contrast to the bilaterality seen in Murray men and its frequency is negligible. All this serves to confirm something Lawrence (1969:143) has already noted from the ethnography:

> *The division of labour in the southeastern coastal areas appears to have differed significantly from the pattern commonly encountered in inland areas and also from that in some other coastal districts. While the collection of vegetable foods and shellfish was still the responsibility of the women and fishing and hunting that of the men, the women did spend a considerable part of their time fishing,....and men were seen collecting shellfish....and vegetable foods....*

Unfortunately, the palaeopathological evidence tells us little about the amount of work done by either sex, the exact form work took and what activities were confined to whom. Taking note of the negative evidence, one might, nevertheless, presume that whatever these were, the tasks of resource procurement were not nearly so rigorous on the coasts as they were in the central Murray.

Overall, the level of aggression, particularly domestic violence, is high. There is, however, less fighting among men on the south coast than on the eastern coasts, but far more than is evident in the central Murray. Initially, it could be argued that these results would make more sense if they were reversed. It would be easier to explain aggression between and within groups obviously under substantial stress, such as those in the central Murray. At this point I can only suggest that if there were social mechanisms to cope with antagonism resulting from demographic and/or environmental stresses, substantial aggression might have been avoided. These mechanisms could have taken the form of very large gatherings, within which mock battles and fights could take place and disputes between groups or individuals settled using a ceremonial form of redress. If such mechanisms were in place in the central Murray but not to the same extent along coastal New South Wales, this might explain the differences in patterns of trauma between these two areas.

In summary, it is reasonable to conclude that the eastern seaboard imposes a regime of persistent stress similar to the central Murray but environmental circumstances allow relief in such a way that the overall palaeopathological picture is somewhat better. Death rates among young women living along the Queensland and New South Wales coasts are similar and high, but they are very high for coastal Victoria (the second highest anywhere). On the other hand, deaths among young women from Adelaide and Swanport are low. Judging by the pattern of trauma on females from coastal New South Wales and Queensland, death by violence, domestic or otherwise, might be proposed as one cause of their demise, together with the effects of an accompanying persistent nutritional or physiological stress, while the early death of young females in coastal Victoria originates from the accumulated effects of seasonal stress, which probably reflect dietary insufficiencies. This may be contrary to the belief that: *...it is unlikely that Aboriginals in any part of Victoria died from malnutrition or starvation....* (Coutts *et al.* 1978:39), but it is worth considering in the light of these results.

Rufus River

As with the central Murray region, a many children from Rufus River suffered anaemia but adult relief was greater (24.4%), particularly for women. One reason for the latter finding may be that these people experienced a greater component of nomadism among their groups. This might have led to irregular or infrequent menstruation which provided additional relief in terms of their general physiological situation which came on top of that derived from reaching adulthood. Small children seem to be generally healthier than those in other regions. Some disparity arises between older girls and boys, but overall anaemic stress in this group is far less than in the same group from the central Murray and rivals the situation of east coast children.

The pattern of Harris lines in the Rufus River area actually tells us very little about whether seasonal or continual stress is taking place. Both these can be invoked, depending upon the interpretation of the index of morbidity (1.3). With some of the highest frequencies of anaemia, it could be argued that as metabolic and/or nutritional stress is high and constant, the number of Harris lines will be low. A similar argument was proposed for the central Murray. Frequencies of dental hypoplasia, however, show that stress persistence is not a feature of this area as it is further up river in the central Murray, meaning there should be no barrier to the formation of Harris lines. However, less than

46% of individuals are affected, which is similar to the east coast but contrasts with over 70% in the desert. This can be taken as an indication that seasonal stress is not as great a factor in life at Rufus River as it is in the desert.

I conclude that the population of this rather arid area was larger than would be expected normally in a region of lower rainfall, with a generally arid area ecology and subsequent extremes of nutritional variability. This relatively high population or, more correctly, high aggregation, was probably due to the effects of demographic compaction around riverine and lacustrine resources, such as the Murray and lower Darling Rivers, along smaller tributaries and creeks like the Lindsay and Rufus Rivers and around Lake Victoria. A dependence on these water sources, of which the Murray was probably the most reliable, was fraught with stress at times of low river and lake levels when food resources became scarce. Kefous (1983:167) argues that at such times people would have resorted to the 'hinterland' and 'ecotone' (between river and hinterland) in the north and the Mallee in the south (Ross 1981). While the ecotone may have been exploited on a more regular basis, movement into the hinterland was probably a last resort because of the poor availability of water. Kefous says that *...in times of drought the hinterland would have been more intensively utilized* (1983:167), but it is hard to believe that people resorted to the hinterland during drought. Surely this area would be the worst place to be in terms of food resources rather than close to the main river systems, in which, even at times of low water level, resource extraction is easier. Moreover, even limited water availability would entice people to remain near rivers, while the uncertainty of the hinterland at such times would tend to attract and concentrate game to and dissuade humans from moving away from it. Furthermore, group contraction or aggregation is usually the response to drought, where permanent residency around dwindling water holes, soaks and springs becomes a necessity (Cane 1984). Here is what Long (1983:72, cited in Nathan and Jabanangka) has to say about these times: *Here [at permanent water].... tribes congregate in dry times. It is to these that the Pintubi retreat in dry periods like the present when the waters further East fail.*

The clustering of people along water courses, particularly during dry times, and some dietary insufficiency could cause the frequencies of anaemia seen in the region while population dispersal, following drought relief, would help prevent the build up of parasites and the creation of a generally infectious environment. Thus the chances of persistent stress occurring among these people are considerably reduced, as in the desert groups. Harris lines at Rufus River seem only to imply some stress of an intermittent nature; they probably formed during a major river failure such as the one that can *...occur at least once in a human generation, and possibly twice* (Kefous 1983:167), or some similar event.

The lifestyle of the people in this area would, by necessity, have had a large component of continuous competition among bands and/or tribes for resources, either to protect those which they had or to relieve others of them. Such a situation would have undoubtedly involved a great deal of animosity and inter- and intragroup belligerence would have been high. This would explain why men here are second only to those from coastal New South Wales for head trauma and the amount of head injury per individual. Deaths among young men are high (the highest on an area basis). While this offers one explanation for the prevalence of violence among these people, the rather low frequency of parrying fractures does not.

The incidence of osteoarthritis points to high activity levels, especially in men. Like most inland groups, these are far higher than those encountered in coastal people and are commensurate with the kind of lifestyle outlined above. Women, on the other hand, have less overall osteoarthritis than coastal females and far less in the elbow than central Murray women. This may be linked to the high death rates among young females which would have prevented many from reaching an age at which they would become susceptible to this disease. This picture is quite opposite to that for the central Murray, however, where the early death among young women there has not prevented them having high rates of arthritis.

The Tropics

As noted earlier, the sample size for this area is small, particularly for post-cranial remains, which are almost non-existent. Therefore very little information concerning osteoarthritis, limb fractures, non-specific infection and Harris lines could be obtained. This situation severely limits what can be said about precontact Aboriginal health in the area and, of course, restricts comparisons with other regions.

Toddlers suffer stress in the tropics to the same degree they do in other areas. The main difference here is that the incidence among girls is far below that found in boys, something reflected in the desert and Rufus River groups also. This may just be the inherent susceptibility of males at this age to stressful events and the nutritional vagaries of weaning. It is, however, worth comparing female frequencies of canine hypoplasia in this group (about 16%) with those of the central Murray (42%), in which the frequencies for boys and girls are nearly equal. Anaemia is prevalent in young children (50%), but this probably stems from helminthiasis, including hookworm, and the semi-sedentary nature of groups both on Cape York (Thomson

1939; Chase and Sutton 1981; Hynes and Chase 1982) and coastal Northern Territory (Meehan 1982) during the wet season. But compare the reduction in frequency on reaching adulthood (13.7%) with that of the central Murray (33.7%).

The composition of the 'tropics' sample does not allow me to make too much of the results, as it represents the whole of northern Australia. It combines communities, for example, which may have led very different lifestyles and had different sized populations.

The only observation I feel justified in making is that people of the north are, generally, healthier with respect to environmental stress than those in the south. Deaths among young males and females are almost equal, which is something not recorded for any other area, and may reflect survival chances for both sexes are more equal than they are in the south, where they are particularly biased against females. On the other hand, it could be that there are mortality factors independent of gender which balance the population. Only more representative samples from different locations across tropical Australia will provide better data and more precise interpretations.

A Note on Aboriginal Palaeodemography

This survey has shown a diverse pattern of stress and other pathology across the continent. Inter-regional variations in stress and trauma emphasise radical differences in lifestyle and behaviour between groups which are so different that I believe there is a need for a reappraisal of the concept of prehistoric Aboriginal society in terms of its palaeodemography, nutrition, socio-economic status and some aspects of its biology. I want to emphasise that these results point strongly to permanent biological pressures on some Aboriginal populations inconsistent with the comparatively stress-free lifestyle often associated with the hunter-gatherer and which has been demonstrated in this study among desert dwellers (cf. Sahlins 1972).

I have intimated throughout this work that population pressures in some areas of southeastern Australia may have been to blame for the frequency of stress found there. Indeed, these findings have prompted me to explain further how sedentism, population aggregation and large populations in this area may have arisen (Webb 1987b). The model thus developed concentrates on the central Murray region and uses palaeopathological, physiological and reproductive data to show how substantial population growth could have occurred during the last two to three thousand years.

Alternative explanations for the presence of these stresses in the central Murray, and to a lesser extent among coastal New South Wales people, could be that Aboriginal society was maladapted to a hunter-gatherer lifestyle; or that the Murray River and the coasts were not as resource-rich as has been thought; or that environments were lacking essential trace elements and minerals; or it could be a combination of all these. Present evidence, however, does not support such conclusions. The findings of this survey have no adequate explanation outside that already offered.

CONCLUSION

The emphasise on mere description of disease in palaepathological work has long been abandoned. The reporting of an unusual injury is generally recognised as of little value unless it says something also about the society in which it occurred or it is exceptional in its appearance. For the last 20 years at least the role of palaeopathological investigation has become one of learning about the health and wellbeing of past populations in terms of changes to their subsistence strategies and human ecology. Recent improvements in palaeopathological methodology now provide better data than ever before concerning behaviour, demography, nutrition, socio-economic patterns, biological adaptation and the cultural strategies of past societies. This would not be possible using non-biological data. It is hoped that this work helps provide new insights into Australian prehistory on many of these topics.

Small sample sizes have prevented a number of inter-regional comparisons that would have been desirable. Examination of bones from northern, western and central regions of Australia, while providing some insight into the incidence of stress in these areas, are of limited use and reliability because of this, particularly given the large regions they represent. Nevertheless, they have shown that there is less stress in those areas than might otherwise be assumed on general environmental grounds.

Despite the inadequacy of temporal control for many of the samples used here, results can be compared through time even though we do not know exactly what that span of time is. By comparing the Pleistocene and later groups it can be seen that a more severe form of temporomandibular arthritis; a greater incidence of anaemia (cribra orbitalia) and persistent stress appears, together with a greater frequency of bone infection, in the latter. These are basic trends over a very long time, but they are a real reflection of changes taking place during the Holocene, at least in the central Murray. It is suggested that these changes are due, at least in part, to socio-demographic factors consistent with population growth, intensification and sedentism.

The diverse range of results that have emerged from examination of the more recent samples has focussed attention towards reassessment of Australian prehistory, at least in the southeast of the continent. The prominent and unique pattern of stress detected in the central Murray region underlines the need for such a reassessment of the type of Aboriginal society that existed there. The data from this group strongly suggest that sedentism and population aggregation were extensive and the population was much larger than has been suggested until now. There seems no other suitable explanation for the patterns of palaeopathological findings presented here.

Results from coastal New South Wales are, to a large extent, similar to those from the central Murray. Nevertheless, coastal habitation required less work to obtain an adequate diet, it provided all-year-round nutrition and a somewhat lower population density overall. The differences in pattern of stress between the Queensland Victoria on the one hand, and New South Wales on the other, emphasises the need for further investigation of precontact Aboriginal society along the New South Wales coastline. Queensland people have little sign of any stress. Victoria provides the change of emphasis from the persistent forms, seen in the central Murray and coastal New South Wales, to those of an accute nature - seasonal shortage, for example.

Tasmania has not been discussed to any great extent because of the rather small cranial sample available for examination and the total absence of post-cranial remains. All that can be said is that stress among these people was minimal. This probably reflects little or no sedentism and a low population density.

Certainly the biological evidence presented in this study seems to to separate the mainland groups in the southeast from those in the north, west and centre of the continent.

Spatial variation in stress patterns shows that Aboriginal society was heterogeneous, at least in terms of its demography. There is support, for example, for recent arguments that precontact Aboriginal populations in southeastern Australia were large, but variously so. Moreover, the data point to population aggregation and sedentism for centralised areas of the Murray River in particular. The arguments presented must be tied in, inevitably, with those concerning Aboriginal intensification in the late Holocene. Although these results are not conclusive with regard to when and where 'intensification' occurred, in the light of the quantitative and qualitative stress differences between the southeast of the continent and elsewhere and between areas within the southeast, one cannot avoid the conclusion that moves towards an intensified socio-economic system could have been a catalyst for demographic changes which inevitably led to the stress observed there.

Well healed trauma in acient skeletal remains highlights the nursing and surgical skills possessed by people in the past. The examples described here are valuable in as much as they tell us more about the care and concern Aborigines showed for their sick. Moreover, they are important to any discussion concerning the inability of such societies to support the infermed or severely injured or their inability to administer nursing care.

It has been proposed that stress on the jaws and teeth of central Murray River people has probably maintained a selection for the retention of teeth. The exact nature of this stress is yet to be accurately determined, but the use of the mouth for preparation of fibre must have partially contributed to this. This slection hypothesis supports the conclusions of others who have proposed that the persistence of a rugged morphology and big teeth and jaws well into the mid-Holocene among Murray people has been a product of continued selection from an ancestral gene pool expressing a robust morphology. There may now be some evidence for such selection in the form of physical stress on the oro-facial complex.

This survey is the first of its kind in Australia. As a preliminary investigation it raises more questions than it can answer. Much more palaeopathological/ palaeobiological work needs to be undertaken throughout Australia. More detailed research is required on the age at which dental hypoplasia occurs on the canine teeth. Patterns of vertebral arthritis should be examined to provide more information about work patterns in various parts of the continent. New approaches to palaeobiology and palaeonutrition which involve the study of bone remodelling dynamics, bone growth and density, cortical bone thickness, bone histology and other microstructural studies should begin here. The use of chemical and isotopic analysis of skeletal tissue has helped in our understanding of palaeonutrition and these sorts of studies on dental calculus from such areas as the central Murray may also prove invaluable for forming clearer conclusions.

While palaeopathology in any holistic sense is only just beginning in Australia, similar surveys should be conducted for Melanesia and Oceania generally. The results would create a framework of palaeopathological data for the whole region, including Indonesia and Southeast Asia, and more clearly identify Australia's position both regionally and globally. Such an exercise will throw light on the palaeoepidemiology of the region and, possibly, provide insights into past prehistoric migrations, using certain diseases and stresses as markers. What diseases came with and confronted the first people on these shores? Was there unknown, indigenous diseases here? perhaps there were zoonoses among the marsupial fauna which proved hostile to humans? What is the history of disease introduction prior to Europeans and what effect might such introductions have had on the Aboriginal population? These question, and many more, may never be answered, but without palaeopathological and palaeoepidemiological research 'never' will be the only reality.

REFERENCES

Anon 1823 On the Aborigines of New Holland and Van Diemen's Land. *Medical and Physical Journal* (London), 50:180-191.

Anon 1861 Surgery of Australian Aborigines. *British Medical Journal* 2:518.

Abbie, A.A. 1950 Closure of cranial articulations in the skull of the Australian Aborigine. *Journal of Anatomy* 84:1-12.

Abbie, A.A. 1960 Physical changes in Australian Aborigines consequent upon European contact. *Oceania* 31:140-144.

Abbie, A.A. 1966 Physical characteristics. In *Aboriginal man in South and Central Australia.* pp9-45. Government Printer, Adelaide.

Abbie, A.A. 1969 *The Original Australians.* A.H. and A.W. Reed, Sydney.

Acheson, R.M. 1959 Effects of starvation, septicaemia and chronic illness on the growth cartilage plate and metaphysis of the immature rat. *Journal of Anatomy* 93:123-130.

Acheson, R.M. 1960 Effects of nutrition and disease on human growth. In *Human Growth.* ed., J.M. Tanner. Pergamon Press: New York, pp.73-92.

Ackerknecht, E.H. 1953 Palaeopathology. In *Anthropology Today*, ed. Kroeber, A.L.. University of Chicago Press. pp120-126.

Ackerknecht, E.H. 1967 Primitive surgery. In *Diseases in Antiquity*, eds. Brothwell, D.R. and A.T. Sandison. Charles C. Thomas: Springfield, USA.

Acsadi, G. Y. and J. Nemeskeri 1970 *History of Human Life Span and Mortality.* Ackademiai Kiado: Budapest.

Adachi, B. 1904 Die Orbitae und die Hauptmasse des Schadels der Japaner. *Z. Morphol. Anthropol.* 7:379.

Adams, J.C. 1976 *Outline of Orthopaedics.* Churchill Livingstone: London, New York.

Adams, J.C. 1978 *Outline of Fractures.* Churchill Livingstone: London, New York.

Aegerter, E. and Kirkpatrick, J.A. 1975 *Orthopedic Diseases.* W.B. Saunders Company, Philadelphia.

Aksoy, M., Camli, N. and Erdem, S. 1966 Roentgenographic bore changes in chronic iron deficiency anemia. A study of twelve patients. *Blood* Vol.27, No.5:677-686, May 1966.

Alexandersen, V. 1967 The evidence for injuries to the jaws. In *Diseases in Antiquity* , eds. Brothwell D.R. and A.T. Sandison, Charles C. Thomas: Springfield, Illinois. pp623-629

Allen, F.J. 1969 *Archaeology and the History of Port Essington.* Unpublished PhD thesis, ANU.

Allen, H.R. 1972 *Where the Crow Flies Backwards.* Unpublished PhD thesis, ANU.

Allison, M.J., D. Mendoza and A. Pezzie 1974 A radiographic approach to childhood illness in precolumbian inhabitants of southern Peru. *American Journal of Physical Anthropology* 40:409-416.

Alpagut, B. 1979 Some palaeopathological cases of the ancient anatolian mandibles. *Journal of Human Evolution* 8:571-574.

Anderson, J.A.D., J.J.R. Duthie and B.P. Moody 1962 Social and economic effects of rheumatic diseases in a mining population. *Annals of Rheumatic Diseases* 21:342-352.

Anderson, R.M. and R.M. May 1979 Population biology of infectious disease. Part I *Nature* 280:361-367, Part II Nature 280:455-461.

Angel, J.L. 1964 Osteoporosis: Thalassemia? *American Journal of Physical Anthropology* 22:369-371.

Angel, J.L. 1966 Early skeletons from Tranquillity, California. *Smithsonian Contributions to Anthropology* 2:1-15.

Angel, J.L. 1967 Porotic hyperostosis or osteoporosis symmetrica. In *Disease in Antiquity* , eds. Brothwell, D.R. and A.T. Sandison Thomas: Illinois, pp.378-389.

Angel, J.L. 1974 Patterns of fractures from Neolithic to modern times. *Anthropoligiai Kozlemenyek* 18:9-18.

Angel, J.L. 1979 Osteoarthritis in prehistoric Turkey and medieval Byzantium. *Henry Ford Hospital Medical Journal* 27(1):38-43.

Arundel, J.H., I.K. Barker and I. Beveridge 1977 Disease of marsupials. In *Biology of Marsupials* , eds. B. Stonehouse and D. Gilmore. University Park Press.

Asada, T. 1924 Uber die Entstehung und pathologische Bedentung der im Rontgenbild des Rohrenknochers am Diaphysenende zum Vorschein kommenden 'parallelen Querbinienbildung'. *Mitt. Med. Fak. Univ.* Kyushee Inknoka 9:43-95.

Attenbrow, V. 1982 The archaeology of Upper Mangrove Creek catchment: research in progress. In *Coastal Archaeology in Eastern Australia*, ed. S. Bowdler. Proceedings of the 1980 Valla Conference on Australian prehistory.

Aufdermaur, M. 1957 The morbid anatomy of ankylosing spondylitis. *Documenta Rheumatologica* January 1957.

Bahr, P. 1914 Notes on yaws in Ceylon, with special references to its distribution in that island and its tertiary manifestations. *Annals of Tropical Medical Parasitology* VIII:675-680.

Barrett, M.J. 1957 Dental observations on Australian Aborigines: tooth eruption sequence. *Australian Dental Journal* 2:217-227.

Barrett, M.J. 1969 Functioning occlusion. *Annals of the Australian College of Dental Surgeons* 2:68.

Barrett, M.J. 1977 Masticatory and non-masticatory uses of teeth. In *Stone Tools as Cultural Markers* , ed. R.V.S. Wright, Prehistory and Material Culture Series No.12,pp.18-23. Australian Institute of Aboriginal Studies: Canberra.

Barrett, M.J. and T. Brown 1966 Eruption of deciduous teeth in Australian Aborigines. *Australian Dental Journal* 11(1):43-50.

Barwick, D.E. 1971 Changes in the Aboriginal population of Victoria, 1863-1966. In *Aboriginal Man and Environment in Australia*, eds. Mulvaney, D.J. and J. Golsen. ANU Press: Canberra. pp218-315

Basedow, H. 1904 Anthropological notes made on the South Australian Government north-west prospecting expedition, 1903. *Transactions of the Royal Society of South Australia* 28:12-51.

Basedow, H. 1932 Diseases of the Australian Aborigines. *Journal of Tropical Medicine and Hygiene* 35:1-32.

Bates, D.M. n.d. *Diseases, remedies, death and burial and disease remedies.* Typescript copy of manuscript Section 10, 2a, 2a(1), 2a. (a) entitled: Diseases of natives before white contact, idiocy, deformities, and names for diseases.

Beaton, J.M. 1982 Fire and water: aspects of Australian Aboriginal management of cycads. *Archaeology Oceania* 17:59-67.

Berndt, R.M. and C.H. 1981 *The World of the First Australians.* Revised edition, Lansdowne Press: Sydney, London.

Berryman, A. n.d. MA thesis in preparation. Division of Prehistory, La Trobe University, Melbourne.

Beute, G.H. 1979 Arthritis: a radiologic overview. *Henry Ford Hospital Medical Journal* 27(1):24-31.

Beveridge, P. 1883 Of the Aborigines inhabiting the great lacustrine and riverine depression of the lower Murray, lower Murrumbidge, lower Lachlan and lower Darling. *Journal of Proceedings of Royal Society of New South Wales* 17:19-68

Beyron, H. 1964 Occlusal relations and mastication in Australian Aborigines. *Acta Odontologica Scandinavica* 22:597-678.

Bickford, A. 1966 *Traditional Economy of the Aborigines of the Murray Valley.* Unpublished BA (Hons) thesis, University of Sydney.

Billington, B.P. 1960 The health and nutritional status of the Aborigines. In *Records of the American-Australian scientific Expedition to Arnhem Land.* Vol.II:27-59, Anthropology and Nutrition.

Binns, R.T. 1945 A study of diseases of Australian natives in the Northern Territory. *Medical Journal of Australia* (1):421-426.

Birdsell, J.B. 1957 Some population problems involving Pleistocene man. *Cold Spring Harbour Symposia on Quantitative Biology* 27:47-69.

Black, E.C. and J.B. Cleland 1938 Pathological lesions in Australian Aborigines, central Australia (granites) and Flinders Range. *Journal of Tropical Medicine and Hygiene.* March 1 1938:1-15.

Black, F.D. 1975 Infectious diseases in primitive society. *Science* 187:515-518.

Blackwood, H.J.J. 1963 Arthritis of the mandibular joint. *British Dental Journal* 115:317-326.

Blackwood, Sir R. and K.N.G. Simpson 1973 Attitudes of Aboriginal skeletons excavated in the Murray Valley region between Mildura and Renmark, Australia. *Victorian National Museum Memoirs* 34:99-150.

Blakely, R.L. 1980 Sociocultural implications of pathology between the village area and mound C skeletal remains from Etowa, Georgia. *Tennessee Anthropological Association* 5:28-38.

Bluhm, G.B. 1979 Gout and pseudogout. *Henry Ford Hospital Medical Journal* 27(1):14-17.

Blumberg, B.S. (ed.) 1961 *Proceedings of the conference on genetic polymorphisms and geographic variations in disease* . Grune and Stratton: New York.

Blumberg, B.S., K.J. Bloch, R.L. Black and C. Dotier 1961 A study of prevalence of arthritis in Alaskan Eskimos. *Arthritis and Rheumatism* 4(4):325-41, August 1961.

Blumberg, J.M. and E.R. Kerley 1966 A critical consideration of roentgenology and microscopy in palaeopathology. *In* *Human Palaeopathology*, ed. S. Jarcho. Yale University Press: New Haven and London.

Borah, W. 1964 America as a model: the demographic impact of European expansion upon the non-European world. *Actas y Memorias del XXXV Congreso International de Americanists*, Vol.3:173-205. Mexzico: Editorial Libros de Mexico.

Boserup, E. 1965 *The Conditions of Agricultural Growth.* Aldine Press: Chicago.

Boule, M. 1911-13 L'homme fossile de La Chapelle-aux-Saints. *Annals de Paleontologie* 6:109-172.

Boule, M. 1923 *Fossil Men.* Oliver and Boyd: Edinburgh.

Bourke, J.B. 1967 A review of the palaeopathology of the arthritic diseases. In *Diseases of Antiquity* , eds. Brothwell D.R. and A.T. Sandison. Charles C. Thomas: Springfield, Illinois.

Bourke, J.B. 1971 The palaeopathology of the vertebral column in ancient Egypt and Nubia. *Medical History* 15:363-375.

Bowdler, S. 1975 Hook, line and dilly bag: an interpretation of an Australian coastal shell midden. *Mankind* 10:248-58.

Bowdler, S. 1977 Coastal colonisation of Australia. In *Sunda and Sahul*, eds. Allen, J., Golson, J. and Jones R. Academic Press 1977:205-246.

Bowdler, S. 1981 Hunters in the highlands: Aboriginal adaptations in the eastern Australian uplands. *Archaeology in Oceania* 16:99-111.

Bowdler, S. 1983 *Archaeological Investigation of a Threatened Aboriginal Burial Site.* A report to the Victorian Archaeological Survey, March 1983.

Bowers, W.F. 1966 Pathological and functional changes found in 864 pre-Captain Cook contact polynesian burials from the sand dunes at Mokapu, Oahu, Hawaii. *International Surgery* 45:206-217.

Bowler, J.M. and A.G. Thorne 1976 Human remains from Lake Mungo. In *Origin of the Australians*, eds. Kirk, R.L. and A.G. Thorne. AIAS Publications, pp.127-138.

Brace, C.L. 1980 Australian tooth-size clines and the death of a stereotype. *Current Anthropology* 21(2):141-146.

Breinl, A. 1912 Report on the health and disease in the Northern Territory. *Bulletin of the Northern Territory*, No 1.

Breinl, A. and M.J. Holmes 1915 Medical report on data collected during a journey through some districts of the Northern Territory. *Bulletin of the Northern Territory*, No.15.

Breinl, A. and H. Priestley 1915 Boomerang leg. *Journal of Tropical and Medical Hygiene* XVIII:217.

Britton, C.J,C. 1963 *Disorders of the Blood.* J. and A. Churchill Ltd.: London.

Britton, H.A., J.P. Canby and C.M. Kohler 1960 Iron deficiency anemia producing evidence of marrow hyperplasia in the calvarium. *Pediatrics* April 1960, pp.621-628.

Brodelius, A. 1961 Osteoarthritis of the talar joints in footballers and ballet dancers. *Acta. Orthopedica Scandinavica* 30:309-14.

Brothwell, D.R. 1963a The macroscopic dental pathology of some earlier human populations. In *Dental Anthropology* , ed. D.R. Brothwell. Pergamon Press: Oxford. pp271-288.

Brothwell, D.R. 1963b *Digging Up Bones.* British Museum: London.

Brothwell, D.R. 1970 The real history of Syphilis. *Science Journal* 6-7(1):27-32.

Brothwell, D.R. 1976 Further evidence of treponematosis in pre- European population from Oceania. *Bulletin Historical Medicine* 50:435-442.

Brothwell, D.R. 1977 Human population history, from Afro-Asia to the new world, and its possible relationship to economic change. In *Hunters, Gatherers and First Farmers Beyond Europe*, ed. J.V.S. Megaw. Leicester University Press. pp15-23.

Brothwell, D.R. 1978 Possible evidence of the parasitisation of early Mexican communities by the micro-organism treponema. *Bulletin of the Institute of Archaeology* 15:113-30.

Brown, A.R. 1930 Former numbers and distribution of the Australian Aborigines. *Official Year Book of the Commonwealth of Australia* 23:687-696.

Brown, P. 1981a Sex determination of Aboriginal crania from the Murray River Valley: a reassessment of the Larnach and Freedman technique. *Archaeology in Oceania* 16:53-63.

Brown, P. 1981b Artificial cranial deformation: a component in the variation in Pleistocene Australian Aboriginal crania. *Archaeology in Oceania* 16:156-67.

Brown, P. 1982 *Coobool Creek: a prehistoric Australian hominid population.* Unpublished PhD thesis, ANU.

Butler, C.S. 1928 Presidential address. *American Journal of Tropical Medicine* 7:363-370.

Butlin, N.G. 1983 *Our Original Aggression.* George Allen and Unwin, Sydney and Boston.

Cadell, F. 1868 Report of an exploring expedition to the Northern Territory. *South Australian Parliamentary Papers* March 31.

Caffey, J. 1931 Clinical and experimental lead poisoning: some roentgenologic and anatomic changes in growing bones. *Radiology* 17:957-983.

Campbell, J. 1983 Smallpox in Australia. *Historical Studies* 20:536-556.

Campbell, T.D. 1925 *Dentition and palate of the Australian Aboriginal.* University of Adelaide Publications 1:72-81.

Campbell, T.D. 1937 Observations on the teeth of Australian Aborigines, Hermannsburg, central Australia. *The Australian Journal of Dentistry* 41:1-6, January 1937.

Campbell, T.D. 1939 Food, food values and food habits of the Australian Aborigines in relation to their dental condition. *Australian Journal of Dentistry* 43(1):1-198.

Campbell, T.D. and M.J. Barrett 1953 Dental observations on Australian Aborigines: a changing environment and food pattern. *Australian Journal of Dentistry* 57:1-6.

Cane, S.B. 1984 *Desert Camps.* Unpublished PhD thesis, Australian National University.

Cannefax, G.R., L.C. Norrins and E.J. Gillespie 1967 Immunology of syphilis. *Annual Review of Medicine* 18:471-482.

Carlson, D.S., G.J. Armelagos and D.P. van Gerven 1974 Factors influencing the etiology of cribra orbitalia in prehistoric Nubia. *Journal of Human Evolution* (3):405-410.

Cassidy, C.M. 1979 Arthritis in dry bones: diagnostic problems. *Henry Ford Hospital Medical Journal* 27(1):68-69.

Chapman, F.H. 1962 Incidence of arthritis in a prehistoric middle Mississippian Indian population. *Proceedings of the Indiana Academy of Science* 72:59-62.

Chapman, F.H. 1968 Comparison of osteoarthritis in three Aboriginal populations. *Proceedings of the Indiana Academy of Science* 74:84-86.

Chase, A. and P. Sutton 1981 Hunter-gatherers in a rich environment: Aboriginal coastal exploitation in Cape York peninsula. In *Ecological Biogeography of Australia* , ed. A. Keast. Junk: The Hague.

Cilento, R. 1942 *Tropical Diseases in Australia.* Smith and Paterson: Brisbane.

Chesterman, C.C. 1927 The relation of yaws and goundou. *Transactions of the Royal Society of Tropical Medicine and Hygiene* 20:554.

Clabeaux, M.S. 1976 Health and disease in the population of an Iroquois ossuary. *Yearbook of Physical Anthropology* 19:359-370.

Clark, G.A. and J.A. Delmond 1979 Vertebral osteophytosis in Dickson mound populations: a biomechanical interpretation. *Henry Ford Hospital Medical Journal* 1:54-58.

Clarke, S.K. 1980 Early childhood morbidity trends in prehistoric populations. *Human Biology* 52(1):79-85.

Clarke, S.K. 1982 The association of early childhood enamel hypoplasias and radiopaque transverse lines in a culturally diverse prehistoric skeletal sample. *Human Biology* 53(1):77-84.

Cleland, J.B. 1928 Disease amongst the Australian Aborigines. *Journal of Tropical Medicine and Hygiene* 31:125-130.

Clement, A.J. 1963 Variations in the microstructure and biochemistry of human teeth. In *Dental Anthropology,* ed. D.R. Brothwell. Pergamon Press: Oxford. pp254-264.

Cockburn, E. (ed.) 1977 Porotic hyperostosis: an enquiry . *Paleopathology Association Monograph No.2.* Detroit: Michigan.

Cockburn, T.A. 1971 Infectious Diseases in Ancient Populations. *Current Anthropology* 12:45-62.

Cockburn, T.A., H. Duncan and J.M. Riddle 1979 Arthritis, ancient and modern: guidelines for field workers. *Henry Ford Hospital Medical Journal* 27(1):74-79.

Cohen, H.J. and H. Diner 1970 The significance of developmental dental enamel defects in neurological diagnosis. *Pediatrics* 46(5):737-747.

Coleman, J. 1982 A new look at the north coast: fish traps and 'villages'. In *Coastal Archaeology in Eastern Australia,* ed. S. Bowdler. Proceedings of the 1980 Valla conference on Australian prehistory.

Collins, D. 1798 *An account of the English colony in New South Wales.* A Royal Australian Historical Society facsimile edition (1975). A.H. and A.W. Reed: Sydney, London.

Cook, C.E. 1970 Notable changes in the incidence of disease in Northern Territory Aborigines. In *Diprotodon to Detribalization*, eds. Pilling A.R. and R.A. Waterman. Michigan State University Press.

Cook, D.C. and J.E. Buikstra 1979 Health and differential survival in prehistoric populations: prenatal dental defects. *American Journal of Physical Anthropology* 51:649-664.

Cook, S.F. 1976 *The Population of the California Indians 1769-1970.* University of Californin Press, Berkley.

Copeman, W.S.C. 1955 *Textbook of the Rheumatic Diseases.* E.S. Livingstone Ltd.: Edinburgh and London.

Courville, C.B. 1942 Cranial injuries of the pre-columbian Incas. *Los Angeles Neurological Society* 7:107-142.

Courville, C.B. 1951 Injuries to the skull and brain in Oceania. *Los Angeles Neurological Society* 16:14-70.

Courville, C.B. 1952 Cranial injuries among the early Indians of California. *Los Angeles Neurological Society* 17:137-162.

Courville, C.B. 1967 Cranial injuries in prehistoric man. In *Disease in Antiquity,* eds. Brothwell D.R. and A.T. Sandison. Charles, C. Thomas: Springfield, USA.

Coutts, P.J.F., R.K. Frank and P. Hughes 1978 Aboriginal engineers of the western district, Victoria. *Records of the Victorian Archaeological Survey* No.7 June.

Coutts, P.J.F., P. Henderson and R.L.K. Fullagar 1979 A preliminary investigation of Aboriginal mounds in northwestern Victoria. *Records of the Victorian Archaeological Survey* No.9 August.

Coutts, P.J.F., D.C. Witter, M. McIlwraith and R. Frank 1976 The mound people of western Victoria: a preliminary statement. *Records of the Victorian Archaeological Survey* No.1:1-54.

Cowlishaw, G. 1981 The determinants of fertility among Australian Aborigines. *Mankind* 13:37-55.

Cran, J.A. 1955 Notes on the teeth and gingival of central Australian Aborigines. *Australian Journal of Dentistry* 59:356-361.

Cunningham, P. 1827 *Two years in New South Wales.* Henry Colburn: London.

Curr, E.M. 1886 *The Australian race* (4 vols). John Ferris, Government Printer: Melbourne.

Cybulski, J.S. 1977 Cribra orbitalia, a possible sign of anemia in early historic native populations of the British Columbia coast. *American Journal of Physical Anthropology* 47:31-40.

Cybulski, J.S. 1980 Skeletal remains from Lillooet, British Columbia, with observations for a possible diagnosis of skull trephination. *Syesis* 13:53-59.

Darling, A.I. 1956 Some observations on amelogenesis imperfecta and calcification of the dental enamel. *Proceedings of the Royal Society of Medicine* 49:759-765.

Dastugue, J. 1980 Possibilities, limits and prospects in paleopathology of the human skeleton. *Journal of Human Evolution* 9:3-8.

Davivongs, V. 1963a The femur of the Australian Aborigine. *American Journal of Physical Anthropology* 21:457-467.

Davivongs, V. 1963b The pelvic girdle of the Australian Aborigine; sex difference and sex determination. *American Journal of Physical Anthropology* 21:443-456.

Dawson, J. 1881 *Australian Aborigines: the languages and customs of several tribes in the western district of Victoria.* George Robertson: Melbourne, Sydney and Adelaide.

Dinning, T.A.R. 1949 A study of healed fractures in the Australian Aboriginal. *Medical Journal of Australia* November 12 (2):712-713.

Dongen, R. van 1963 The shoulder girdle and humerus of the Australian Aborigine. *American Journal of Physical Anthropology* 21:469-488.

Dreizen, S., C.N. Spirakis and R.E. Stone 1964 The influence of age and nutritional status on 'bone scar' formation in the distal end of the growing radius. *American Journal of Physical Anthropology* 22:295-306.

Duke-Elder, Sir Stewart 1961 *The Anatomy of the Visual System,* Vol.II. Henry Kimpton: London.

Duncan, H. 1979 Osteoarthritis. *Henry Ford Hospital Medical Journal* 27(1):6-9.

Dunn, F.L. 1968 Epidemiological factors: health and disease in hunter-gatherers. In *Man the Hunter*, eds. Lee R.B. and I. Devore. Aldine Press: Chicago, pp.221-228.

Edynak, G.J. 1976 Life-styles from skeletal material: a medieval Yugoslav example. In *The Measures of Man*, eds. Giles, E. and Friedlaender. Peabody Museum Press: Cambridge, Mass.. pp408-432.

Elkin, A.P. 1979 *The Australian Aborigines* . Revised edition. Angus and Robertson: Sydney and London.

Elkington, J.S.C. 1912 Report on an inspection of the Torres Straits Islands and vaccination of the islanders. *Annual Report of the Commonwealth Public Health, Queensland*. pp21-27.

El-Najjar, M. 1976 Maize, malaria and the anemias in the pre- columbian new world. *Yearbook of Physical Anthropology* 20:329-337.

El-Najjar, M. 1979 Human treponematosis and tuberculosis: evidence from the new world. *American Journal of Physical Anthropology* 51:599-618.

El-Najjar, M.Y., M.V. DeSanti and L. Ozebek 1978 Prevalence and possible etiology of dental enamel hypoplasia. *American Journal of Physical Anthropology* 48:185-192.

El-Najjar, M.Y., D.J. Ryan, C.G. Turner and B. Lozoff 1976 The etiology of porotic hyperostosis among the prehistoric and historic Anasazi Indians of southwestern United States. *American Journal of Physical Anthropology* 44:477-488.

Elphinstone, J.J. 1971 The health of Australian Aborigines with no previous association with Europeans. *Medical Journal of Australia* (2):293-301.

Enwonwu, C.O. 1973 Influence of socio-economic conditions on dental development in Nigerian children. *Archives of Oral Biology* 18:95-107.

Eylmann, E. 1908 *Die Engeborenen der Kolonie Sudaustralien.* Dietrich Riemer: Berlin.

Eyre, E.J. 1845 *Journals of Expeditions of Discovery into Central Australia.* Australiana Facsimile edition no.7 (1964). Libraries Board of South Australia: Adelaide.

Falin, L.I. 1961 Histological and histochemical studies of human teeth of the Bronze and Stone ages. *Archives of Oral Biology* 5:5- 13.

Fenner, F.J. 1938 Some Australian Aboriginal scaphocephalic skulls. *Records of South Australian Museum* 6(2):144-157.

Flood, J. 1980 *The Moth Hunters.* Australian Institute of Aboriginal Studies: Canberra.

Foelsche, P. 1882 Notes on the Aborigines of north Australia. *Transactions and Proceedings of the Royal Society of South Australia* 5:1-18

Foelsche, P. 1886 Port Darwin: the Larrakia tribe. In *The Australian Race* , ed. E.M. Curr Vol.1, pp250-259.

Follis, R.H. and E.A. Park 1952 Some observations on bone growth with particular reference to zones and transverse lines of increased density in the metaphysis. *American Journal of Roentgenology* 68:709-724.

Freedman, L. and M. Lofgren 1981 Odontometrics of Western Australian Aborigines. *Archaeology in Oceania* 16:87-93.

Garn, S.M. and R.S. Baby 1969 Bilateral symmetry in finer lines of increased density. *American Journal of Physical Anthropology* 31:89- 92.

Garn, S.M. and P.M. Schwager 1967 Age dynamics of persistent transverse lines in the tibia. *American Journal of Physical Anthropology* 27:375-378.

Garn, S.M., F.N. Silverman, K.P. Hertzog and C.G. Rohmann 1968 Lines and bands of increased density. *Medical Radiography and Photography* 44(3):58-89.

Garner, M.F. and R.W. Hornabrook 1973 Treponematosis in Papua New Guinea - a review of surveys undertaken between 1964 and 1972. *Papua New Guinea Medical Journal* 16:189-193.

Garner, M.F., J.L. Backhouse, P.M. Moodie and G.J. Tibbs 1972 Treponemal infection in the Australian Northern Territory Aborigines. *Bulletin of World Health Organisation* 42:285-293.

Gejvall, N-G and F. Henschen 1971 Anatomical evidence of pre- columbian syphilis in the West Indian islands. *Beitrag Pathologie Bd.* 144:138-157.

Giles, E. 1889 *Australia Twice Traversed.* Sampson Low, Marston, Searle and Rivington: London.

Gill, E. 1968 Examination of Harris' lines in recent and fossil Aboriginal bones. *Current Anthropology* 9(2-3):215.

Gindhart, P.S. 1969 The frequency of appearance of transverse lines in the tibia in relation to childhood illnesses. *American Journal of Physical Anthropology* 31:17-22.

Giro, C.M. 1947 Enamel hypoplasia in human teeth: an examination of its causes. *Journal of American Dental Association* 34:309-317.

Goff, C.W. 1967 Syphilis. In *Diseases in Antiquity*, eds. Brothwell D.R. and A.T. Sandison. Charles C. Thomas: Illinois, USA.

Goldstein, M.S. 1957 Skeletal pathology of early Indians in Texas. *American Journal of Physical Anthropology* 15:299-307.

Goldstein, M.S., B. Arensberg and H. Nathan 1976 Pathology of Bedouin skeletal remains from two sites in Israel. *American Journal of Physical Anthropology* 45:621-640.

Goodman, A.H., G.J. Armelagos and J.C. Rose 1980 Enamel hypoplasia as indicators of stress in three prehistoric populations from Illinois. *Human Biology* 52(3):515-528.

Goodman, A.H., J. Lallo, G.J. Armelagos and J.C. Rose 1983 Health changes at Dickson mounds, Illinois (AD 950-1300). In *Paleopathology and the Origins of Agriculture*, eds. Cohen M.N. and G.J. Armelagos. Academic Press: New York.

Gott, B. 1982 Ecology and root use by the Aborigines of southern Australia. *Archaeology in Oceania* 17:59-67.

Green, M.K. 1982 *A Review of Enamel Hypoplasia and its Application to Australian Prehistory.* Unpublished BA (Hons) thesis, ANU.

Green, N. 1981 Aborigines and white settlers in the nineteenth century. In *A new history of Western Australia*, ed. C.T. Stannage. University of Western Australia Press.

Grin, E.I. 1953 Endemic syphilis in Bosnia: clinical and epidemiological observations on a successful mass-treatment campaign. *Bulletin of World Health Organisation* 7:1-74.

Hackett, C.J. 1936a *Boomerang leg and yaws in Australian Aborigines Monograph I.* September 1936. Royal Society of Tropical Medicine and Hygiene. London.

Hackett, C.J. 1936b Boomerang legs and yaws in Australian Aborigine. *Transactions Royal Society of Tropical Medicine and Hygiene* 30:137-150.

Hackett, C.J. 1936c A critical survey of some references to syphilis and yaws among the Australian Aborigine. *Medical Journal of Australia* 1:773.

Hackett, C.J. 1947 Incidence of yaws and of venereal diseases in Lango (Uganda). *British Medical Journal* January 18, pp88-90.

Hackett, C.J. 1951 *Bone Lesions of Yaws in Uganda.* Blackwell: Oxford.

Hackett, C.J. 1963 On the origin of human treponematoses. *Bulletin of World Health Organisation* 29:1-41.

Hackett, C.J. 1968 Some aspects of treponematoses in past populaions. *Journal of the Anthropological Society of South Australia* 6(9):5-14. Minutes of 1967 Annual General Meeting.

Hackett, C.J. 1974 Possible treponemal changes in a Tasmanian skull. *Man* 9:436-443.

Hackett, C.J. 1975 An introduction to diagnostic criteria of Syphilis, Treponarid and Yaws (Treponematoses) in dry bones, and some implications. *Virchows Archives of Pathology, Anatomy and Histology* 368:229-241

Hackett, C.J. 1976 *Diagnostic criteria of Syphilis, Yaws and Treponarid (Treponematoses) and of some other diseases in dry bones.* Springer-Verlag: Berlin, Heidelberg and New York.

Haglund-Calley, L. 1968a The Aboriginal burial ground at Broadbeach, Queensland: excavation report. *Mankind* 6:676-680.

Haglund-Calley, L. 1968b *The Relation Between the Broadbeach Burials and the Cultures of Eastern Australia.* Unpublished MA thesis, University of Queensland.

Haglund, L. 1976 *The Broadbeach Aboriginal Burial Ground.* University of Queensland Press: St Lucia, Queensland.

Hall, J. 1982 Sitting on the crop of the bay: and historical and archaeological sketch of Aboriginal settlement and subsistence in Morton Bay. In *Coastal archaeology in eastern Australia,* ed. S. Bowdler. Proceedings of the 1980 Valla conference on Australian prehistory.

Hamilton, A. 1981 *Nature and Nurture: Aboriginal child-rearing in north-central Arnhem Land.* Australian Institute of Aboriginal Studies: Canberra.

Harpending, H.C. 1974 Demographic changes accompanying sedentarization in Kung hunter-gatherers. *American Journal of Physical Anthropology* 40:139.

Harris, H.A. 1926 The growth of the long bones in childhood, with special reference to certain bony striations of the metaphysis and to the role of the vitamins. *Archives of Internal Medicine* 38:785- 806.

Harris, H.A. 1931 Lines of attested growth in the long bones in childhood. Correlation of histological and radiographic appearances in clinical and experimental conditions. *British Journal of Radiology* 4:561-588.

Harris, H.A. 1933 *Bone Growth in Health and Disease.* Oxford University Press: London.

Harrison, G.A., J.S. Weiner, J.M. Tanner and N.A. Barnicot 1977 *Human Biology.* Oxford University Press, 2nd edition.

Hartney, P.C. 1978 *Palaeopathology of Archaeological Aboriginal Populations from Southern Ontario and Adjacent Region.* Unpublished PhD thesis, Department of Anthropology, University of Toronto.

Head, L. 1983 Environment as artefact: a geographic perspective on the Holocene occupation of southwestern Victoria. *Archaeology in Oceania* 18:73-80.

Hengen, O.P. 1971 Cribra orbitalia: pathogenesis and probable etiology. *Homo* XXII (2):57-76.

Henschen, F. 1961 Cribra cranii, a skull condition said to be of racial or geographical nature. 7th Conference International Society Geographical Patholology London 1960. *Pathology and Microbiology* 24:734-729.

Henschen, F. 1966 *The History of Diseases.* Longmans: London.

Hochberg, M.C., D.G. Borenstein and F. Arnett 1978 The absence of back pain in classical ankylosing spondylitis. *Johns Hopkins Medical Journal* 143:181-183.

Hoets, J. 1949 A study of healed fractures in the Australian Aboriginals. *Medical Journal of Australia* (Correspondence), November 26 (2):794.

Hogan, M.J. and L.E. Zimmerman 1962 *Opthalmic Pathology.* Second edition. Saunders: Philadelphia.

Hollander, J.L. 1962 *Arthritis and Allied Conditions.* Seventh edition, Lea and Febiger: Philadelphia.

Holmes, M.J. 1913 Report of the medical officer for the year ending 31 December 1912. *Bulletin of the Northern Territory* No.6.

Hooton, E.A. 1930 *Indians of Pecos Pueblo.* Yale University Press: New Haven. pp306-33

Houghton, P. 1980 *The First New Zealanders.* Hodder and Stoughton: Auckland, London and Sydney.

Howells, N. 1976 *The population of the Dobe area !Kung.* In Kalahari Hunter-Gatherers, eds. Lee R.B. and I. DeVore. Harvard University Press. pp138-151.

Hoyme, E. St 1969 On the origins of new world paleopathology. *American Journal of Physical Anthropology* 31:295-302.

Hudson, C., R. Butler and D. Sikes 1975 Arthritis in the prehistoric southeastern United States: biological and cultural variables. *American Journal of Physical Anthropology* 43:57-62.

Hudson, E.H. 1958a *Non-venereal syphilis: a sociological and medical study of Bejel.* Livingstone: London.

Hudson, E.H. 1958b The treponematoses - or treponematosis? *British Journal of Venereal Diseases* 34:22-23.

Hudson, E.H. 1961 Endemic syphilis - heir of the syphiloids. *Archives of Internal Medicine* 108:1-4.

Hudson, E.H. 1963 Treponematosis and anthropology. *Annals of Internal Medicine* 58:1037-1048.

Hudson, E.H. 1965 Treponematosis in perspective. *World Health Organisation Bulletin* 32:735-748.

Hughes, P.J. and R.J. Lampert 1982 Prehistoric population change in southern coastal New South Wales. In *Coastal archaeology in eastern Australia*, ed. S. Bowdler. Proceedings of the 1980 Valla conference on Australian prehistory.

Hunt, E.E. Jr. and J.W. Hatch 1981 The estimation of age at death and ages of formation of transverse lines from measurements of human long bones. *American Journal of Physical Anthropology* 54:461-469.

Huss-Ashmore, R., A.H. Goodman and G.J. Armelagos 1982 Nutritional inference from paleopathology. *Advances in Archaeological Method and Theory* 5:395-474.

Hynes, R.A. and A.K. Chase 1982 Plants, sites and domiculture: Aboriginal influence upon plant communities in Cape York Peninsula. *Archaeology Oceania* 17:38-50.

Infante, P.F. 1974 Enamel hypoplasia in Apache Indian children. Ecology of Food and Nutrition 2:155-164.

Infante, P.F. and G.M. Gillespie 1974 An epidemiologic study of linear enamel hydroplasia of deciduous anterior teeth in Guatemalan children. Archives of Oral Biology 19:1055-1061.

Infante, P.F. and G.M. Gillespie 1977 Enamel hypoplasia in relation to caries in Guatemalan children. Journal of Dental Research 56:493-498.

Jacobs, D.S. 1978 Syphilis in Australian Aborigines in the Northern Territory. *Medical Journal of Australia* (1):10-12.

Jaffe, H.L. 1975 *Metabolic, Degenertive and Inflammatory Diseases of Bones and Joints.* Lea and Febiger: Philadelphia.

Jannsens, P.A. 1970 *Palaeopathology.* John Baker: London.

Jarcho, S. 1966 *Human Palaeopathology*, ed. S. Jarcho. Yale University Press: New Haven.

Jelliffe, D.B. and E.F.P. Jelliffe 1971 Linear hypoplasia of deciduous incisor teeth in malnourished children. *American Journal of Clinical Nutrition* 24:893.

Johnston, T.H. and J.B. Cleland 1937 A survey of the literature relating to the occurrence in Australia of helminth parasites of man. *Transactions and Proceedings of the Royal Society of South Australia* 61:250-277

Jones, P.R.M. and R.F.A. Dean 1959 The effects of kwashiorkor on the development of the bones of the knee. *Journal of Pediatrics* 54:176-184.

Jose, D.G. and J.S. Welch 1970 Growth retardation, anaemia and infection, with malabsorption and infestation of the bowel. The syndrome of PCM in Australian Aboriginal children. *Medical Journal of Australia* (1):349-356.

Jurisich, M. and D. Davies 1976 The palaeopathology of prehistoric Aboriginal skeletal remains excavated by the Victoria Archaeological Survey. *The Artefact* 1(4):194-218.

Jurmain, R.D. 1977a Paleoepidemiology of degenerative knee disease. *Medical Anthropology* 1:1-23.

Jurmain, R.D. 1977b Stress and the etiology of osteoarthritis. *American Journal of Physical Anthropology* 46:353-366.

Karsh, R.S. and J.D. McCarthy 1960 Archaeology and arthritis. *Archives of Internal Medicine* 105:640-644.

Kefous, K. 1983 *Riveraine*. Unpublished MA Thesis, Australian National University.

Kelley, M.A. 1979 A survey of Joint Disease at the Libben Site, Ottawa County, Ohio. *Henry Ford Hospital Medical Journal* 27(1):64-67

Kellgren, J.H. and J.S. Lawrence 1958 Osteoarthritis and disc degeneration in an urban population. *Annals of Rheumatic Disease* 17:388-397

King, A.J. and R.D. Catterall 1959 Syphilis of Bones British. *Journal Venereal Disease* 35:116-128

Kirk, R.L. 1981 *Aboriginal Man Adapting*. Research monographs on human population biology. Clarendon Press, Oxford

Klepinger L.L 1979 Paleopathologic Evidence for the Evolution of Rheumatoid Arthritis. *American Journal Physical Anthropology* 50:119-122.

Klepinger, L.L. 1983 Differential Diagnosis in Palaeopathology and the Concept of Disease Evolution. *Medical Anthropology* 7:73-77.

Koganei, Y. 1912 Cribra cranii and cribra Orbitalia. *Mitt. med. Falk University of Tokyo*. 10:113.

Krefft, G. 1862 On the manners and customs of the aborigines of the Lower Murray and Darling. *Transactions of the Philosophical Society of New South Wales* 1862-65:357-74.

Krogman, W.M. 1962 *The Human Skeleton in Forensic Medicine*. Charles C. Thomas publisher, Springfield. Ill.

Krzywicki, L. 1934 *Primitive Society and its Vital Statistics*. Macmillan, London.

Kuhl, I. 1980 Harris's lines and their occurrence also in bones of prehistoric cremations. *Ossa* 7:129-171.

Kuzell, W.C. and G.P. Gaudin 1956 Gout. *Documenta Rheumatologica,* September 1956.

Ladurie, E. LeRoy 1975 Famine Amenorrhoea (Seventeenth-Twentieth Centuries). In *Biology of Man in History*, eds. Forster, R. and O. Ranum. Johns Hopkins University Press, Baltimore and London.

Lallo, J.W. and J.C. Rose 1979 Patterns of Stress, Disease and Mortality in Two Prehistoric Populations from North America. *Journal of Human Evolution* 8:323-335

Lampert, R.J. 1966 An excavation at Durras North, New South Wales. *Archaeology and Physical Anthropology in Oceania* 1(2):83-118.

Lampert, R.J. 1971a *Burrill Lake and Currarong: Coastal sites in southern New South Wales.* Terra Australis No.1 Department of Prehistory, Australian National University.

Lampert, R.J. 1971b Coastal Aborigines of Southeastern Australia. In *Aboriginal Man and Environment in Australia*, eds. Mulvaney, D.J. and J. Golson. Australian National University Press, Canberra.

Lampert, R.J. and Hughes, P.J. 1974 Sea level changes and Aboriginal coastal adaptations in southern New South Wales. *Archaeology and Physical Anthropology in Oceania* IX(3):226-235.

Lane, L. 1980 Aboriginal earthen mounds along the Little Murray River in northern Victoria. *Records of the Victorian Archaeological Survey* 10:98-108.

Larnach, S.L. and L. Freedman 1963 Sex determination of Aboriginal crania from coastal New South Wales, Australia. *Records of the Australian Museum* 26(11):295-308.

Larnach, S.L. and N.W.G. Macintosh 1966 *The Craniology of the Aborigines of New South Wales.* Oceania Monographs No.13. University of Sydney.

Larnach, S.L. and N.W.G. Macintosh 1970 *The craniology of the Aborigines of Queensland.* Oceania Monographs No.15. University of Sydney.

Larsen, C.S. 1981 Skeletal and dental adaptations to the shift to agriculture on the Georgia Coast. *Current Anthropology* 22(4):422- 423

Larsen, C.S. 1982 The Anthropology of St. Catherines Island 3. Prehistoric Human Biological Adaptation. *Anthropological Papers American Museum of Natural History* 57:197-270.

Lawrence, J.S. 1955 Rheumatism in coalminers. Part III: Occupational factors. *British Journal of Industrial Medicine* 12:249-261

Lawrence, R. 1969 *Aboriginal Habitat and Economy.* Occasional Paper No.6, Department of Geography, Australian National University.

Leigh, R.W. 1925 Dental pathology of Indian Tribes of varied environmental and food conditions. *American Journal of Physical Anthropology* 8:179-199.

Leigh, R.W. 1929 Dental Pathology of Aboriginal California. *Dental Cosmos* 71:756-767.

Leisen, J.C. and H. Duncan 1979 The Impact of Rheumatic Disease on Society. *Henry Ford Hospital Medical Journal* 27(1):70-72.

Lourandos, H. 1976 Aboriginal settlement and land use in south western Victoria: a report on current fieldwork. *The Artefact* 1:174-93.

Lourandos, H. 1977 Aboriginal spatial organization and population: South Western Victoria reconsidered. *Archaeology and Physical Anthropology in Oceania* X11(3):202-225.

Lourandos, H. 1980 Change or stability?:hydraulics, hunter-gatherers and population in temperate Australia. *World Archaeology* 11:245- 264.

Lourandos, H. 1983 Intensification: a Late Pleistocene-Holocene archaeological sequence from Southwestern Victoria. *Archaeology in Oceania* 18:81-94

Love, J.R.B. 1936 *Stone-Age Bushman of Today.* Blackie & Son Ltd: London and Glasgow.

Lovejoy, C.O. and K.G. Heiple 1981 The analysis of fractures in skeletal populations with an example from the Libben Site, Ottowa County, Ohio. *American Journal of Physical Anthropology* 55:529-541.

Leubbers, R.A. 1978 *Meals and Menus: A study of change in prehistoric coastal settlements in South Australia.* Unpublished PhD Thesis, Australian National University.

McDevitt, H.O. and E.G. Engleman 1976 Association between genes in the major histocompatibility complex and disease susceptibility. In *Arthritis: a quarter century of research.* Proceedings of the Conference Commemorating the 25th Anniversary of the Arthritis Foundation Research Fellowship Program. Proceedings editors E.M. Tan and M. Ziff.

McHenry, H.M. 1968 Transverse lines in long bones of prehistoric California Indians. *American Journal of Physical Anthropology* 29:1- 18

McHenry, H.M. and P.D. Schulz 1976 The Association between Harris Lines and Enamel Hypoplasia in Prehistoric Californian Indians. *American Journal of Physical Anthropology* 44:507-512

Macconaill, M.A. 1969 On the Harris Lines in Bone. *Current Anthropology* 10(2-3):224

Macintosh, N.W.G. 1949 Crania in the MacLeay Museum. *Proceedings of the Linnean Society New South Wales* Vol.1XXXIV.3-4:188-191.

Macintosh, N.W.G. 1952a The Cohuna Cranium: teeth and palate. *Oceania* 23(2):95-105.

Macintosh, N.W.G. 1952b The Cohuna Cranium: history and commentary from November, 1925, to November, 1951. *Mankind* 4(8):307-329.

Macintosh, N.W.G. 1953 The Cohuna Cranium: physiography and chemical analysis. *Oceania* 23(4):277-296.

Macintosh, N.W.G. 1965 The physical aspect of man in Australia. In *Aboriginal Man in Australia*, eds. R.C. and C.H. Berndt. Angus and Robertson, Sydney.

Macintosh, N.W.G. 1967 Fossil Man in Australia. *Australian Journal of Science* 30(3):86-98.

Macintosh, N.W.G. and S.L. Larnach 1976 Aboriginal affinities looked at in a world cntext. In *The Origin of the Australians*, eds. Kirk, R.L. and A.G. Thorne. Human Biology Series No.6:113-126. Australian Institute of Aboriginal Studies, Canberra.

Mackay, C.V. 1938 Some pathological changes in Australian Aboriginal bones. *Medical Journal of Australia* 14(2):537-555.

Mackerras, M.J. 1958 Catalogue of Australian mammals and their recorded internal parasites. Parts I-IV. *Proceedings of the Linnean Society of New South Wales* Vol.83. Part 2.

Mackillop, D. 1893 Anthropological Notes on the aboriginal tribes of the Daly River, North Australia. *Tranactions of the Royal Society of South Australia* 17:254-264.

Macpherson, J. 1902 'Ngarrabul' and other aboriginal tribes. Part I: Medical and Surgical Practices. *Proceedings of the Linnean Society of New South Wales* 27:637-647.

Macumber, P.G. and R. Thorne 1975 The Cohuna Cranium Site - a re- appraisal. *Archaeology and Physical Anthropology in Oceania* 10(1):67-72.

Manson-Bahr, Sir P.H. 1961 *Manson's Tropical Diseases.* 15th ed. Cassell, London.

Martin, D.L., G.J. Armelagos and J.R. King 1979 Degenerative joint disease of the long bones in Dickson mounds. *Henry Ford Hospital Medical Journal* 27(1):60-63.

Masters, G. 1877 Exhibition of, and remarks upon, the skull of an Aboriginal Female. *Proceedings of the Linnean Society of New South Wales* 2.

Meehan, B. 1982 *Shell Bed to Shell Midden.* Australian Institute of Aboriginal Studies, Canberra.

Meggitt, M.J. 1962 *Desert People.* Angus and Robertson, Sydney.

Menken, J., J. Trussell and S. Watkins 1981 The Nutrition Fertility Link: An evaluation of the evidence. *Journal of Interdisciplinary History* 11:425-441.

Mensforth, R.P., C.O. Lovejoy, J.W. Lallo and G.J. Armelagos 1978 The role of constitutional factors, diet, and infectious disease in the etiology of porotic hyperostosis and periosteal reactions in prehistoric infants and children. *Medical Anthropology* 2(1):1-59.

Merrill, E.M. 1875 "One-eyed Billy - the Aboriginal". In *Wagga Wagga Advertiser* August 7th, 1875.

Middelton, M.R. and S.H. Francis 1967 *Yuendumu and its Children.* Australian Government Publishing Service, Canberra.

Miller, J.D. and W.B. Jennett 1968 Complications of depressed skull fracture. *Lancet.* (2):991-995.

Milne, N., L.H. Schmitt, and L. Freedman 1983 Discrete trait variation in Western Australia Aboriginal skulls. *Journal of Human Evolution* 12:157-158.

Moller-Christensen, V. 1961 *Bone changes in Leprosy.* Munksgaard, Copenhagen.

Moller-Christensen, V. 1978 *Leprosy changes of the skull.* Odense University Press, Denmark.

Moller-Christensen, V. and R.G. Inkster 1965 Cases of leprosy and syphilis in the osteological collection of the Department of Anatomy, University of Edinburgh. *Danish Medical Bulletin* 12:11-18.

Molnar, S. 1972 Tooth wear and culture: a survey of tooth functions among some Prehistoric populations. *Current Anthropology* 13:511- 526.

Molnar, S. and S.C. Ward 1975 Mineral metabolism and microstructural defects in primate teeth. *American Journal of Physical Anthropology* 43:3-18.

Moodie, P.M. 1971 *The Health of Australian Aborigines: an annotated bibliography.* Australian Government Publishing Service, Canberra.

Moodie, R.L. 1923 *Paleopathology: an introduction to the study of ancient evidences of disease.* Urbana, University of Illinois Press.

Moody, J.E.H. 1960 The dental and periodontal conditions of Aborigines at settlements in Arnhem Land and adjacent areas. In *Records of the American-Australian Scientific Expedition to Arnhem Land, Vol. 2: Anthropology and Nutrition*, ed. C.R. Mountford, pp.60-71. Melbourne University Press.

Moorrees, C.F.A., E.A. Fanning and E.E. Hunt Jr. 1963 Age variation of formation stages for ten permanent teeth. *Journal of Dental Research* 42:1490-1502.

Morton, R.S. 1967 The Sibbens of Scotland. *Medical History* 11:374-380.

Morton, S.G. 1845 Remarks on the skulls of two natives of New Holland. *Proceedings of the Academy of Natural Science* 2:292-293.

Moseley, J.E. 1966 Radiographic studies in hematologic bone disease: implications for palaeopathology. In *Human Palaeopathology*, ed. Jarcho, S.. Yale University Press. New Haven and London. pp.121-130.

Moskowitz, R.W. 1974 Clinical and laboratory findings in Osteoarthritis. In *Arthritis and Allied Conditions*, eds. Hollander J.L. and D.J. McCarthy Jr. 8th edition. Lea and Febiger, Philadelphia.

Mulvaney, D.J. 1960 Archaeological excavations at Fromm's Landing on the Lower Murray River, S.A. *Proceedings of the Royal Society of Victoria* 72:53-85.

Nathan, H. 1962 Osteophytes of the Vertebral Column. *Journal of Bone and Joint Surgery* 44(A):243-268.

Nathan, H. and N. Haas 1966 Cribra Orbitalia. A bone conditions of the orbit of unknown nature. *Israel Journal of Medical Science* 2:171-191.

Nathan, P. and D.L. Japanangka 1983 *Settle Down Country.* Kibble Books, Central Australian Aboriginal Congress.

Newland, S. 1926 *Memoirs.* F.W. Preece, Adelaide.

Nie, N.H., C.H. Hull, J.G. Jenkins, K. Steinbrenner and D.H. Brent 1975 *Statistical package for the Social Sciences (S.P.S.S.)* 2nd edition. McGraw-Hill Book Company, New York, London.

Oberg, T., G.E. Carlsson and C-M. Fajers 1971 The Temporomandibular Joint. A morphologic study on a human autopsy material. *Acta Odontologica Scandinavica* 29:349-384.

Oetteking, B. 1909 Kraniologische Studien an Altagypten. *Archaeological Anthropology* 8:1-90.

Ortner, D.J. 1968 Description and classification of degenerative bone changes in the distal joint surfaces of the humerus. *American Journal of Physical Anthropology* 28:139-156.

Ortner, D.J. and W.G.J. Putschar 1981 Identification of Pathological Conditions in Human Skeletal Remains. *Smithsonian Contributions to Anthropology No.28.* Smithsonian Institution Press, Washington.

Pales, L. 1930 *Paleopathologie.* Masson and Cie, Paris.

Park, E.A. 1954 Bone growth in health and disease. *Archives of Disease in Children* 29:269-281.

Park, E.A. 1964 The imprinting of nutritional disturbances on the growing bone. *Pediatrics* 33:815-862.

Park, E.A. and C.P. Richter 1953 Transverse lines in bone: the mechanism of their development. *Bulletin of Johns Hopkins Medical School* 93:234-248.

Park, E.A., D. Jackson, T.C. Goodwin and L. Kajdi 1933 Shadows in growing bones produced by lead: their characteristics, cause, anatomical counterpart in bone and differentiation. *Journal of Pediatrics* 3:265-298.

Parker, K.L. 1905 *The Euahlayi Tribe.* Archibald Constable, London.

Parry, T.W. 1914 Prehistoric man and his early efforts to combat disease. *The Lancet* 1914(1):1699-1702.

Parry, T.W. 1931 Neolithic Man and penetration of living human skulls. *Lancet* 2:1388-1390.

Peterson, N. 1972 Totemism yesterday: sentiment and local organisation and the Australian Aborigines. *Man* 7:12-32.

Pickering, R.B. 1979 Hunter-Gatherer/Agriculturalist Arthritic Patterns: A Preliminary Investigation. *Henry Ford Hospital Medican Journal* 27(1):50-53.

Pietrusewsky, M. 1969 An osteological study of cranial and infracranial remains from Tonga. *Records of the Auckland Institute Museum* 6(4-6):287-402.

Pietrusewsky, M. 1979 Craniometric variation in pleistocene [sic] Australian and more recent Australian and New Guinea populations studied by multivariate procedures. *Occasional papers in Human Biology.* Australian Institute of Aboriginal Studies, Canberra, June 1979.

Platt, B.S. and R.J.C. Stewart 1962 Transverse trabeculae and osteoporosis in bones in experimental protein-calorie deficiency. *British Journal of Nutrition* 16(4):483-495.

Poiner, G. 1976 The process of the year among Aborigines of the Central and South Coast of New South Wales. *Archaeology and Physical Anthropology in Oceania* 11(3):186-206.

Polgar, S. 1975 *Population, Ecology, and Social Evolution.* Mouton Publishers, The Hague.

Presland, G. 1977a Journals of George Augustus Robinson: January- March 1840. *Records of the Victorian Archaeological Survey No.5.*

Presland, G. 1977b Journals of George Augustus Robinson: March-May 1841. *Records of the Victorian Archaeological Survey No.6.*

Presland, G. 1980 Journals of G.A. Robinson: May to August 1841. *Records of the Victorian Archaeological Survey No.11.*

Pretty, G.L. 1977 The cultural chronology of the Roonka Flat. In: *Stone Tools as Cultural Markers*, ed. R.V.S. Wright, Australian Institute of Aboriginal Studies, Canberra. pp.288-331.

Prokopec, M. 1975 *Analysis of Human Remains from Roonka.* Doc.No. 76/2021. Ms 669 in Australian Institute of Aboriginal Studies Library. (Records of the South Australian Museum 1976) - June 1976.

Prokopec, M. 1979 Demographical and morphological aspects of the Roonka population. *Archaeology and Physical Anthropology in Oceania* 14:11-26.

Rabkin, S. 1942 Dental conditions among Prehistoric Indians of Northern Alabama. *Journal of Dental Research* 21:211-222.

Rabkin, S. 1943 Dental conditions among Prehistoric Indians of Kentucky. The Indian Knoll Collection. *Journal of Dental Research* 22:355-366.

Reyman, T.A. 1979 The search for "Arthritis" in Antiquity: paleoarthritis workshop. *Henry Ford Hospital Medical Journal* 27(1):32-37.

Richards, L.C. and T. Brown 1981 Dental attrition and degenerative arthritis of the temporomandibular joint. *Journal of Oral Rehabilitation* 8:293-307

Riddle J.M. 1979 Rheumatoid Arthritis. *Henry Ford Hospital Medical Journal* 27(1):18-23.

Robbins, S.L. and R.S. Cotran 1979 *Pathologic Basis of Disease.* W.B. Saunders Company: London and Toronto.

Robinson, J.T. 1956 The dentition of the Australopithecinae. *Transvaal Museum Memoirs No.9.*

Rogers, S. 1964 The healing of trephined wounds from pre-Columbian America. *American Journal of Physical Anthropology* 23:321.

Roheim, G. 1933 Women and their life in Central Australia. *Journal of the Royal Anthropological Institute* 63:207-265.

Rose, J.C. 1977 Defective enamel histology of Prehistoric teeth from Illinois. *American Journal of Physical Anthropology* 46:439-446.

Rose, J.C., G.J. Armelagos and J.W. Lallo 1978 Histological enamel indicator of childhood stress in prehistoric skeletal samples. *American Journal of Physical Anthropology* 49:511-516.

Ross, A. 1981 Holocene environments and prehistoric site patterning in the Victorian Mallee. *Archaeology in Oceania* 16:145-154.

Roth, W.E. 1897 *Ethnological studies amongst the North-west Central Queensland Aborigines.* E. Gregory, Government Printer: Brisbane.

Roth, W.E. 1903 Superstition, Magic and Medicine. In *North Queensland Ethnography Bulletin No.5.* E. Gregory, Government Printer: Brisbane.

Rothschild, B.M. and W. Turnbull 1987 Treponemal infection in a Pleistocene cave bear. *Nature* 329:59-62

Roydhouse, R.H. and B.O. Simonsen 1975 Attrition of Teeth. *Syesis* 8:263-273.

Ruffer, M.A. and A. Rietti 1912 On Osseous Lesions in Ancient Egyptians. *Journal of Pathology and Bacteriology* 16:439-466.

Rushton, M.A. 1964 Hereditary Enamel Defects. *Proceedings of the Royal Society of Medicine* 57:53-58.

Sahlins, M 1972 *Stone Age Economics.* Aldine, Chicago

Sandford, M.K., D.P. van Gerven and R.R. Meglan 1983 Elemental Hair Analysis: new evidence on the Etiology of Cribra Orbitalia in Sudanese Nubia. *Human Biology* 55:831-844.

Sandison, A.T. 1968 Pathological changes in the skeletons of earlier populations due to acquired disease, and difficulties in their interpretation. In *The Skeletal Biology of Earlier Human Populations*, ed. D. Brothwell. Pergamon Press, Oxford pp.205-243

Sandison, A.T. 1972 Evidence of infective disease. *Journal of Human Evolution* (1972)1:213-224.

Sandison, A.T. 1973a Disease changes in Australian Aboriginal skeletons. *Australian Institute of Aboriginal Studies Newsletter* 1970-73(3):20-23.

Sandison, A.T. 1973b Palaeopathology of human bones from Murray River Region between Mildura and Renmark, Australia. *Memoirs of National Museum of Victoria* 34:173-74.

Sandison, A.T. 1980 Notes on some skeletal changes in Pre-European contact Australian Aborigines. *Journal of Human Evolution* 9:45-47.

Sarnat, B.G. and I. Schour 1941 Enamel hypoplasia (chronologic enamel aplasia) in relation to systemic disease: a chronologic, morphologic and etiologic classification. *Journal of American Dental Association* 28:1989-2000.

Sarnat, B.G. and I. Schour 1942 Enamel Hypoplasia (Chronologic Enamel Aplasia) in relation to systemic disease: a chronologic, morphologic and etiologic classification. *Journal of American Dental Association* 29:67-75.

Saul, F.P. 1972 The human skeletal remains of Altar de Sacrificios. *Papers of the Peabody Museum (Harvard)* 63(2):34-75.

Schamschula, R.G., M.H. Cooper, B.L. Adkins, D.E. Barmed and W.M. Agus 1980 Oral conditions in Australian children of Aboriginal and caucasian descent. *Community Dentistry and Oral Epidemiology* 8:365-369.

Schrire, C. 1982 The Alligator Rivers: prehistory and ecology in western Arnhem Land. *Terra Australis No.7.* Department of Prehistory, Research School of Pacific Studies, Australian National University.

Schuman, E.L. and R.F. Sognnaes 1956 Developmental microscopic defects in the teeth of subhuman primates. *American Journal of Physical Anthropology* 14:193-198.

Schultz, A.H. 1939 Notes on diseases and healed fractures of wild apes. *Bulletin of the History of Medicine* 7:571-582.

Schulze, L. 1891 The aborigines of the Upper and Middle Finke River: their habits and customs, with introductory notes on the physical and natural-history features of the country. *Royal Society of South Australia, Transactions* 14:210-246.

Schulz, P.D. and H.M. McHenry 1975 Age distribution of enamel hypoplasia in prehistoric California Indians. *Journal of Dental Research* 54:913.

Sciulli, P.W. 1977 A descriptive and comparative study of the deciduous dentition of prehistoric Ohio Valley Amerindians. *American Journal of Physical Anthropology* 47:71-80.

Sciulli, P.W. 1978 Developmental abnormalities of the permanent dentition in prehistoric Ohio Valley Amerindians. *American Journal of Physical Anthropology* 48:193-198.

Scott, E.C. and B.R. DeWalt 1980 Subsistence and dental pathology etiologies from prehistoric coastal Peru. *Medical Anthropology* 4:263-290.

Scott, J.H. and N.B.B. Symonds 1977 *Introduction to Dental Anatomy.* 8th ed. Churchill Livingstone, Edinburgh, London.

Scrimshaw, N.S. 1964 Ecological factors in nutritional disease. *American Journal Clinical Nutrition* 14:112-122.

Scrimshaw, N.S., C.E. Taylor and J.E. Gordon 1968 Interactions of Nutrition and Infection. *World Health Organisation (W.H.O) Monograph Series No.57.*

Seward, F.S. 1976 Tooth attrition and the temporomandibular joint. *Angle Orthodontics* 46(2):162-170.

Shermis, S. 1975 Common types of paleopathological lesions and some cultural inferences. *Pacific Coast Archaeological Society Quarterly* 11(3):33-58.

Short, C.L. 1974 The antiquity of rheumatoid arthritis. *Arthritis and Rheumatism* 17(3):193-205

Smith, A. 1976 *The Body.* Penguin. England

Smith, G.E. and W.R. Dawson 1924 *Egyptian Mummies.* Allen and Unwin, London.

Smith, L.R. 1980 *The Aboriginal Population of Australia.* ANU Press, Canberra.

Smith, P., T. Brown and W.B. Wood 1981 Tooth size and morphology in a recent Australian Aboriginal population from Broadbeach, South East Queensland. *American Journal of Physical Anthropology* 55:423-432.

Smyth, R.B. 1878 *Aborigines of Victoria.* 2 vols. John Ferres, Government Printer: Melbourne, London.

Sognnaes, R.F. 1956 Histological evidence of developmental lesions in teeth originating from paleolithic, prehistoric, and ancient man. *American Journal of Pathology* 32(1):547-577.

Sokoloff, L. 1974 The pathology and pathogenesis of osteoarthritis. In *Arthritis and Allied Conditions*, eds. Hollander J.L. and D.J. McCarthy Jr. 8th edition. Lea and Febiger: Philadelphia.

Solonen, K.A. 1966 The joints of the lower extremities of football players. *Arm. Chir. Gyraec. Fenn.* 55:176-80.

Spencer, B. 1928 *Wanderings in Wild Australia.* MacMillan & Co. Ltd, London.

Spencer, B. and F.J. Gillen 1899 *The Native Tribes of Central Australia.* MacMillan and Co. Ltd., London.

Steinbock, R.T. 1976 *Paleopathological Diagnosis and Interpretation.* Charles C. Thomas, U.S.A.

Stewart, R.J.C. and B.S. Platt 1958 Arrested growth lines in the bones of pigs on low-protein diets. *Proceedings of the Nutrition Society* 17:v.

Stewart, T.D. 1958 Stone Age skull surgery: a general review, with emphasis on the New World. *Annual Report of the Smithsonian Institute* (1957) pp.469-91.

Stewart, T.D. 1966 Some problems in Human Palaeopathology. In *Human Palaeopathology*, ed. S. Jarcho. Yale University Press, New Haven and London.

Stewart, T.D. and A. Spoehr 1952 Evidence on the Paleopathology of Yaws. *Bulletin of the History of Medicine* 26:538-53.

Stini, W.A. 1971 Evolutionary implications of changing nutritional patterns in human populations. *American Anthropologist* 73:1019-1030.

Stirling, E.C. 1894 Notes from Central Australia. *Intercolonial Quarterly Journal of Medicine and Surgery* 1:221.

Stirling, E.C. 1911 Preliminary report on the discovery of native remains at Swanport, River Muray; with an Inquiry into the alleged occurrence of a Pandemic among the Australian Aborigines. *Transactions of the Royal Society of South Australia* 35:4-46.

Stone, A.C. 1911 The Aborigines of Lake Boga, Victoria. *Proceedings of the Royal Society of Victoria* 23(2):433-468.

Straus, W.L. and A.J.E. Cave 1957 Pathology and the posture of Neanderthal Man. *Quarterly Review of Biology* 32:348-63.

Sturt, C. 1833 *Two Expeditions into the Interior of Southern Australia.* 2 vols. Libraries Board of South Australia, Facsimile edition (1963), Adelaide.

Sullivan, M.E. 1982 *Aboriginal Shell Middens in the Coastal Landscape of New South Wales.* Unpublished PhD thesis, Australian National University.

Sullivan, M.E. and R.A. Buchan 1980 Distinguishing between Aboriginal natural mounds in the Murray Valley. *Records of the Victorian Archaeological Survey* 10:87-97.

Sunderland, S. and L.J. Ray 1959 A note on the Murray Black collection of Australian Aboriginal skeletons. *Royal Society of Victoria Proceedings* 71:45-48.

Sutton, D.G. 1979 The prehistoric people of Eastern Palliser Bay. *National Museum Bulletin* 21:185-203.

Sweeney, E.A., A.J. Saffir and R. de Leon 1971 Linear hypoplasia of deciduous incisor teeth in malnourished children. *American Journal of Clinical Nutrition* 24:29-31.

Sweeney, E.A., J. Cabrera, J. Urrutia and L. Mata 1969 Factors associated with linear hypoplasia of human deciduous incisors. *Journal of Dental Research* 48:1275-1279.

Sweeney, G. 1947 Food supplies of a desert tribe. *Oceania* 187(3):289-99.

Sweet, W.C. 1924 The intestinal parasites of man in Australia and its dependencies as found by the Australian Hookworm campaign. *Medical Journal of Australia* (1):405-407.

Taplin, G. 1879 The Narrinyeri. In *The Native Tribes of South Australia, ed.* J.D. Woods, Adelaide.

Tempelaar, H.C.G. and J. van Breeman 1932 Rheumatism and Occupation. *Acta Rheumatology* 4:36-38.

Thomson, D.F. 1939 The seasonal factor in human culture. *Proceedings of the Prehistory Society* 5(2):209-221.

Thomson, D.F. 1975 *Bindibu Country.* Thomas Nelson, Melbourne.

Thorne, A.G. 1969 Preliminary comments on the Kow Swamp skeleton. *Australian Institute of Aboriginal Studies Newsletter* 2:6-7.

Thorne, A.G. 1971a The racial affinities and origins of the Australian Aborigines. In *Aboriginal Man and Environment in Australia*, eds. Mulvaney D.J. and J. Golson. ANU Press, Canberra. pp.316-325.

Thorne, A.G. 1971b Mungo and Kow Swamp: morphological variations in pleistocene Australians. *Mankind* 8:85-89.

Thorne, A.G. 1972 Recent discoveries of Fossil Man in Australia. *Australian Natural History* June pp191-196.

Thorne, A.G. 1975 *Kow Swamp and Lake Mungo: towards an osteology of early man in Australia.* Unpublished PhD thesis, University of Sydney.

Thorne, A.G. 1976 Morphological contrasts in pleistocene Australians. In *Origin of the Australians*, eds. Kirk R.L. and A.G. Thorne. Human Biology Series No.6, Australian Institute of Aboriginal Studies. Humanities Press Inc. New Jersey.

Thorne, A.G. and M.H. Wolpoff 1981 Regional continuity in Australian pleistocene hominid evolution. *American Journal of Physical Anthropology* 55:337-349.

Tindale, N.B. 1974 *Aboriginal Tribes of Australia.* University of California Press, Los Angeles.

Tindale, N.B. 1981 Desert Aborigines and the southern coastal peoples: some comparisons. In *Ecological Biogeography of Australia*, ed. A. Keast. Junk, The Hague.

Tobias, P.V. 1967 The cranium and maxillary dentition of Australopithecus (Zinjanthropus) boisei. *Olduvai Gorge, Volume II.* Cambridge University Press.

Toldt, C. 1886 Uber Welcker's Cribra Orbitalia. *Mitt. d. Anthropol. Gessell Wien* 16:20-24.

Tonkinson, R. 1978 *The Mardudjara Aborigines.* Holt, Rinehart and Winston, New York.

Topinard, P. 1894 *Anthropology.* Chapman and Hall, London.

Townsend, G.C. and T. Brown 1980 Dental Asymetry in Australian Aboriginals. *Human Biology* 52:661-673.

Trinkaus, E. 1977 The Alto Salaverry child: a Case cf anemia from the Peruvian preceramic. *American Journal of Physical Anthropology* 46:25-28.

Trinkaus, E. and M.R. Zimmerman 1982 Trauma among the Shanidar Neanderthals. *Australian Journal of Physical Anthropology* 57:61-76.

Vanderwal, R. 1982 *The Aboriginal Photographs of Baldwin Spencer*. John Currey, O'Neil Pty Ltd, National Museum of Victoria.

Wadsworth, G.R. 1975 Nutritional factors in anaemia. *World Review of Nutrition and Dietetics* 21:75-150.

Walker, A.C. and S.P. Bellmaine 1975 Severe alimentary bleeding associated with hookworm infestation in Aboriginal infants. *Medical Journal of Australia* 1:751-752.

Walker, E.G. 1978 The paleopathology of certain skeletal remains from Saskatchewan. *Na'pao: a Saskatchewan Anthropology Journal* 8:30-47.

Webb, S.G. 1981 *Palaeopathology and its Future Use in Australian Prehistory*. BA (Hons) Thesis, Australian National University, Canberra

Webb, S.G. 1982 Cribra Orbitalia: a possible sign of anaemia in pre- and post-contact crania from Australia and Papua New Guinea. *Archaeology in Oceania* 17(3):148-156.

Webb, S.G. 1987a Reburying Australian skeletons. *Antiquity* 61:292-6

Webb, S.G. 1987b A Palaeodemographic Model of Late Holocene Central Murray Aboriginal Society, Australia. *Human Evolution* 2(5):385-406

Webb, S.G. 1988 Two Possible Cases of Trephination from Australia. *American Journal of Physical Anthropology* 75:541-548

Webb, S.G. 1989 *The Willandra Lakes Hominids*.

Webb, S.G. and A.G. Thorne 1985 A Congenital Meningocoele in Prehistoric Australia. *American Journal of Physical Anthropology* 68:525-533

Webb, T.T. 1933 Aboriginal medical practice in East Arnhem Land. *Oceania* 4:91-98.

Wegner, G. 1874 Ueber das normale und pathologische Wachsthum der Rohrenknochen. Eine knitische Untersuchung auf experimenteller und casiustischer Grindlage. *Archives of Pathological Anatomy* 61:44-76.

Weidenreich, F. 1939 Recent Researches on Early Man in China. *Nature* 144:1054-1056 pt.I and 144:1097-1098 pt.II.

Weinmann, J.P., J.F. Svoboda and K.W. Woods 1945 Hereditary disturbances of enamel formation and calcification. *Journal of American Dental Association* 32:397-418.

Welcker, H. 1885-7 Die Abstammung der Bevolkerung von Scotia. Mitthellung im Geographentage zu Hamburg, Sitzung vom 11 April 1885. *Verhandlungen des funften deutschen geographentages.* Berlin. Dietrich Reimer, 1887:92-94.

Welcker, H. 1888 Cribra Orbitalia, ein ethnologisch diagnostisches Merkmal am Schadel mehrerer Menschrassen. *Archives of Anthropology* 17:1-18.

Wells, C. 1961 A new approach to ancient disease. *Discovery* 22:526-531.

Wells, C. 1963 The radiological examination of human remains. In *Science and Archaeology*, eds. Bròthwell, D. and E. Higgs. Thames and Hudson, London. pp.401-412.

Wells, C. 1964 *Bones and Bodies and Disease.* Thames and Hudson, London.

Wells, C. 1965 Diseases of the knee in Anglo-Saxons. *Medical Biological Illustrations* 15:100-107.

Wells, C. 1967 A new approach to palaeopathology: Harris's Lines. In *Diseases in Antiquity*, eds Brothwell, D. and A.T. Sandison. Charles C. Thomas, Springfield, Illinois. pp390-404.

Wells, C. 1973 The palaeopathology of bone disease. *Practitioner* 210:384-391.

Westgarth, W. 1848 *Australia Felix.* London.

White, T.D. 1978 Early hominid enamel hypoplasia. *American Journal of Physical Anthropology* 49:79-84.

Williams, E. 1985 *Wet Underfoot? Earth Mound Sites and the Recent Prehistory of Southwestern Victoria.* PhD Thesis, Australian National University, Canberra, Australia.

Williams, E. 1988 *Complex Hunter-Gatherers.* BAR International Series 423, Oxford, Great Britain.

Wilson, P.W. and M.S. Mathis 1930 Epidemiology and pathology of yaws. A report based on a study of 1,423 consecutive cases in Haiti. *Journal of the American Medical Association* 94:1289.

Winterbotham, L.P. 1951 Primitive medical art and primitive medicine-men of Australia. *Medical Journal of Australia* 1(13):461- 468.

Wolff, E. 1954 *The anatomy of the eye and orbit. Including the central connections, development and comparative anatomy of the visual apparatus.* 4th edition. H.K. Lewis and Co. Ltd., London.

Wood, W.B. 1968 An Aboriginal burial ground at Broadbeach, Queensland. Skeletal material. *Mankind* 6(12):681-686.

Wood, W.B. 1976 The skeletal material from the Brooloo Range and Rocky Hole Creek Burial Sites. *Archaeology and Physical Anthropology in Oceania* 11(3):175-185.

Wood, W.Q. 1920 The tibia of the Australian Aborigine. *Journal of Anatomy* 54:232-257.

Woodall, J.N. 1968 Growth arrest lines in long bones of the Casas Grandes population. *Plains Anthropologist* 13:152-160.

Worsnop, T. 1897 *The Prehistoric Arts, Manufactures, Works, Weapons etc. of the Aborigines of Australia.* Bristow, Adelaide.

Wright, R.V.S. 1976 Evolutionary process and semantics: Australian prehistoric tooth size as a local adjustment. In *The Origin of the Australians*, eds. Kirk, R.L. and A.G. Thorne. Australian Institute of Aboriginal Studies, Canberra. pp.265- 274.

Zaino, D.E. and E.C. Zaino 1975 Cribra Orbitalia in the Aborigines of Hawaii and Australia. *American Journal of Physical Anthropology* 42:91-94.

Zorab, P.A. 1961 The historical and prehistorical background to Ankylosing Spondylitis. *Proceedings of the Royal Society of Medicine* 54:415.

FIGURES

Fig. 1-2

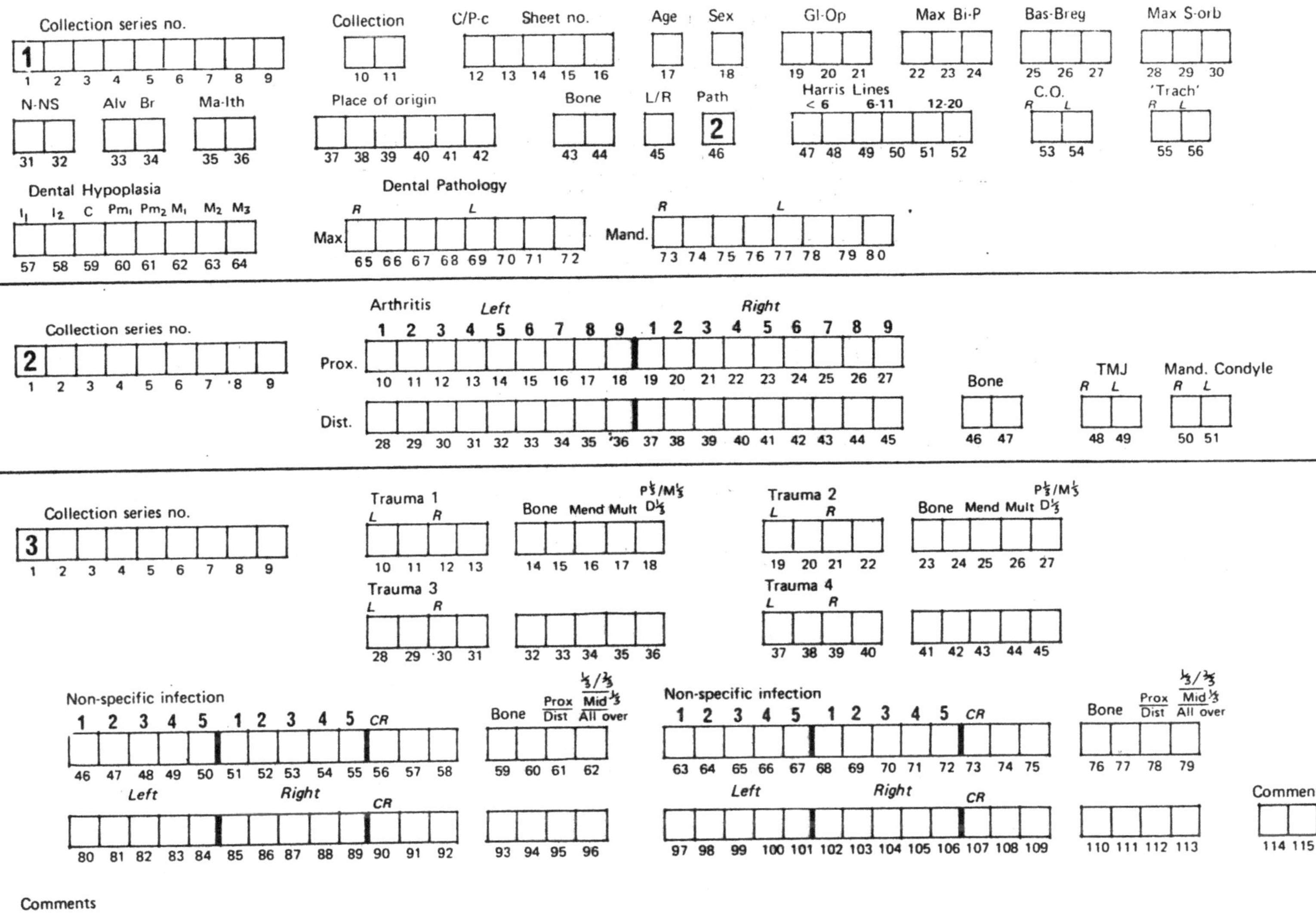

Palaeopathology Data Collection Sheet 1

S.Webb

Fig. 1-1

Collection series no.	Collection	C/P-c	Sheet no.	Age	Sex	Gl-Op	Max Bi-P	Bas-Breg	Max S-orb
1									
1 2 3 4 5 6 7 8 9	10 11	12	13 14 15 16	17	18	19 20 21	22 23 24	25 26 27	28 29 30

N-NS	Alv Br	Ma Ith	Place of origin	Bone	L/R	Path
						1
31 32	33 34	35 36	37 38 39 40 41 42	43 44	45	46

Collection series no.	Collection	C/P-c	Sheet no.	Age	Sex	Gl-Op	Max Bi-P	Bas-Breg	Max S-orb
1									
1 1 2 3 4 5 6 7 8 9	10 11	12	13 14 15 16	17	18	19 20 21	22 23 24	25 26 27	28 29 30

N-NS	Alv Br	Ma-Ith	Place of origin	Bone	L/R	Path
						1
31 32	33 34	35 36	37 38 39 40 41 42	43 44	45	46

Collection series no.	Collection	C/P-c	Sheet no.	Age	Sex	Gl-Op	Max Bi-p	Bas- Breg	Max S-orb
1									
1 2 3 4 5 6 7 8 9	10 11	12	13 14 15 16	17	18	19 20 21	22 23 24	25 26 27	28 29 30

N-Ns	Alv Br	Ma-Ith	Place of origin	Bone	L/R	Path
						1
31 32	33 34	35 36	37 38 39 40 41 42	43 44	45	46

Collection series no.	Collection	C/P-c	Sheet no.	Age	Sex	Gl Op	Max Bi-P	Bas- Breg	Mas S-orb
1									
1 2 3 4 5 6 7 8 9	10 11	12	13 14 15 16	17	18	19 20 21	22 23 24	25 26 27	28 29 30

N-NS	Alv Br	Ma-Ith	Place of origin	Bone	L/R	Path
						1
31 32	33 34	35 36	37 38 39 40 41 42	43 44	45	46

Collection series no.	Collection	C/P-c	Sheet no.	Age	Sex	Gl-Op	Max Bi-P	Bas-Breg	Max S-orb
1									
1 2 3 4 5 6 7 8 9	10 11	12	13 14 15 16	17	18	19 20 21	22 23 24	25 26 27	28 29 30

N-NS	Alv Br	Ma-Ith	Place of origin	Bone	L/R	Path
						1
31 32	33 34	35 36	37 38 39 40 41 42	43 44	45	46

Palaeopathology Data Collection Sheet 2

S.Webb

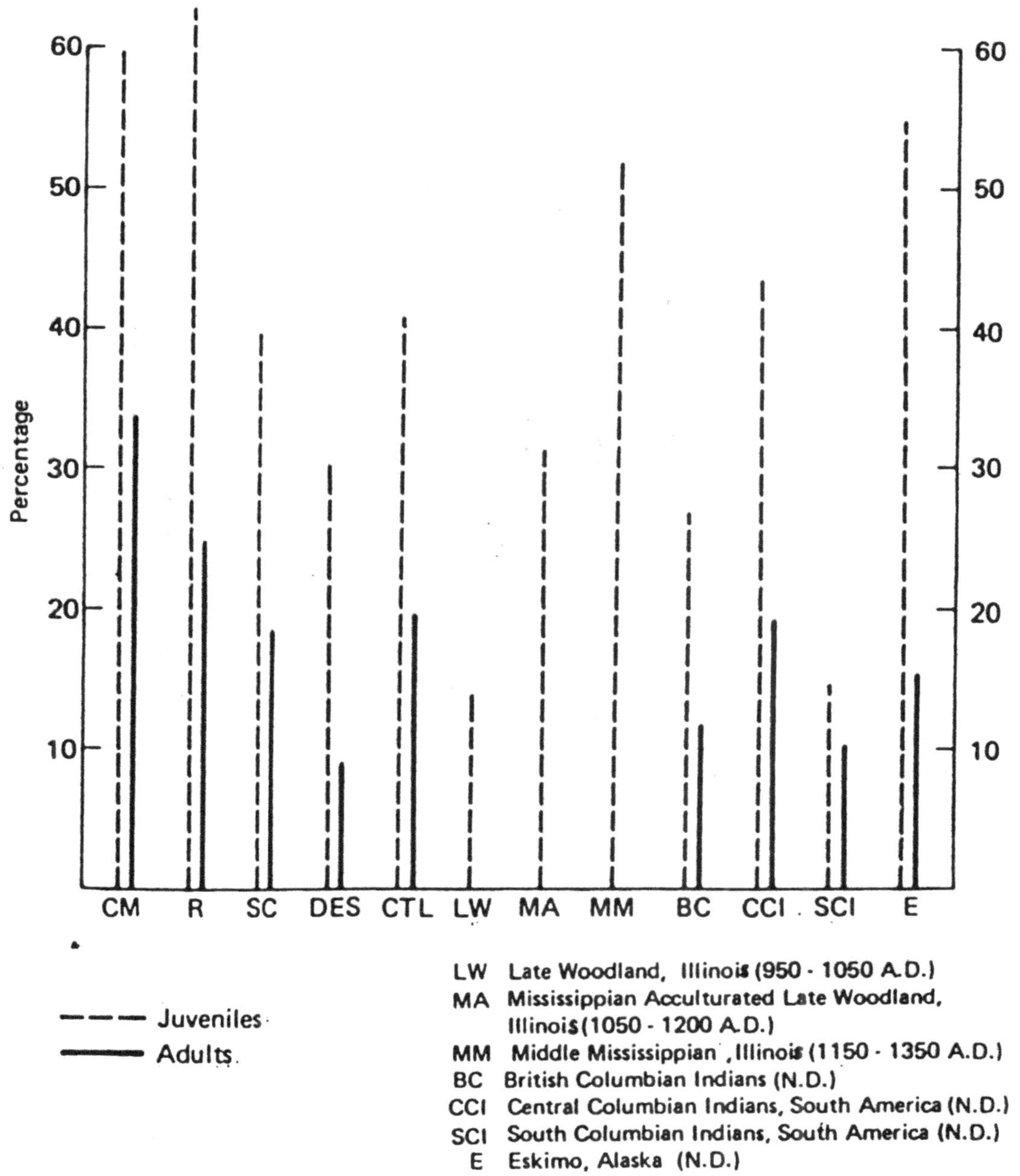

Fig. 2-1 Comparison of cribra orbitalia in Australian Aborigines and various American Indian groups

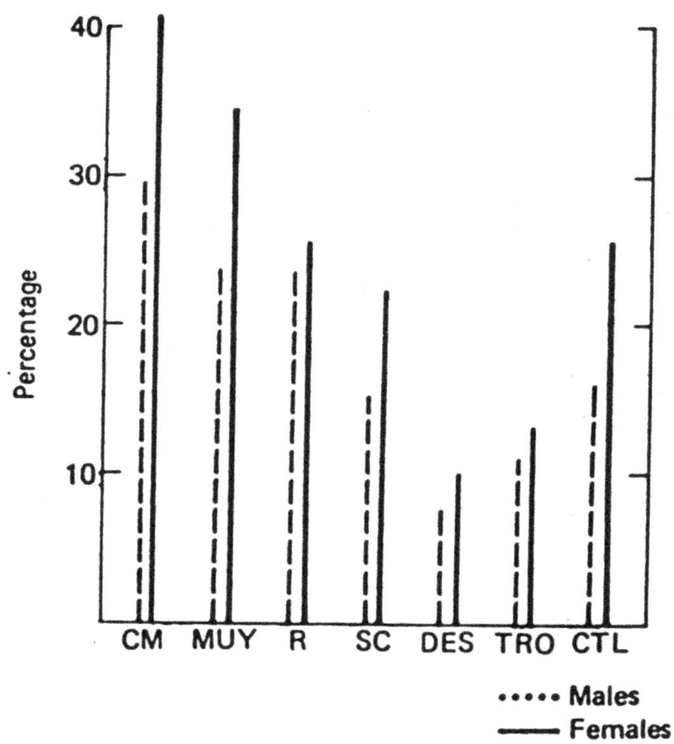

Fig. 2-2 Cribra orbitalia in males and females, by area

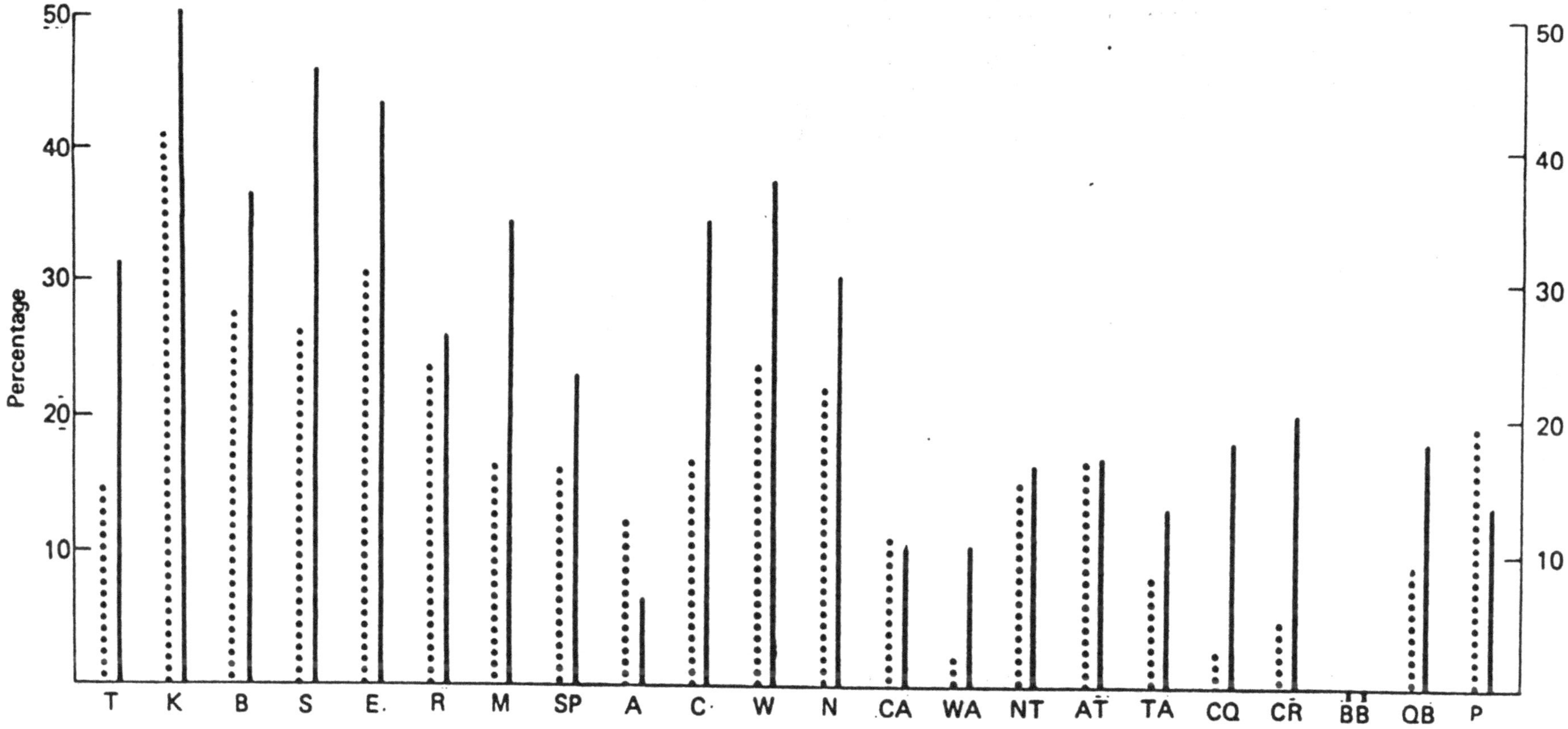

Fig. 2-3 Cribra orbitalia in males and females, by location

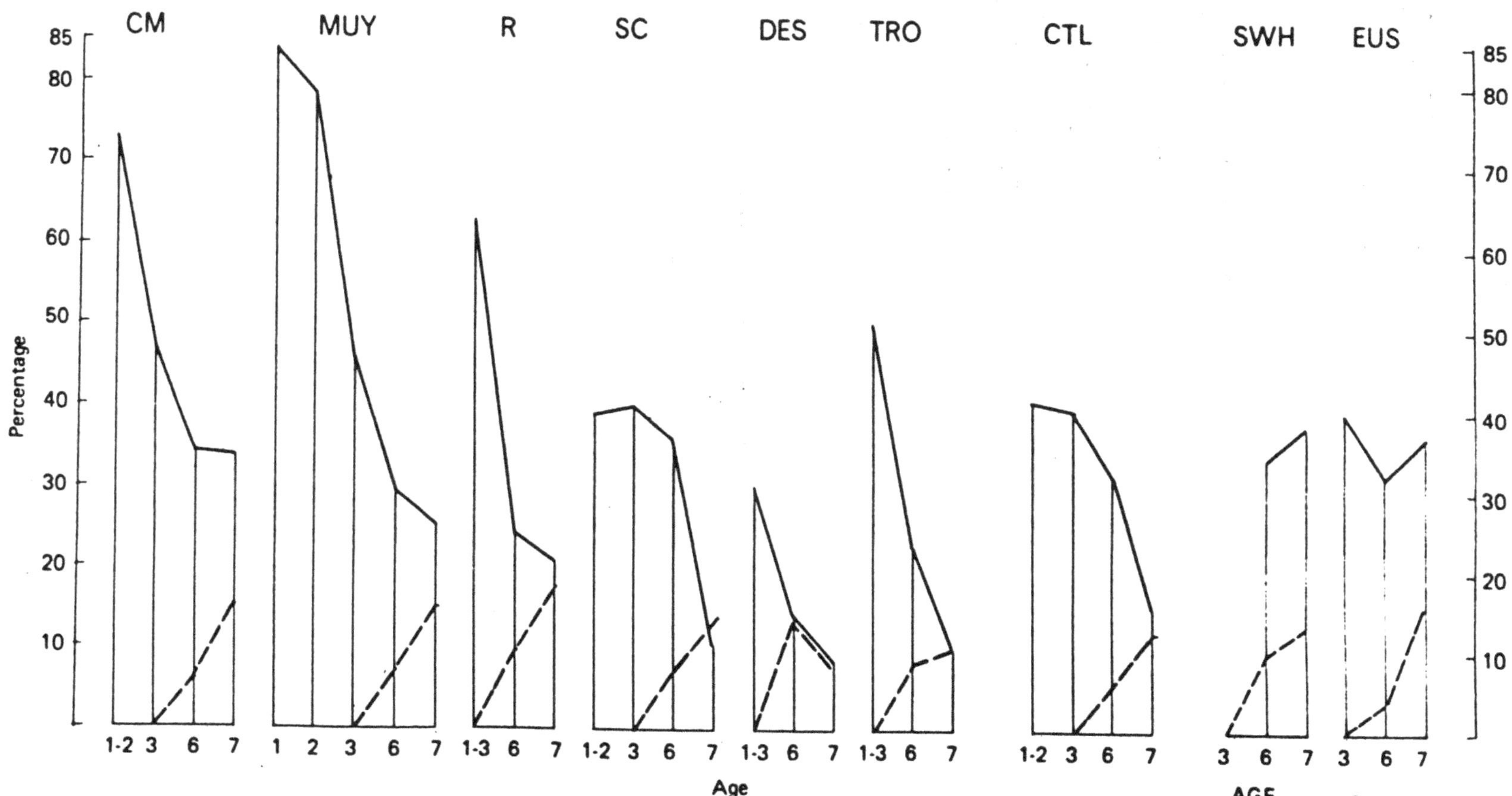

—— Category 1,2,3 combined
– – – Category 4

AGE
1 = 0-5
2 = 6-11
3 = 12-20
6 = Young adults
7 = Old adults

Fig. 2-4 The incidence of anaemia in childhood and adulthood measured through the frequency of cribra orbitalia, also the incidence of the proposed recuperation category (4) is plotted

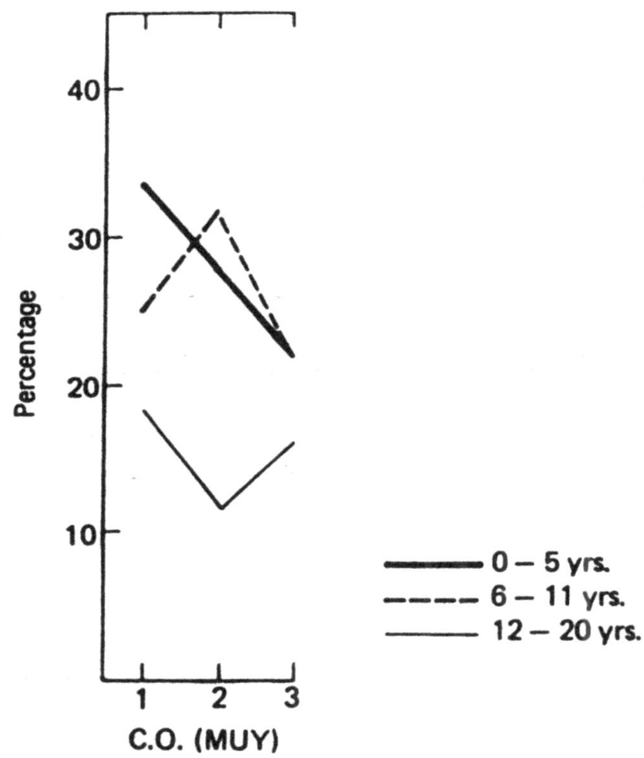

Fig. 2-5 The three cribra orbitalia categories in sub-adult age groups 1, 2 and 3

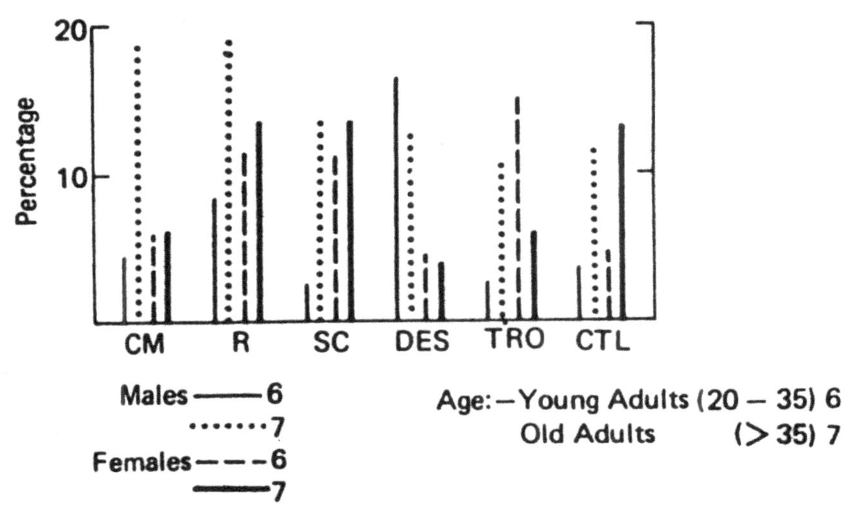

Fig. 2-6 Cribra orbitalia in young (6) and old (7) adults, by area

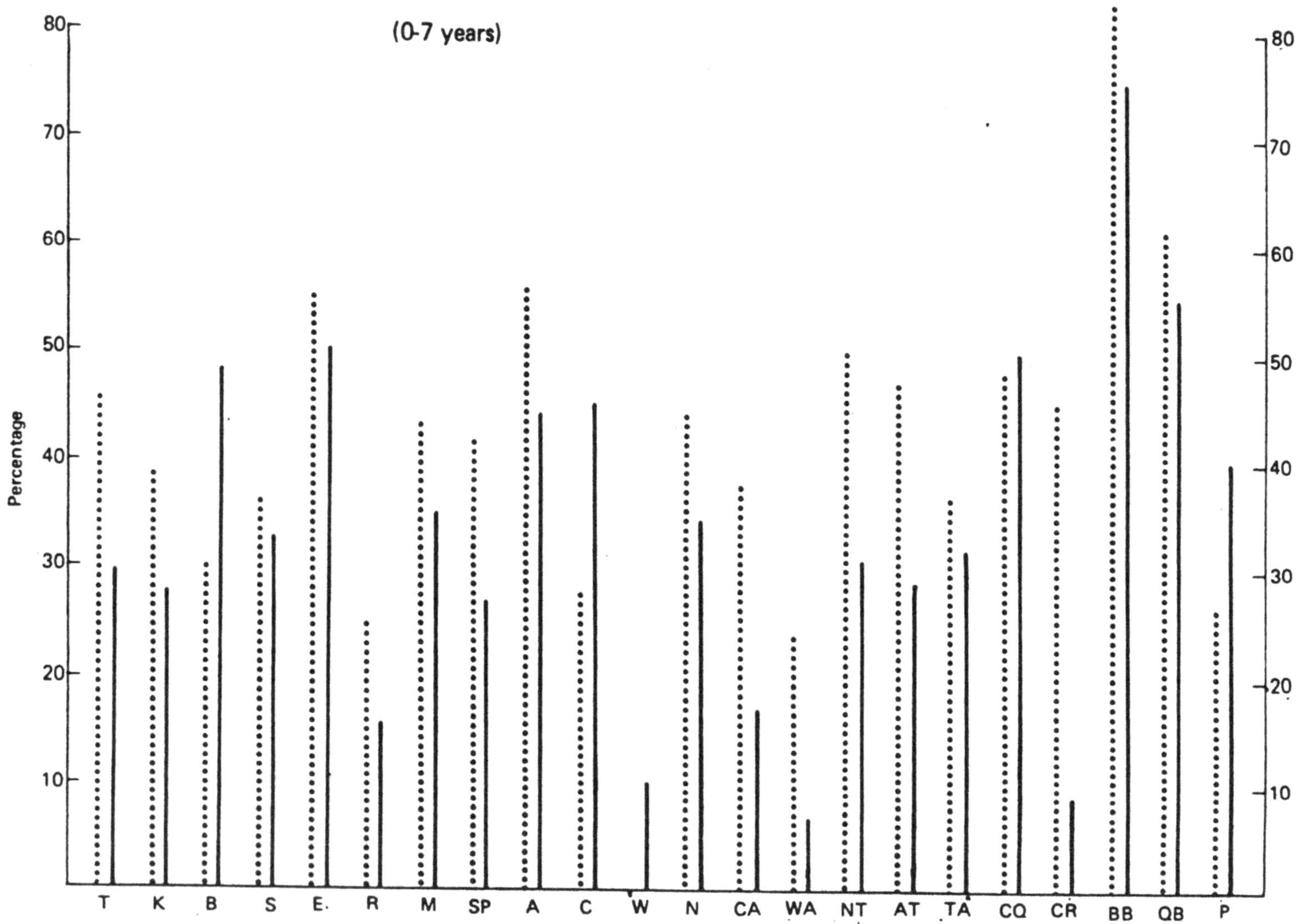

Fig. 3-1 Dental hypoplasia on the canine, by location

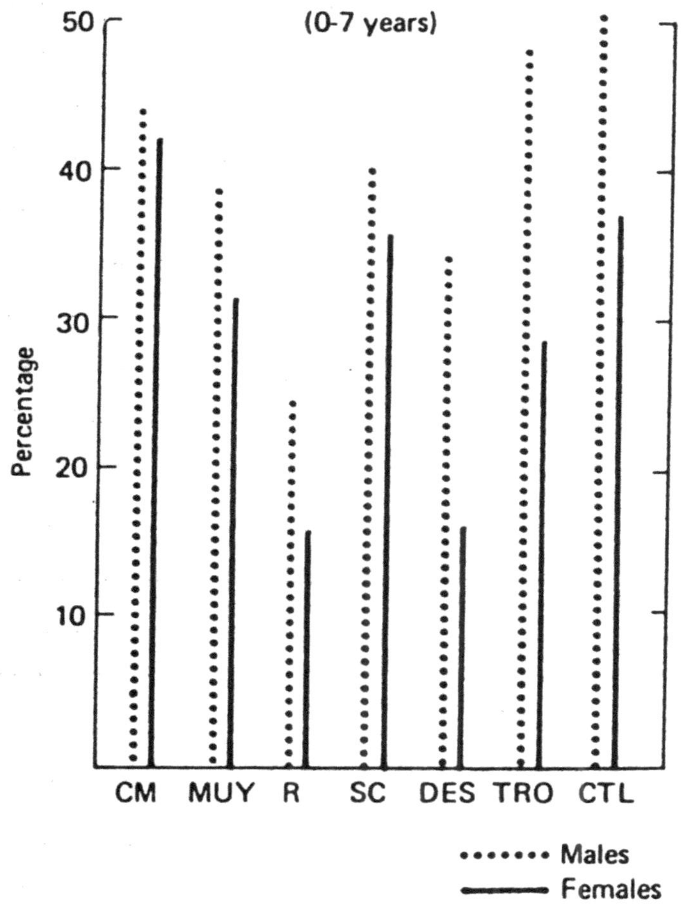

Fig. 3-2 Dental hypoplasia on the canine, by area

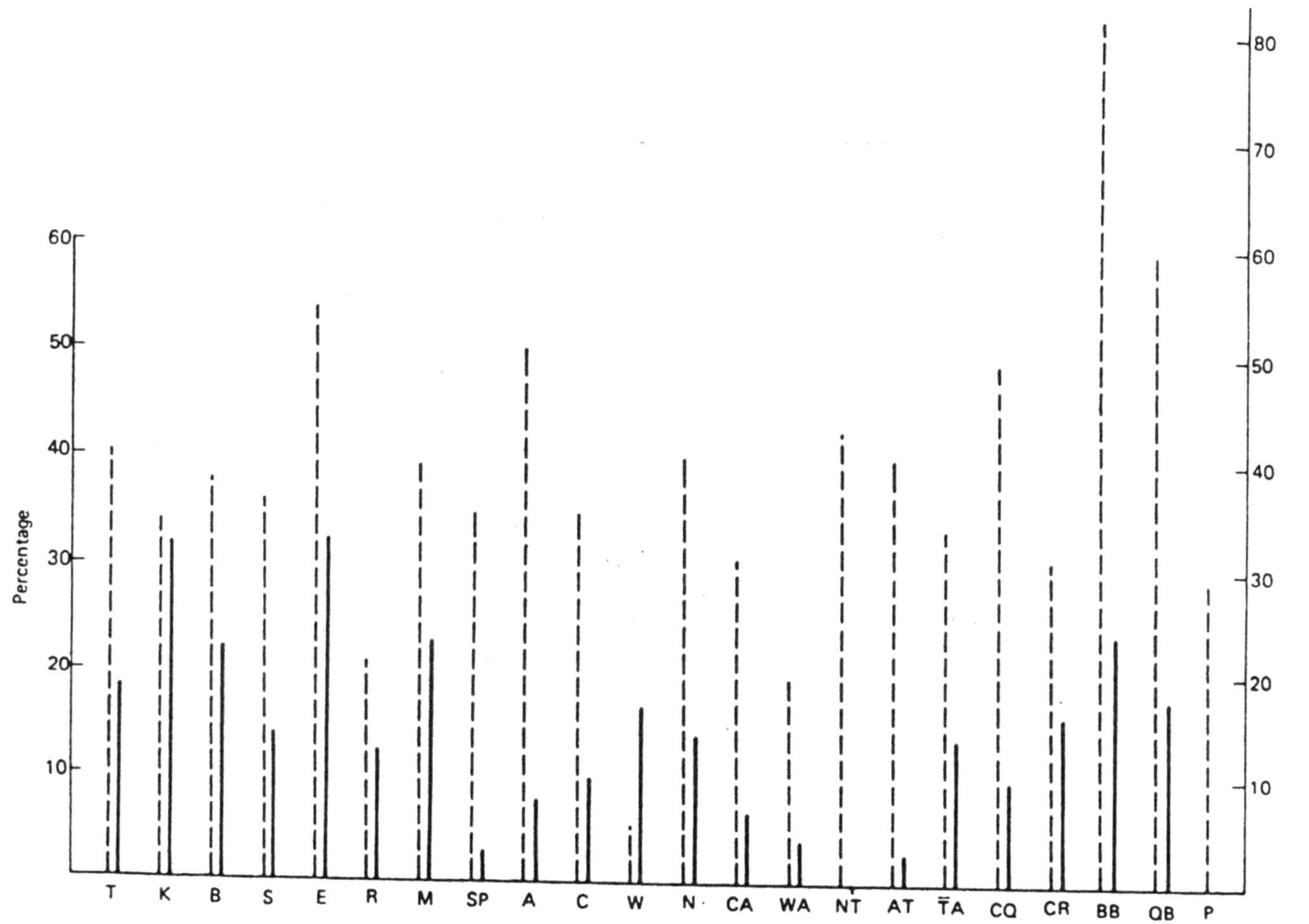

Fig. 3-3 Population frequencies of dental hypoplasia on the canine and molar three, by location

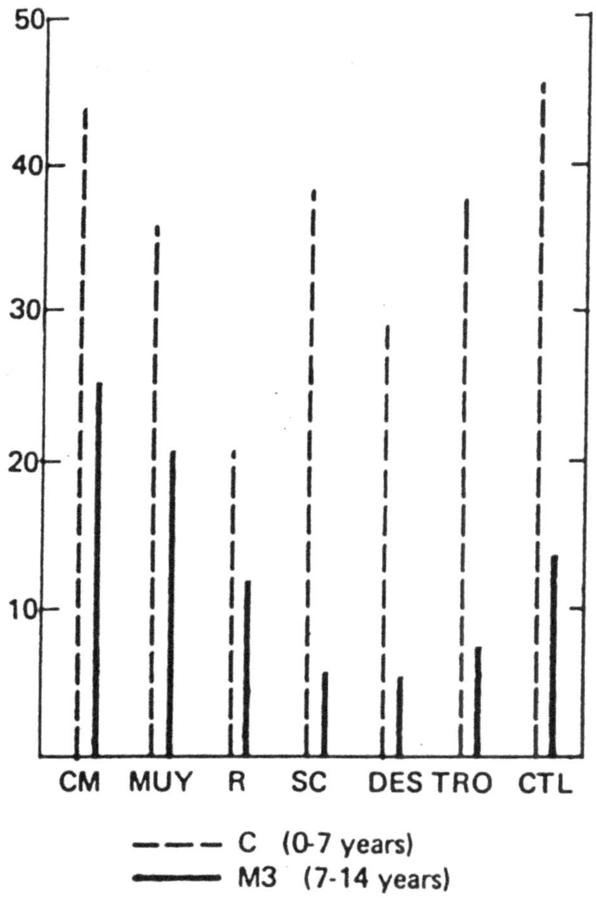

Fig. 3-4 Population frequencies of dental hypoplasia on the canine and molar three, by area

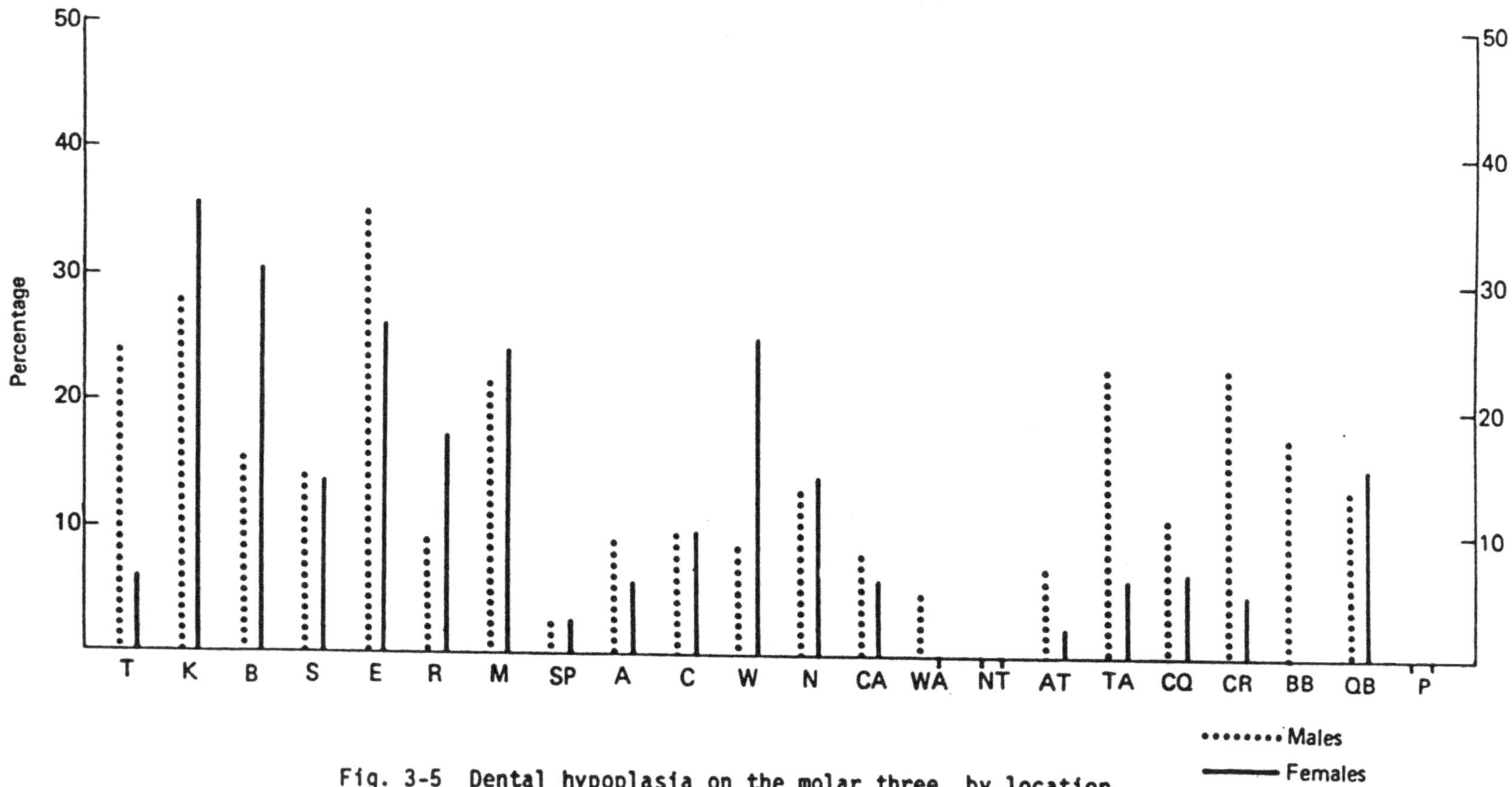

Fig. 3-5 Dental hypoplasia on the molar three, by location

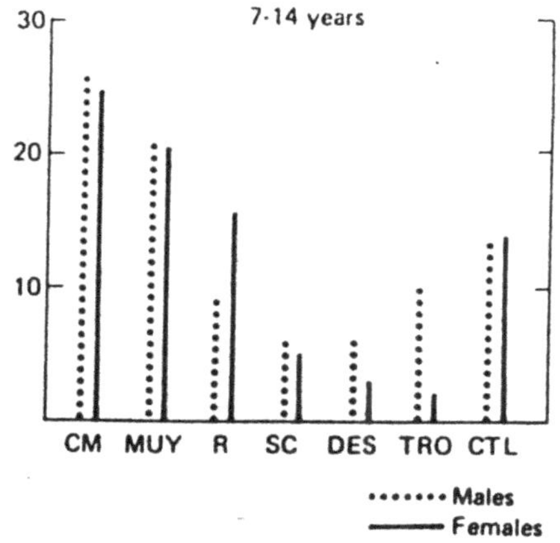

Fig. 3-6 Dental hypoplasia on the molar three, by area

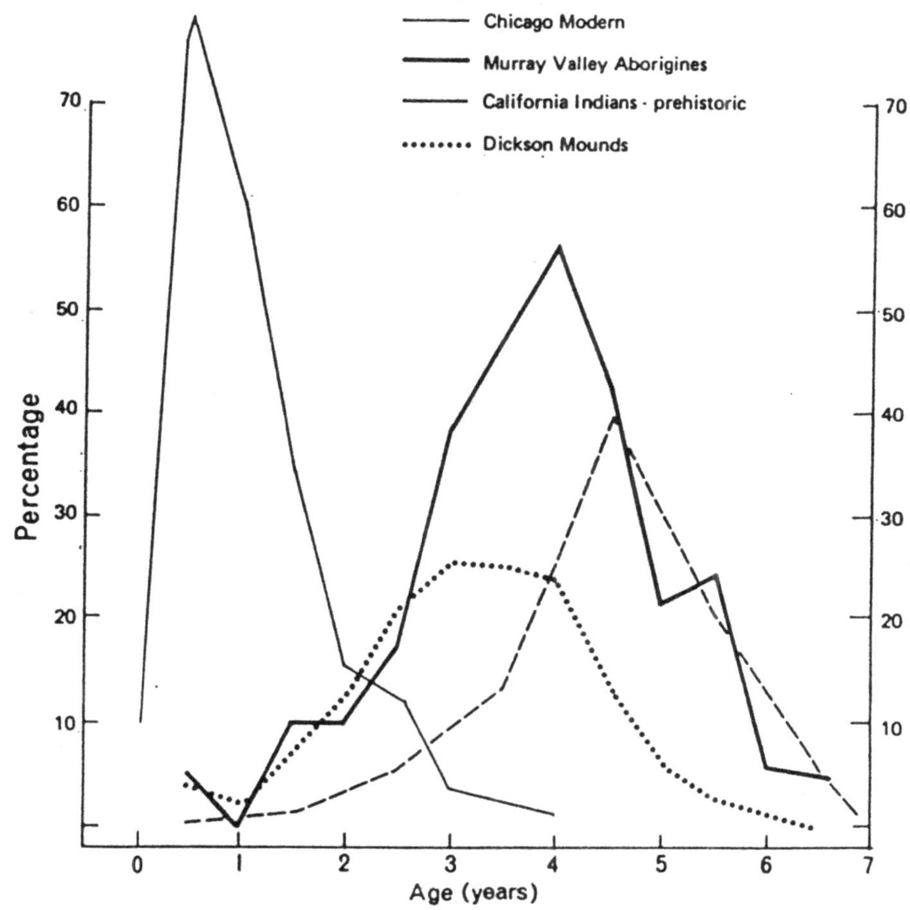

Fig. 3-7 Age distribution of dental hypoplasia in infants (After Goodman and Armelagos 1980, cited in Huss-Ashmore, et al 1982 and Green 1982)

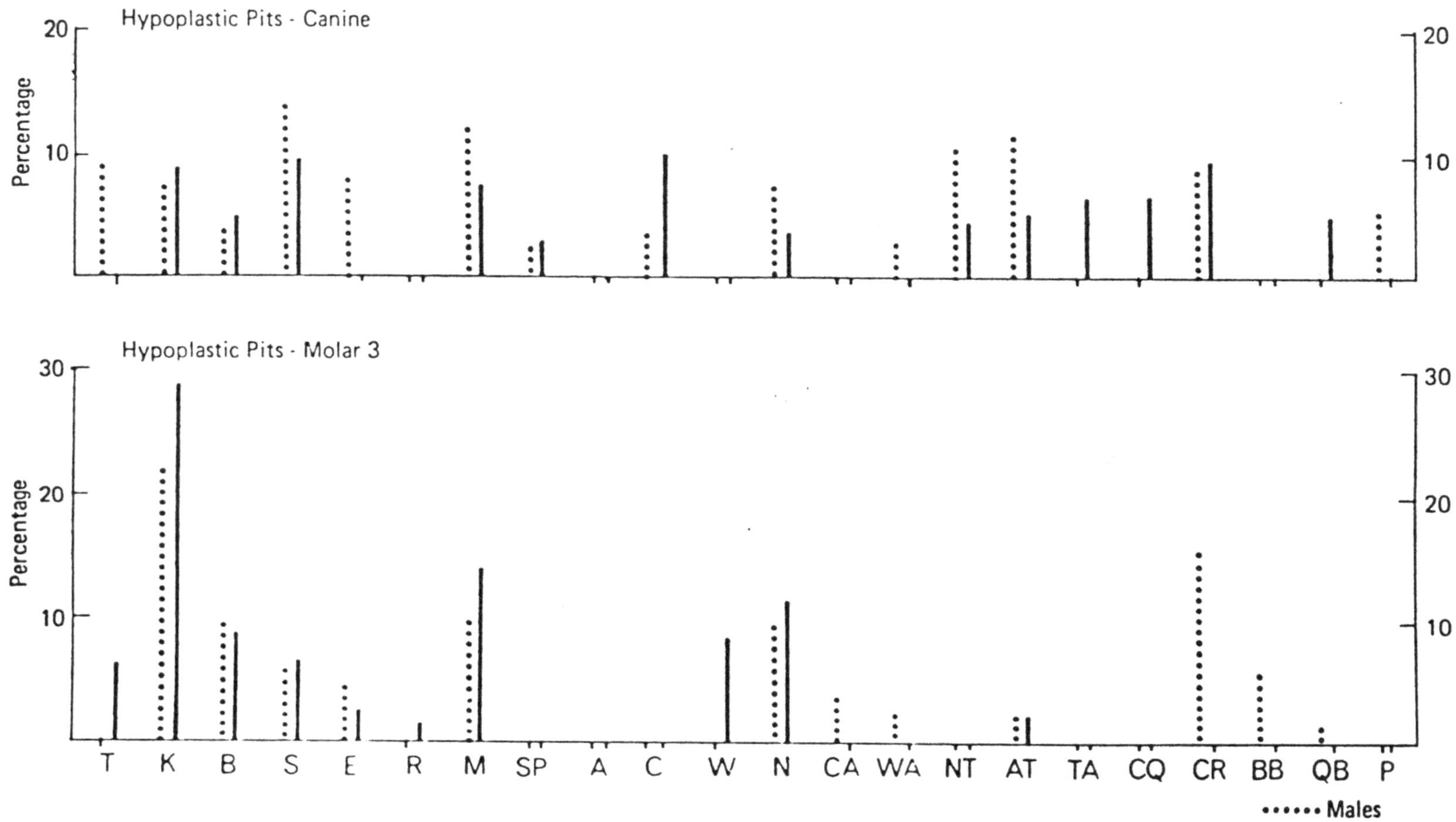

Fig. 3-8 Hypoplastic pitting

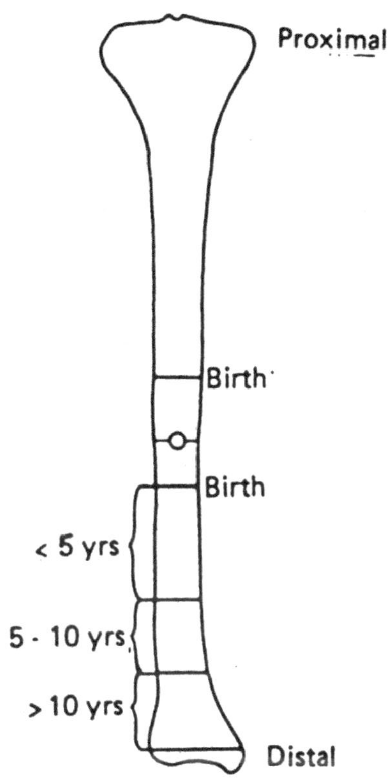

Fig. 4-1 Tibial growth divisions used in this survey (After Hunte and Hatch 1981)

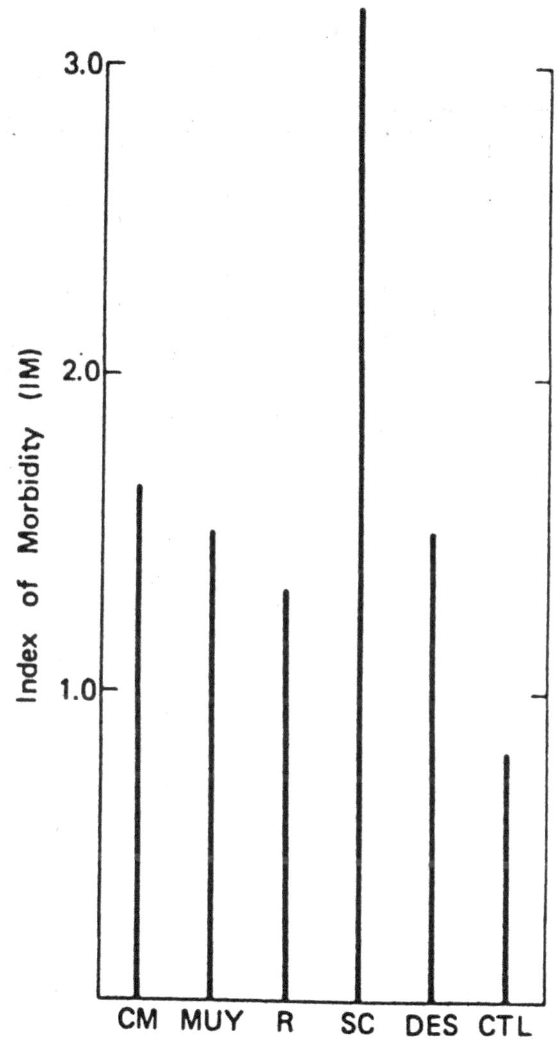

Fig. 4-2 Index of morbidity (IM) for males and females combined, by area

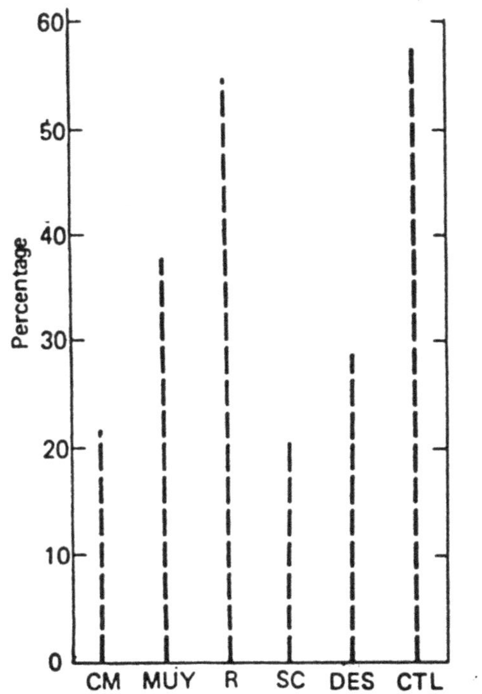

Fig. 4-3 Individuals without Harris lines, by area

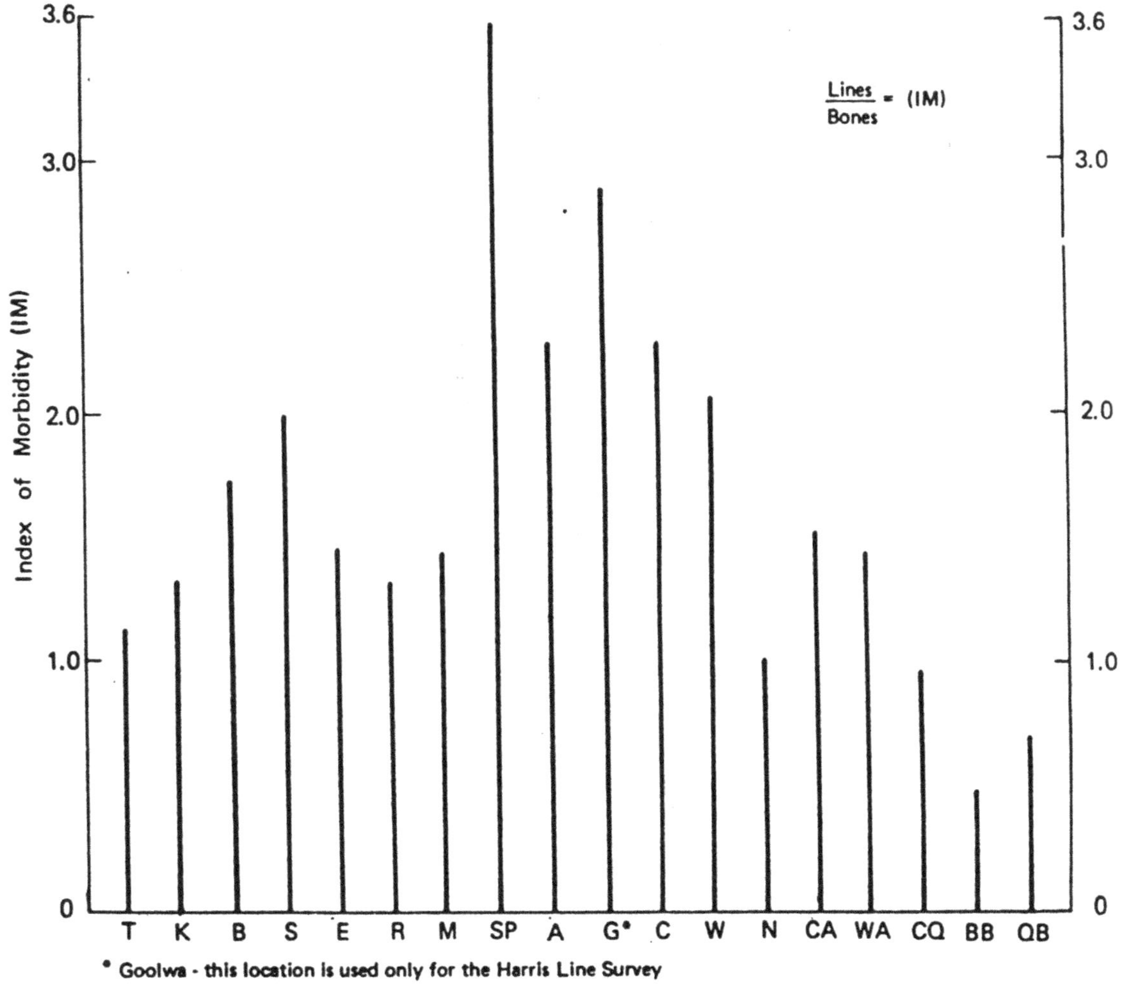

Fig. 4-4 Population morbidity indices, by location

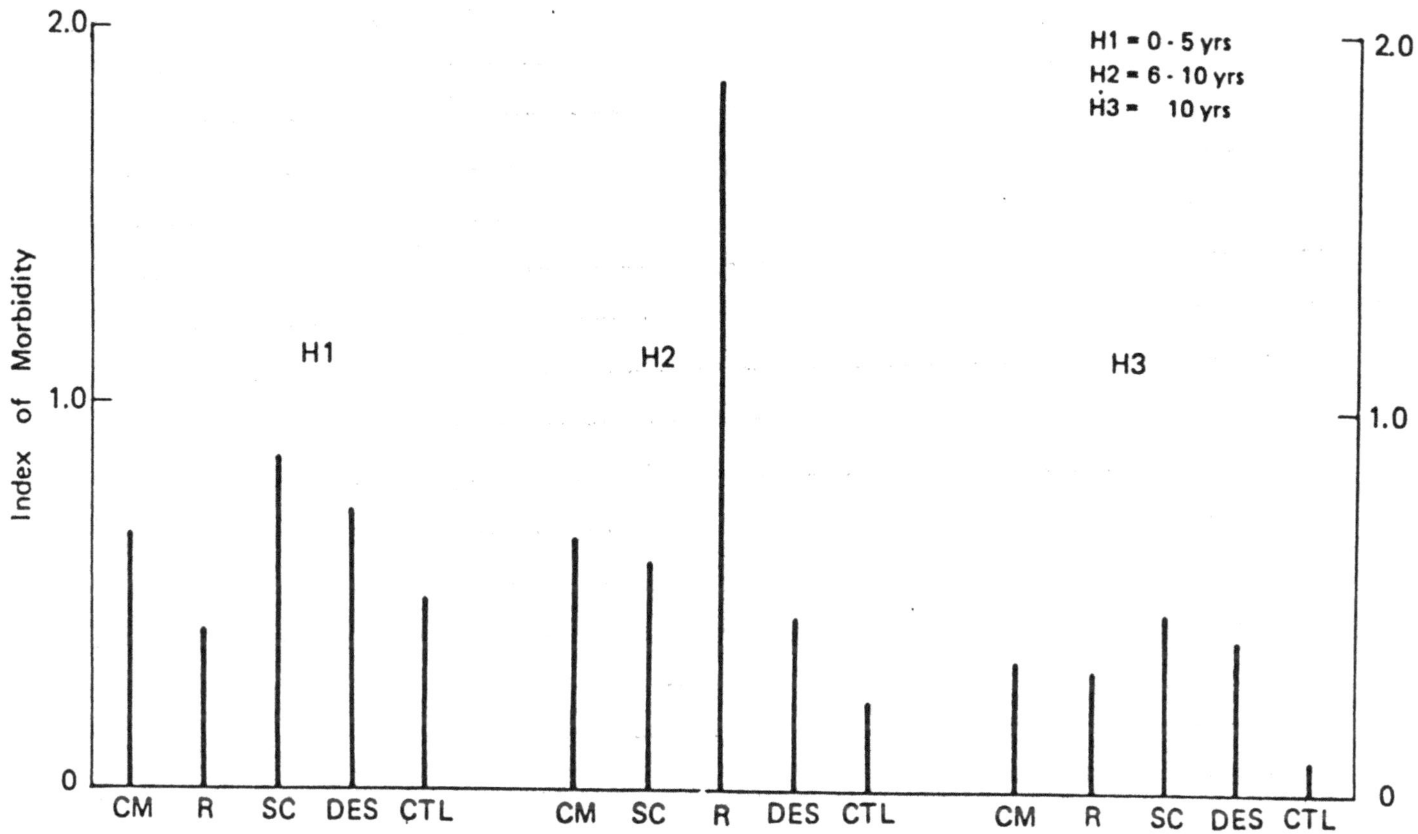

Fig. 4-5 Harris line distribution among three age groups of sub-adults, by area

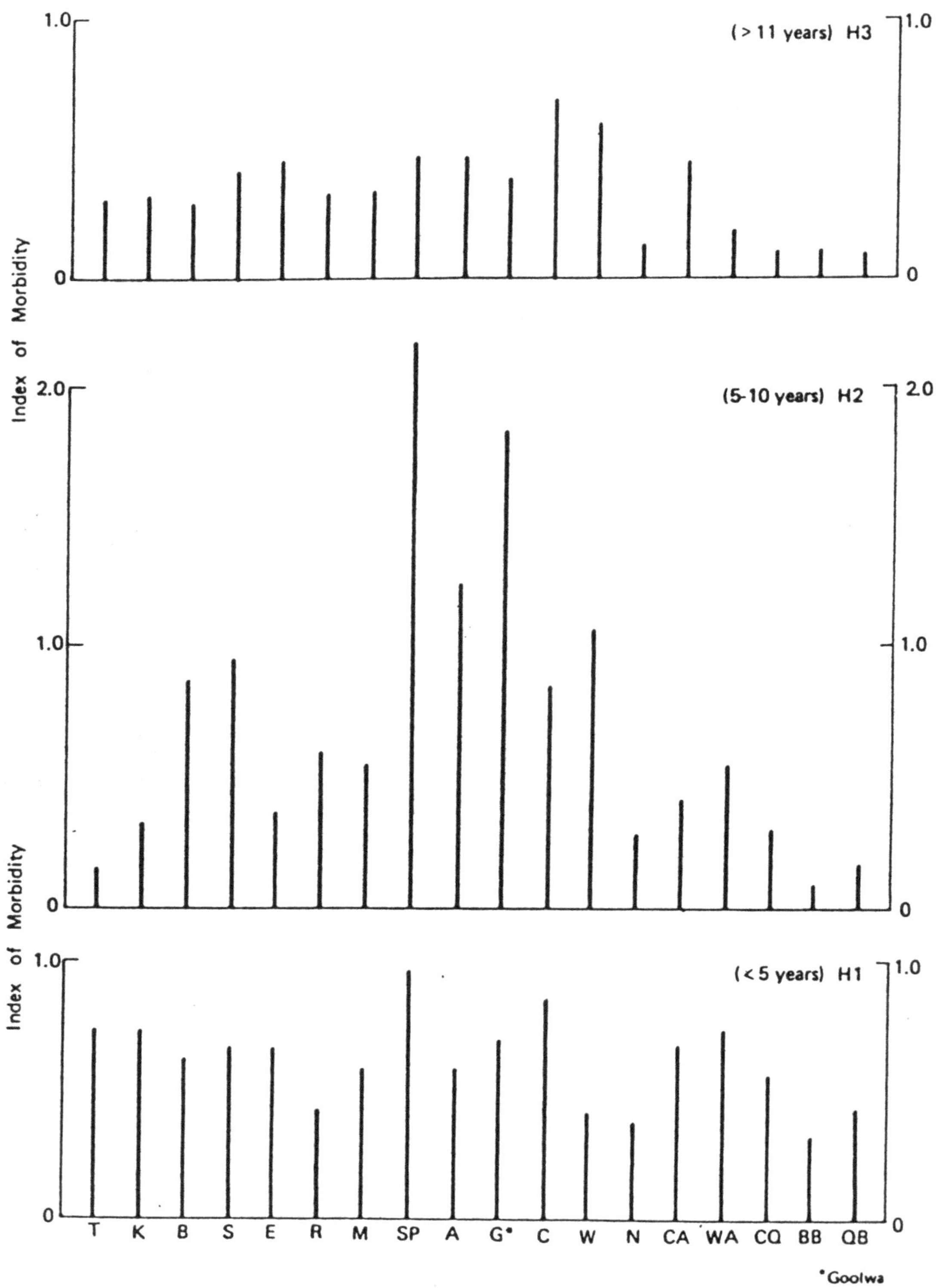

Fig. 4-6 Harris line distribution among the three age groups of sub-adults, by location

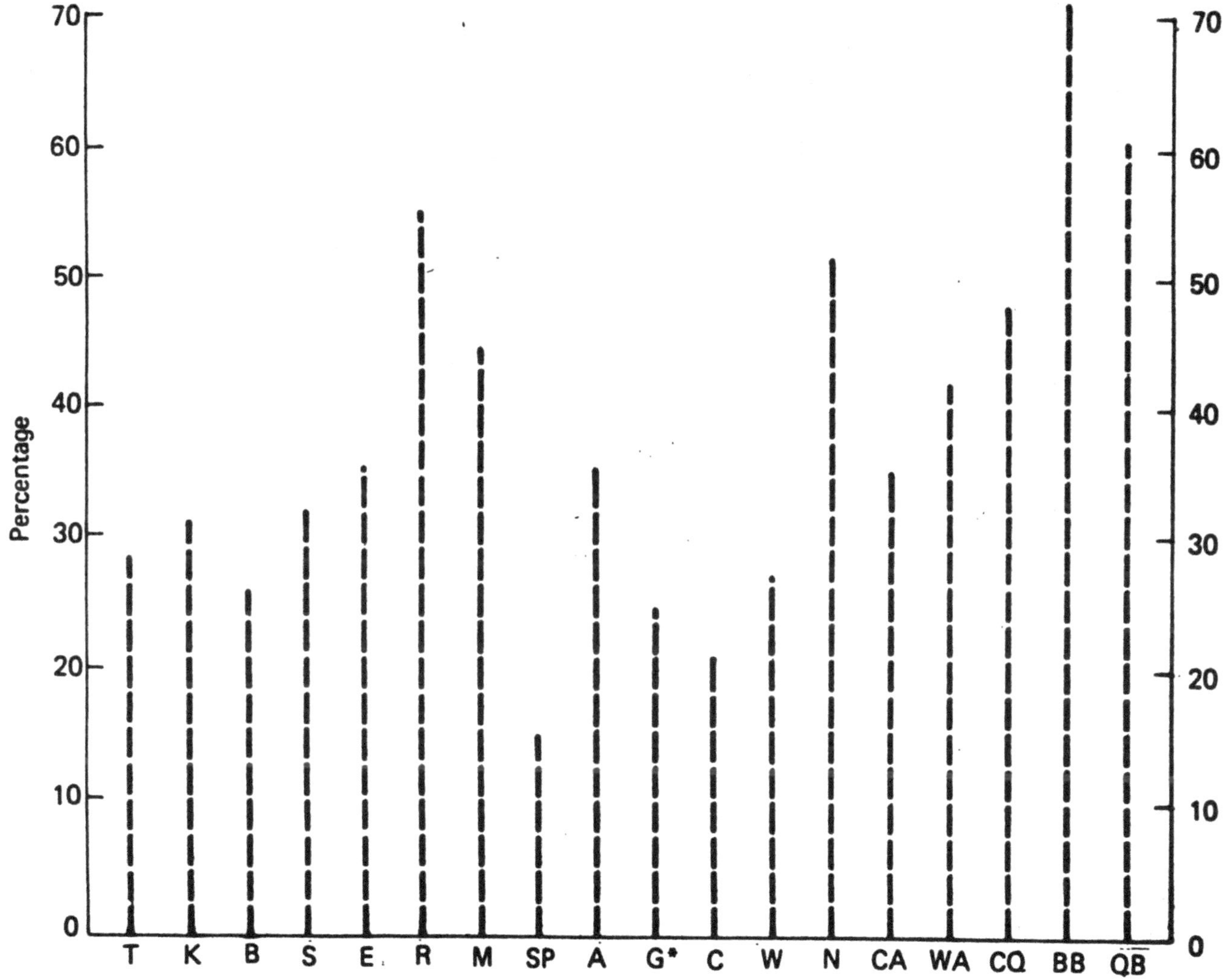

* Goolwa

Fig. 4-7 Individuals without Harris lines, by location

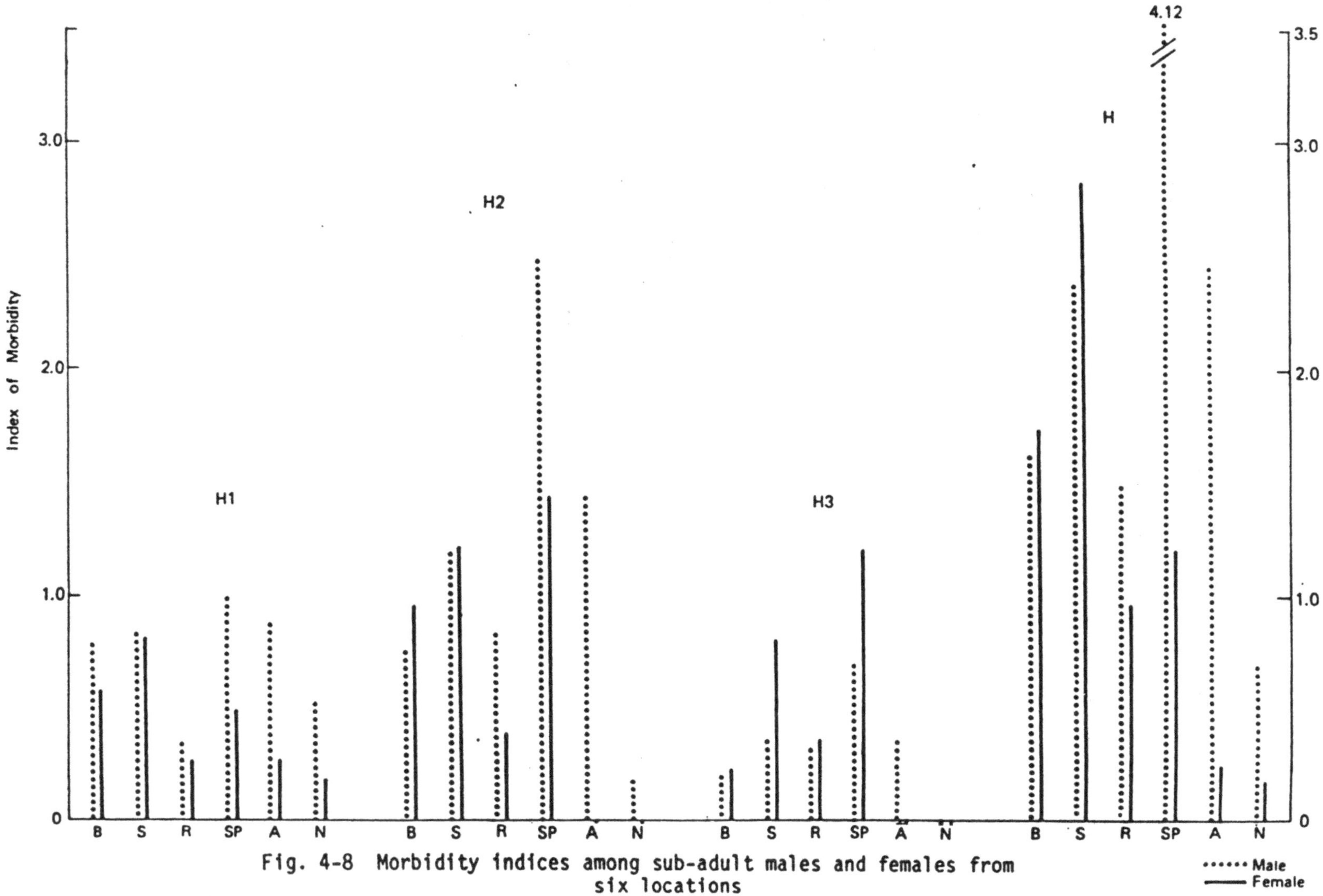

Fig. 4-8 Morbidity indices among sub-adult males and females from six locations

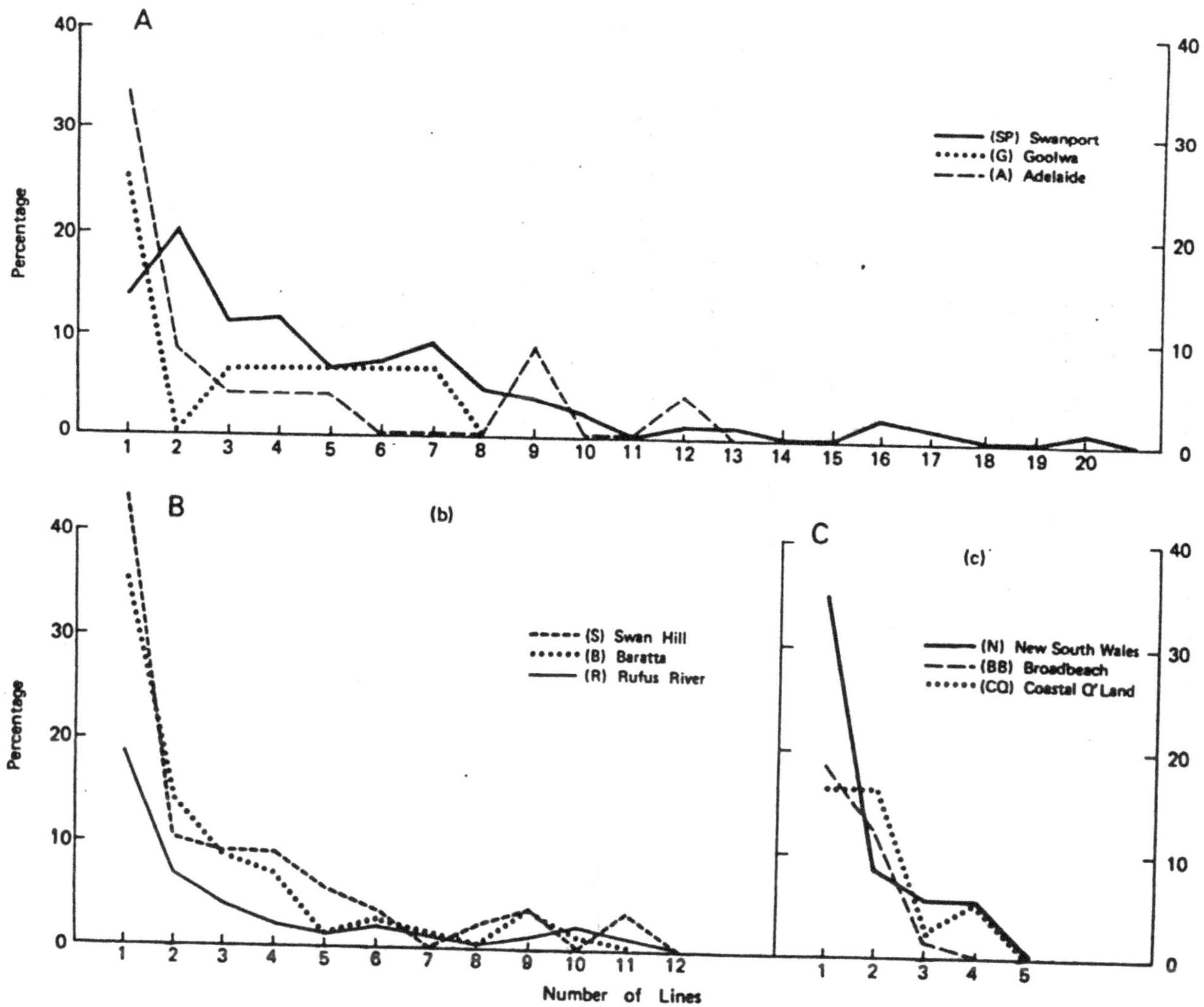

Fig. 4-9 A comparison of individuals from nine locations who have multiple Harris lines

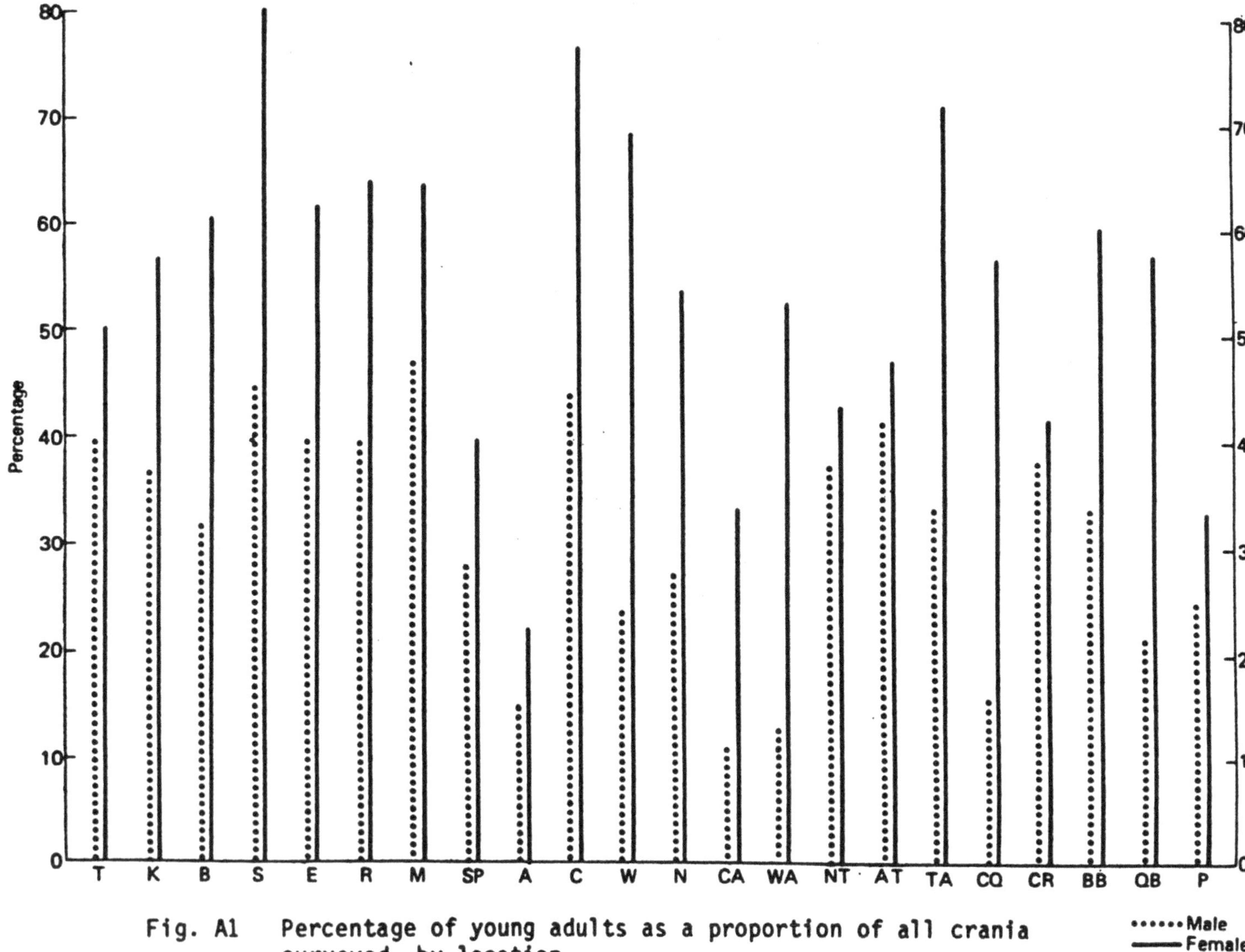

Fig. A1 Percentage of young adults as a proportion of all crania surveyed, by location

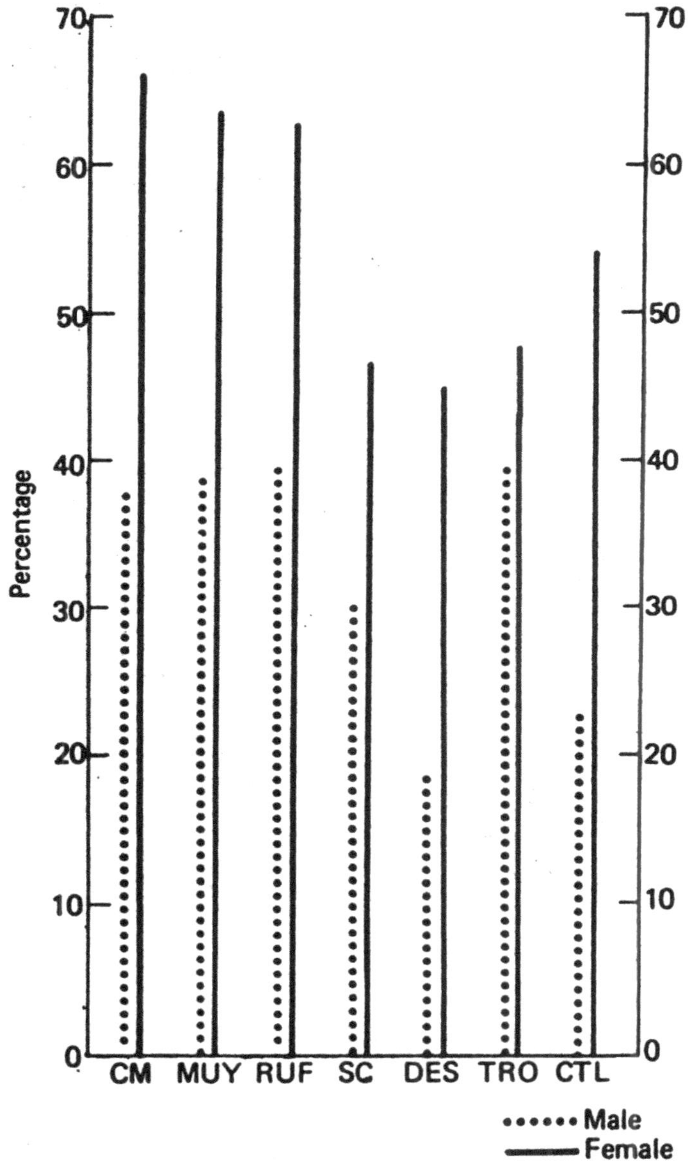

Fig. A2 Percentage of young adults as a proportion of all crania surveyed, by area

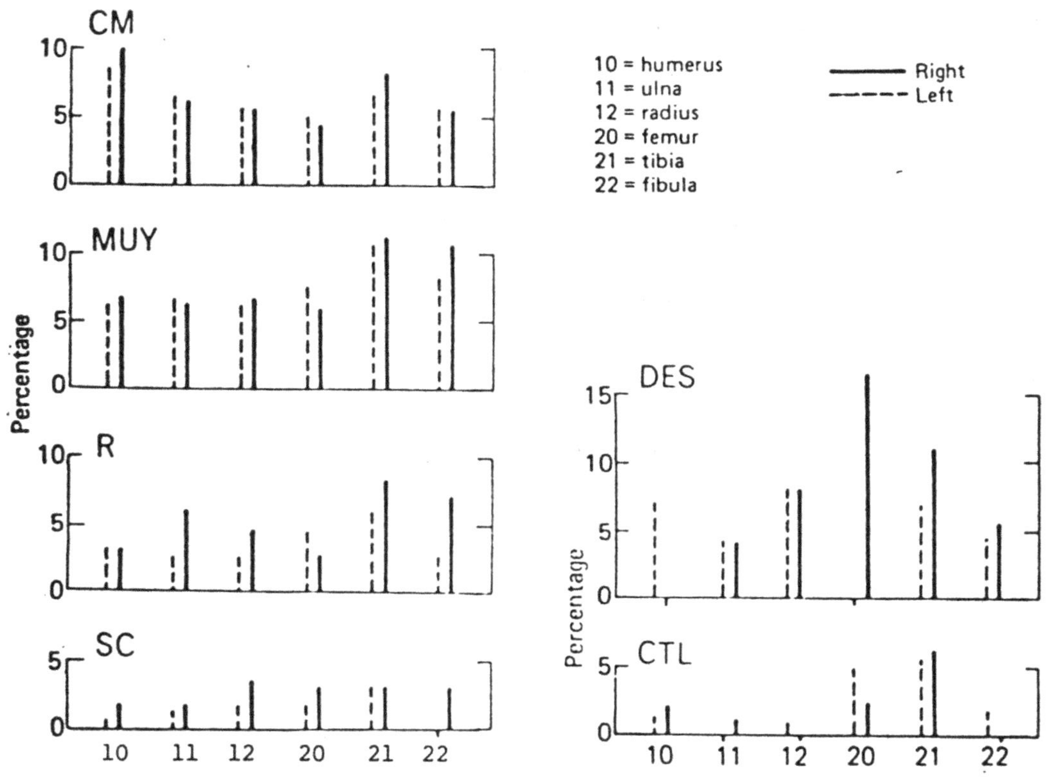

Fig. 5-1 Frequency of non-specific bone infection on lefts and rights of the six long bones

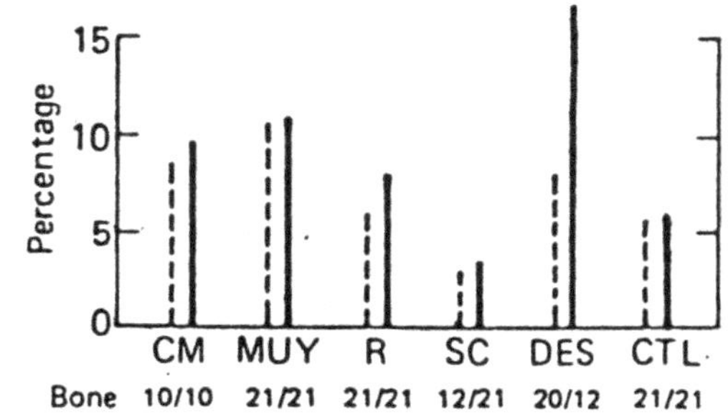

Fig. 5-2 Highest frequencies of non-specific infection from each area. The most infected left and right bones are displayed (ref. Fig. 5-1)

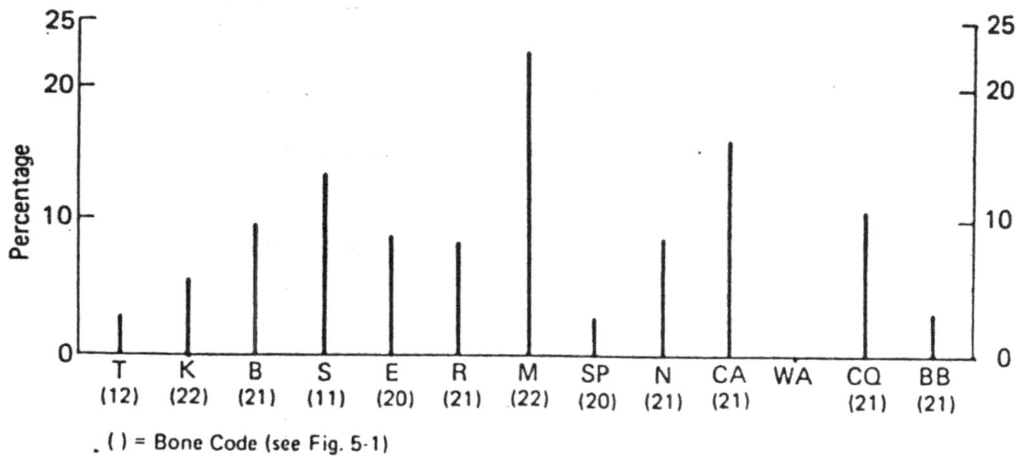

Fig. 5-3 The highest frequencies of non-specific bone infection, by area, irrespective of bone type or side

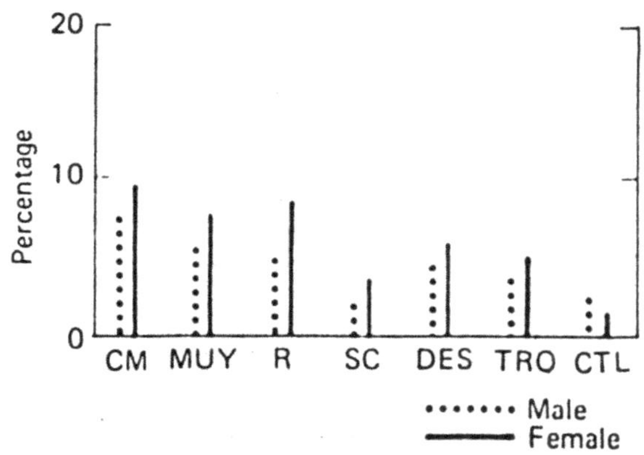

Fig. 5-4 Treponemal infection, by area

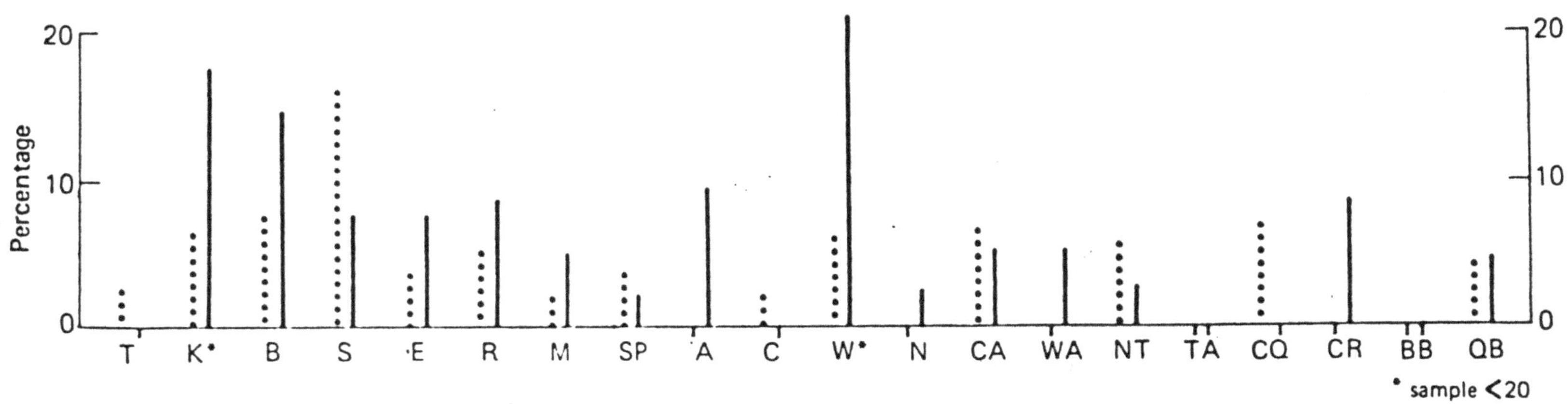

Fig. 5-5 Treponemal infection, by location

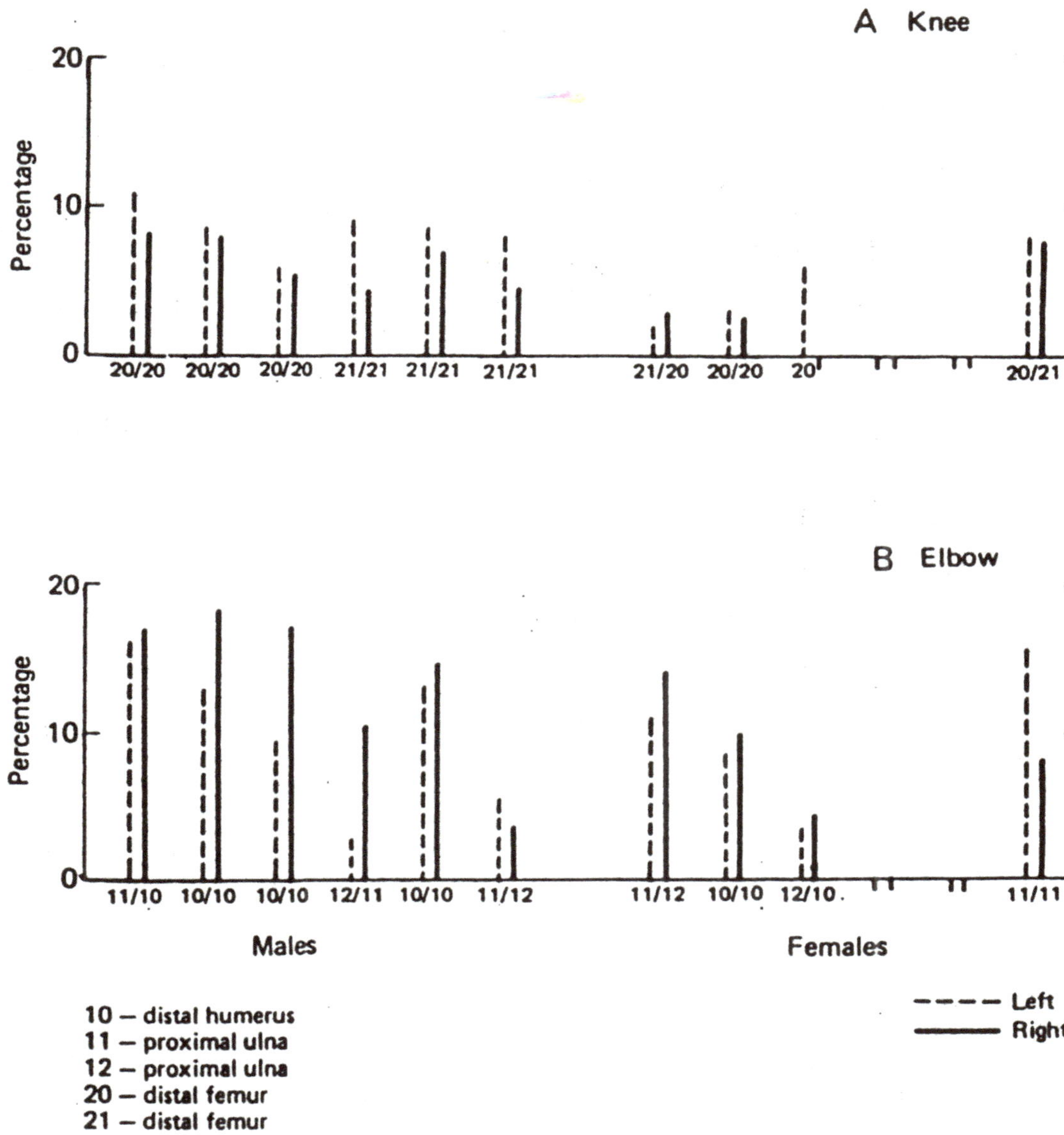

Fig. 6-1 Osteophytosis of the knee and elbow. Each frequency represents the largest for any bone that plays a role in the respective joint

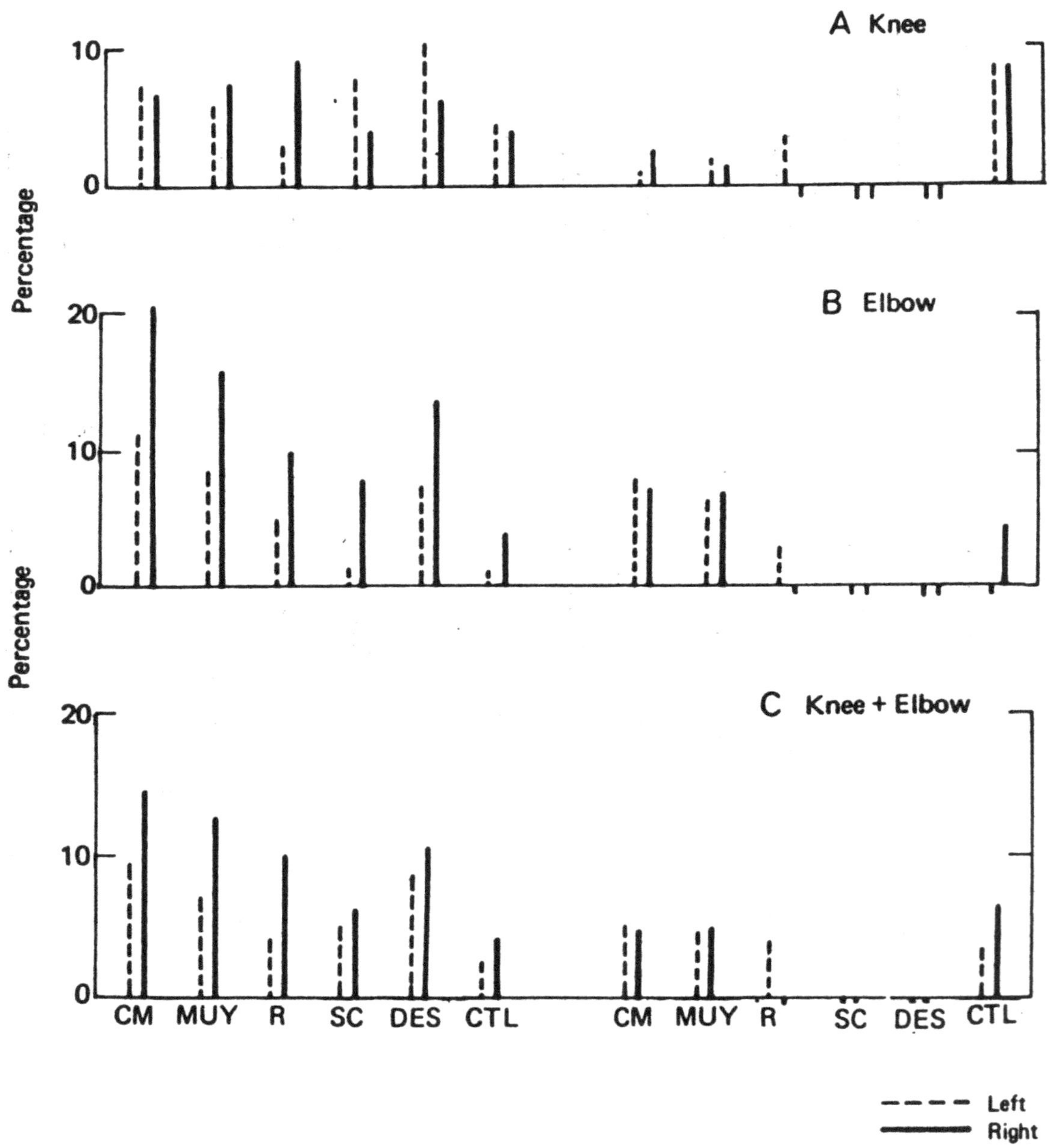

Fig. 6-2 Osteoarthritis (chronic) of the knee, elbow and knee/elbow combined, by area

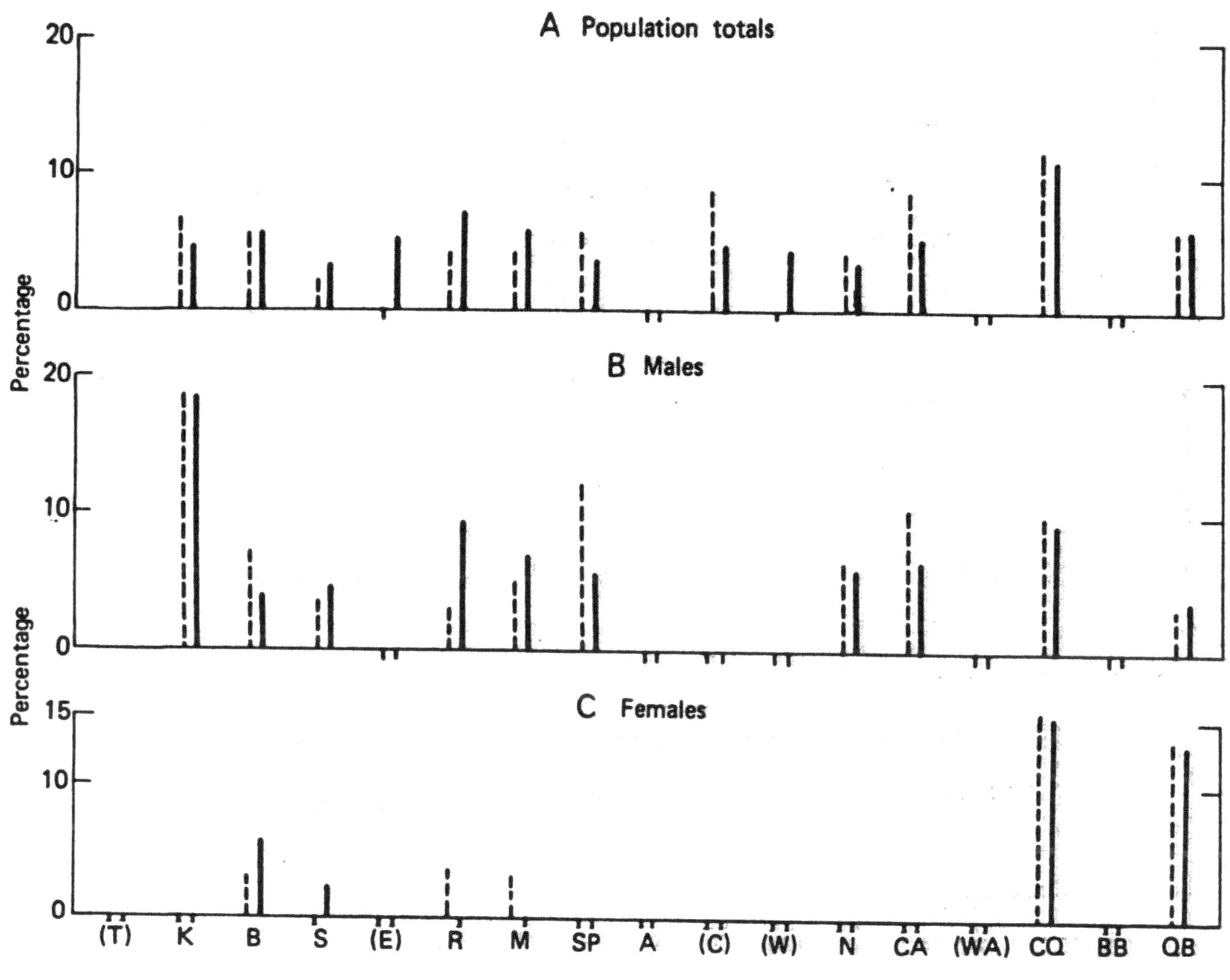

Fig. 6-3 Chronic osteoarthritis of the knee, by location

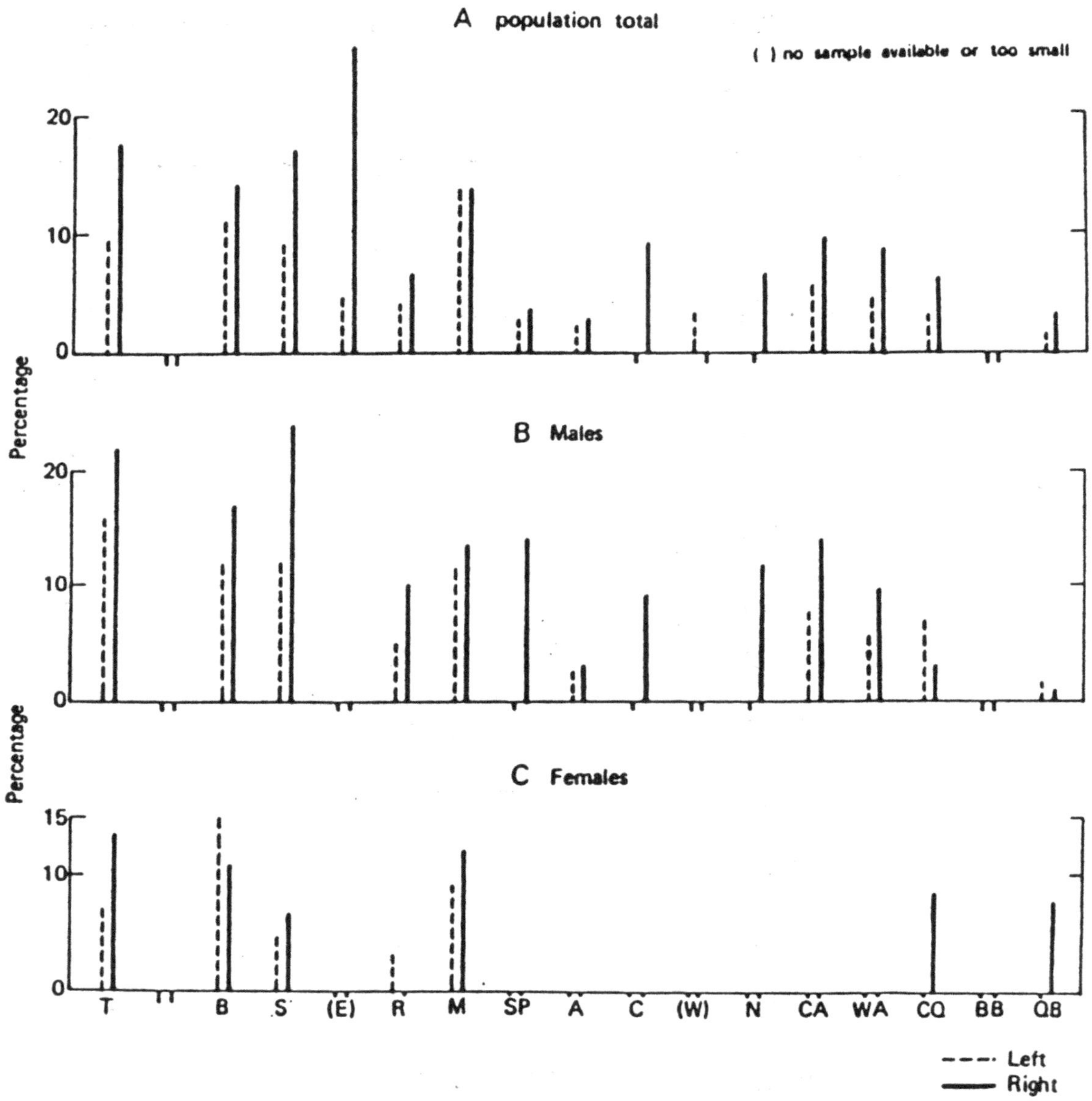

Fig. 6-4 Chronic osteoarthritis of the elbow, by location

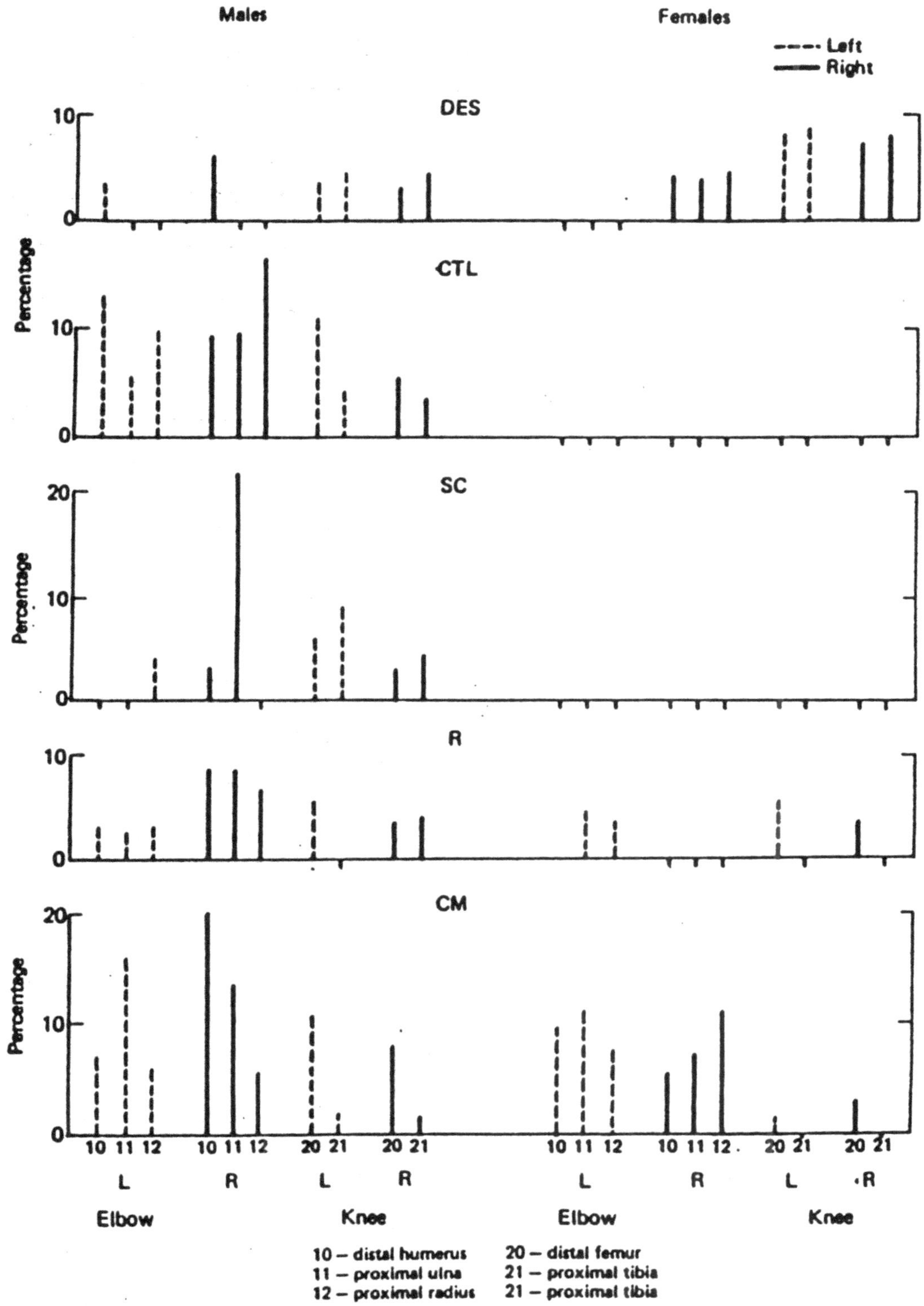

Fig. 6-5 Osteophytosis of bones comprising the elbow and knee in each area

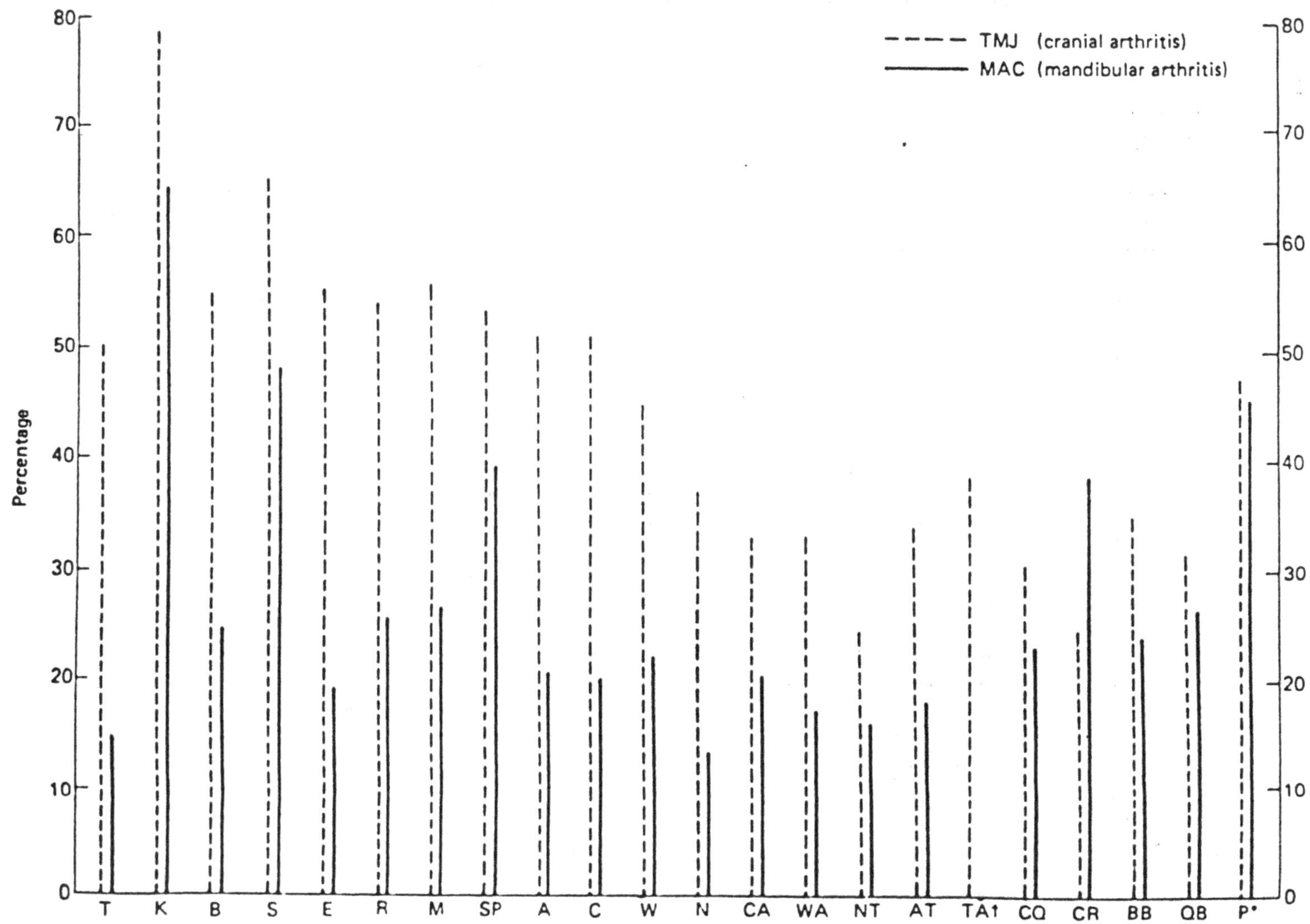

† sample n = 11
* no sample available

Fig. 6-6 Osteoarthritis of the glenoid fossae (TMJ) and the mandibular condyle (MAC). Males and females combined

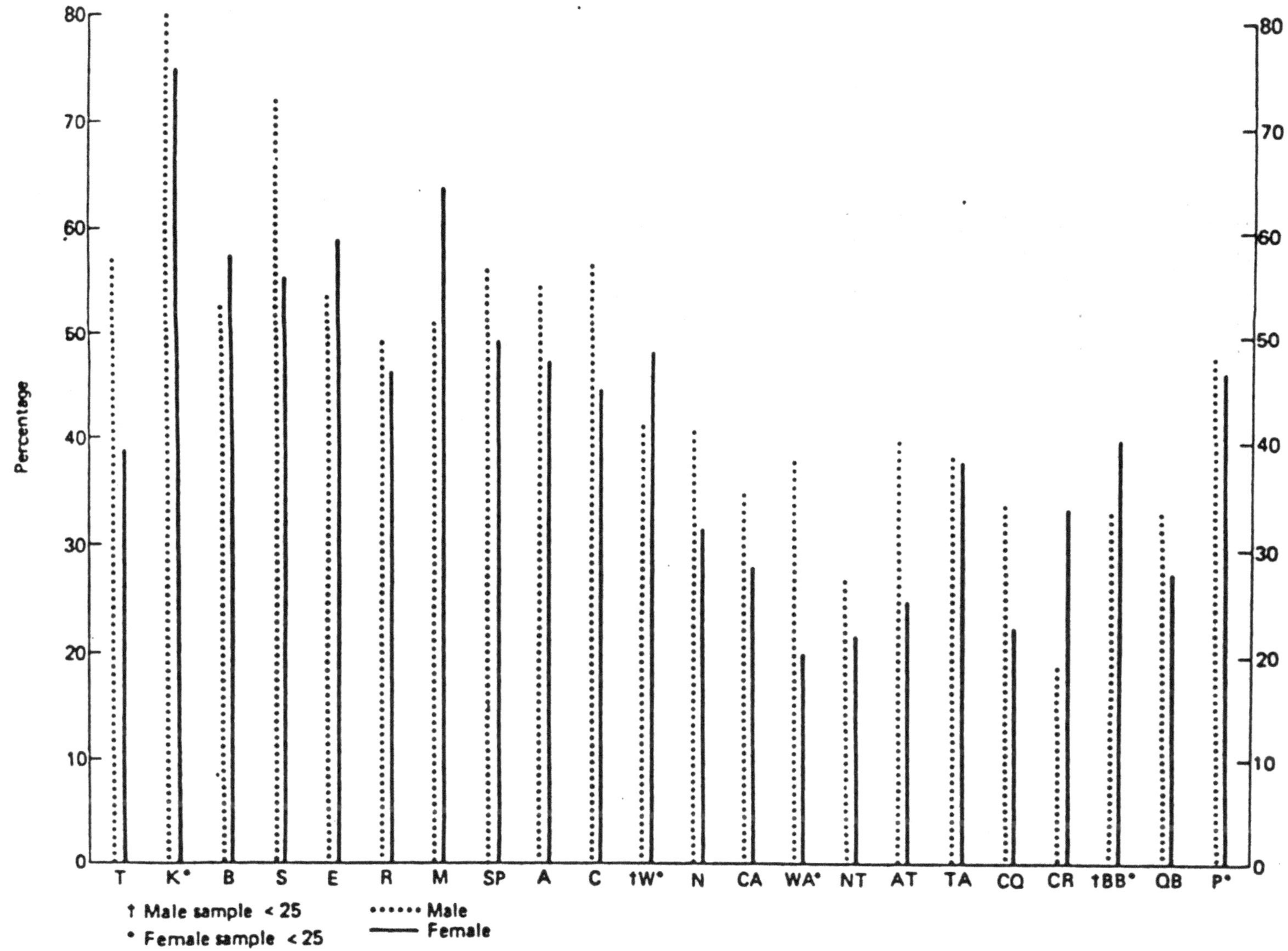

Fig. 6-7 TMJ osteoarthritis (both categories combined), by location

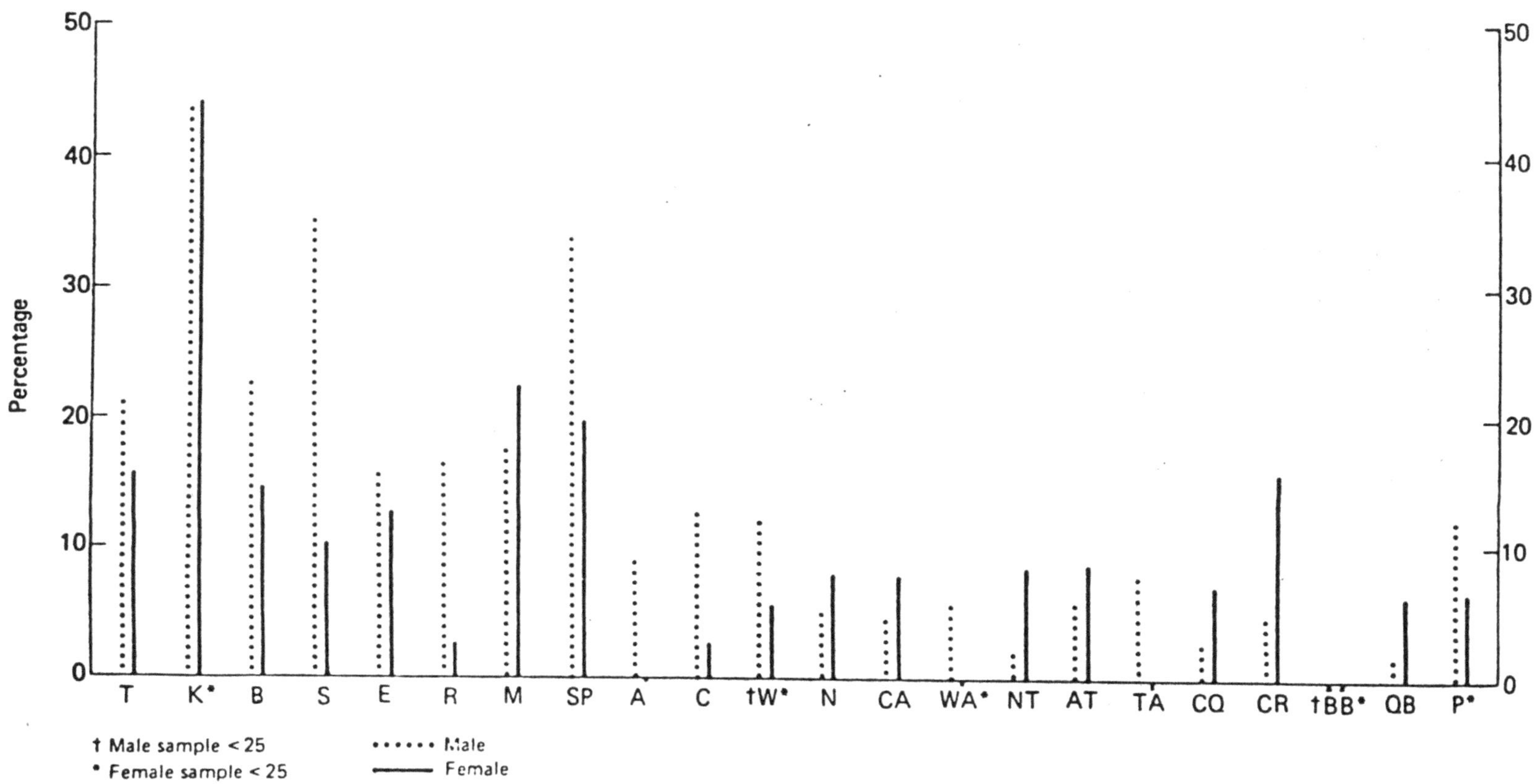

Fig. 6-8 Category 2 TMJ osteoarthritis, by location

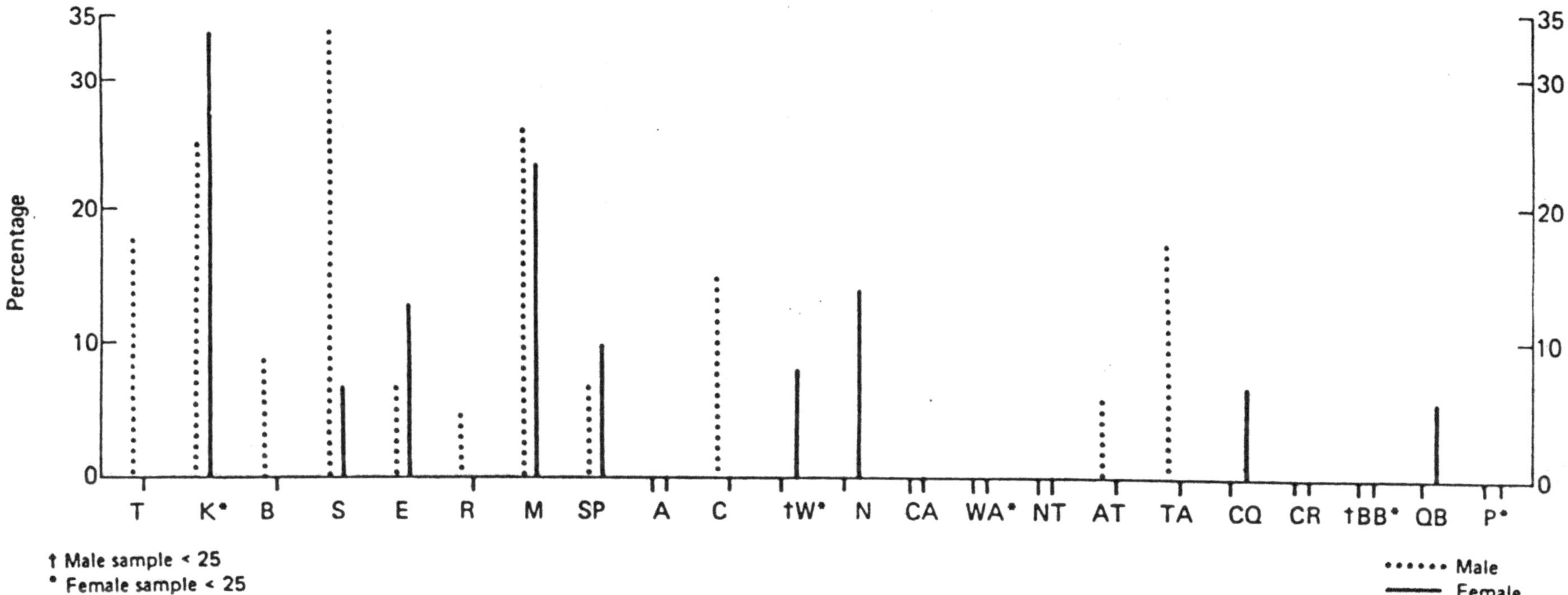

Fig. 6-9A Frequency of category 2 TMJ osteoarthritis in young adults, by location

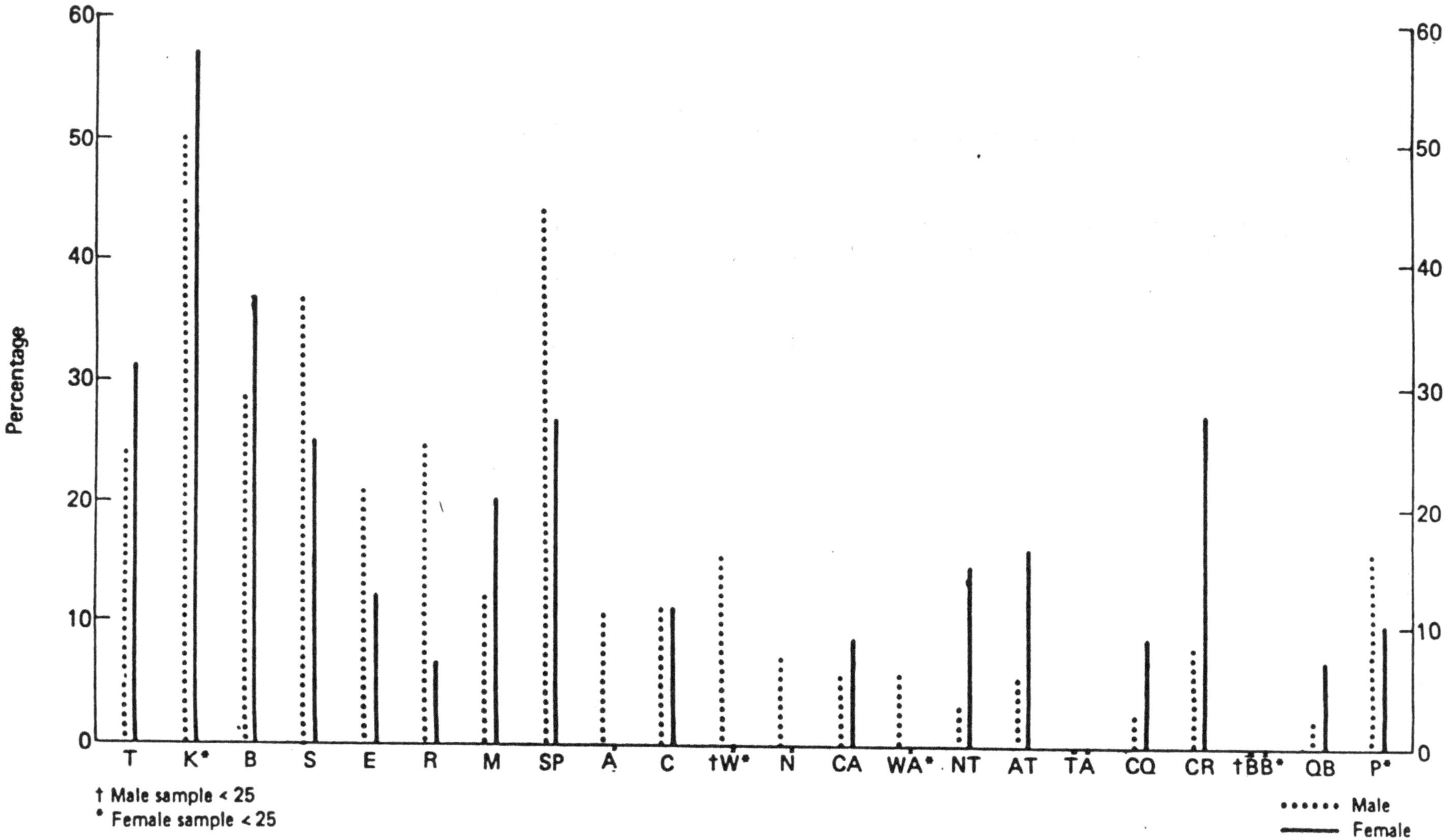

Fig. 6-9B Frequency of category 2 TMJ osteoarthritis in old adults, by location

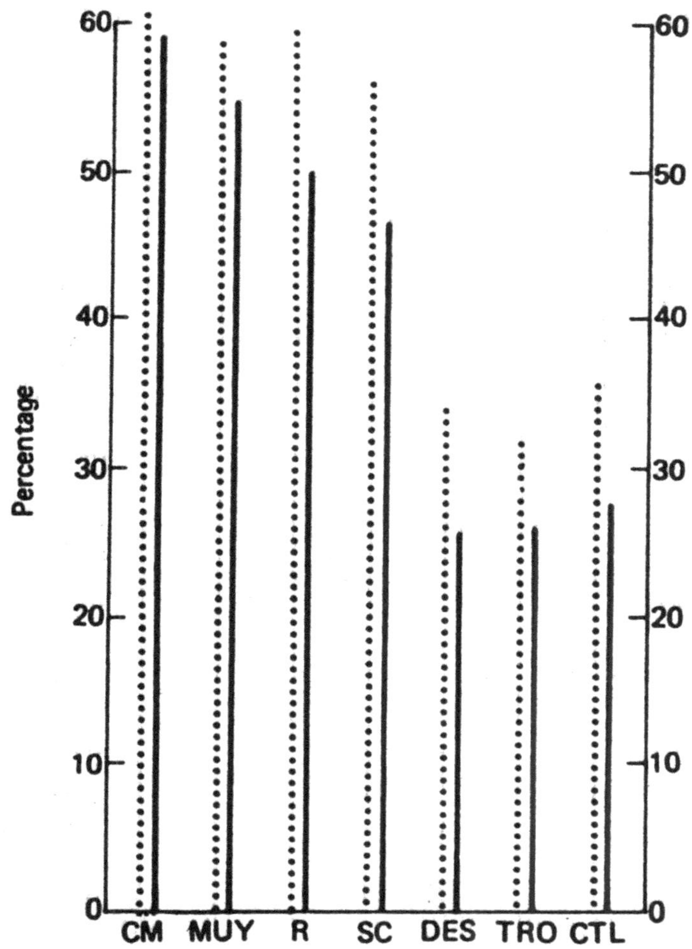

Fig. 6-10A TMJ osteoarthritis, by area

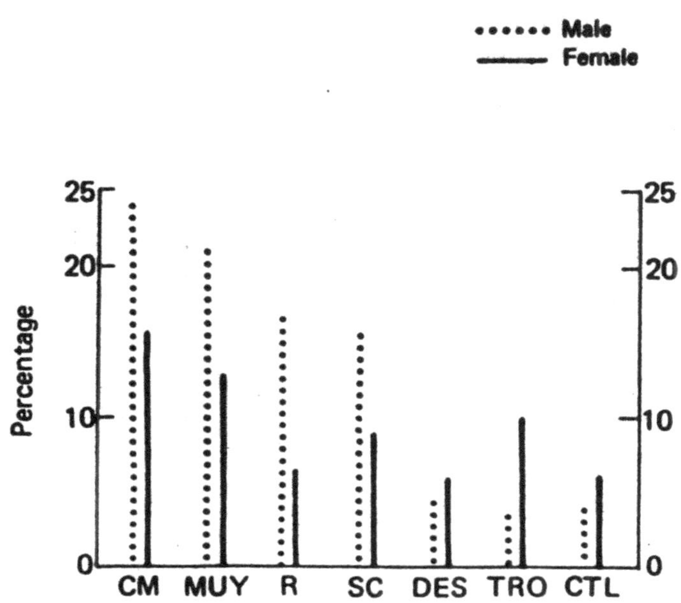

Fig. 6-10B Category 2 TMJ osteoarthritis, by area

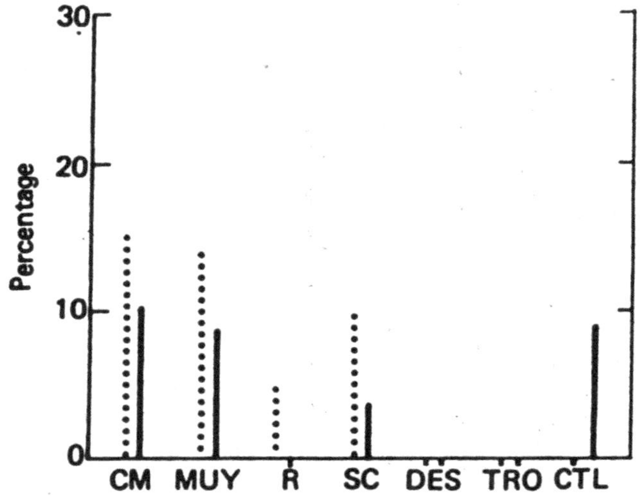

Fig. 6-10C Category 2 TMJ osteoarthritis in young adults, by area

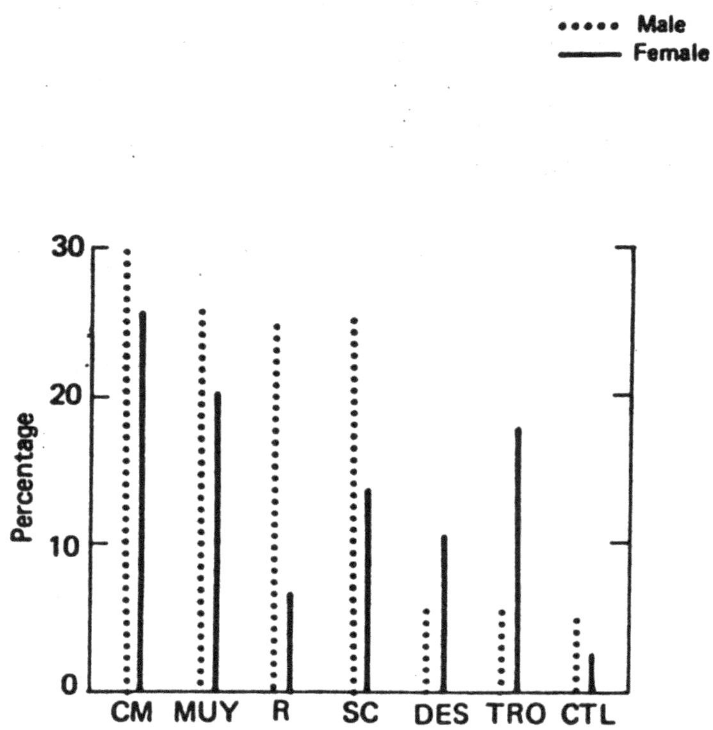

Fig. 6-10D Category 2 TMJ osteoarthritis in old adults, by area

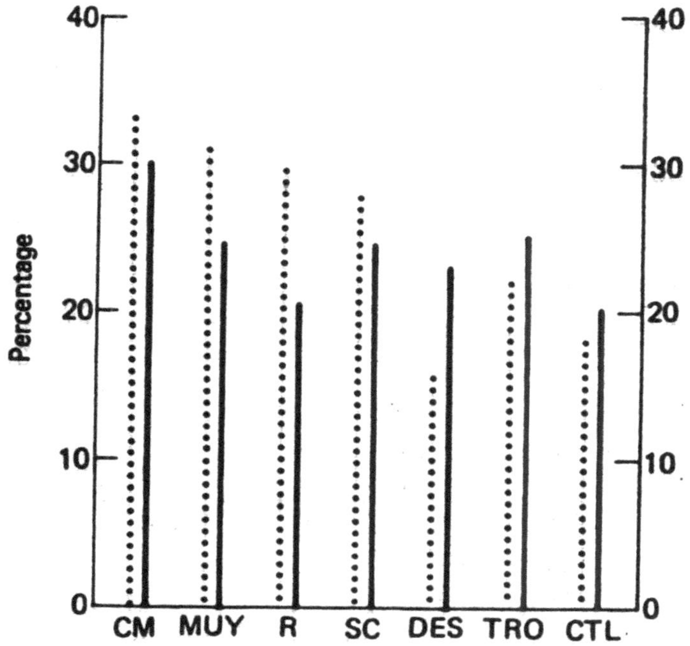

Fig. 6-11A Osteoarthritis of the mandibular condyle (MAC), by area

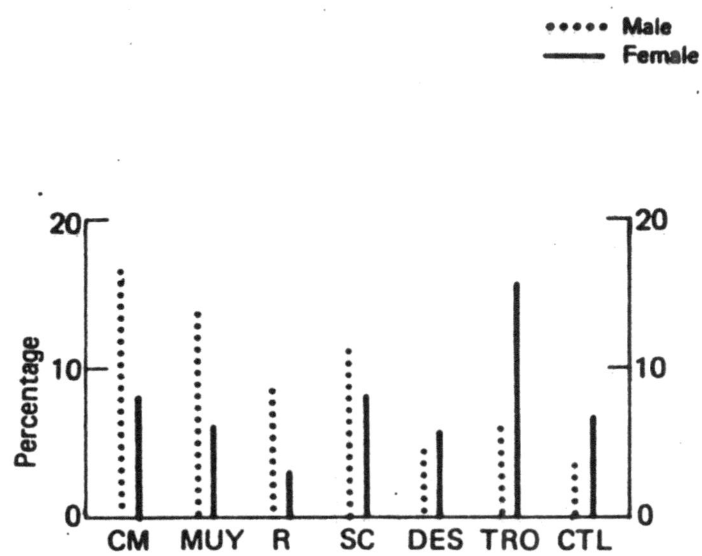

Fig. 6-11B Category 2 MAC osteoarthritis, by area

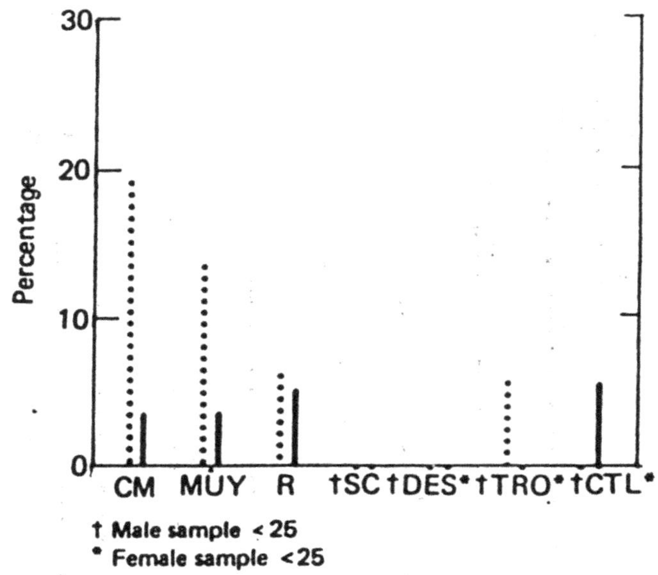

Fig. 6-11C Category 2 MAC osteoarthritis in young adults, by area

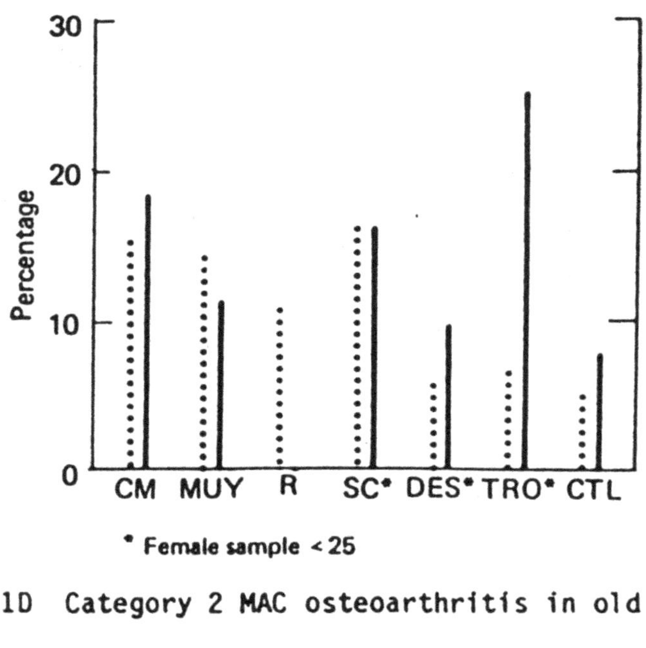

Fig. 6-11D Category 2 MAC osteoarthritis in old adults, by area

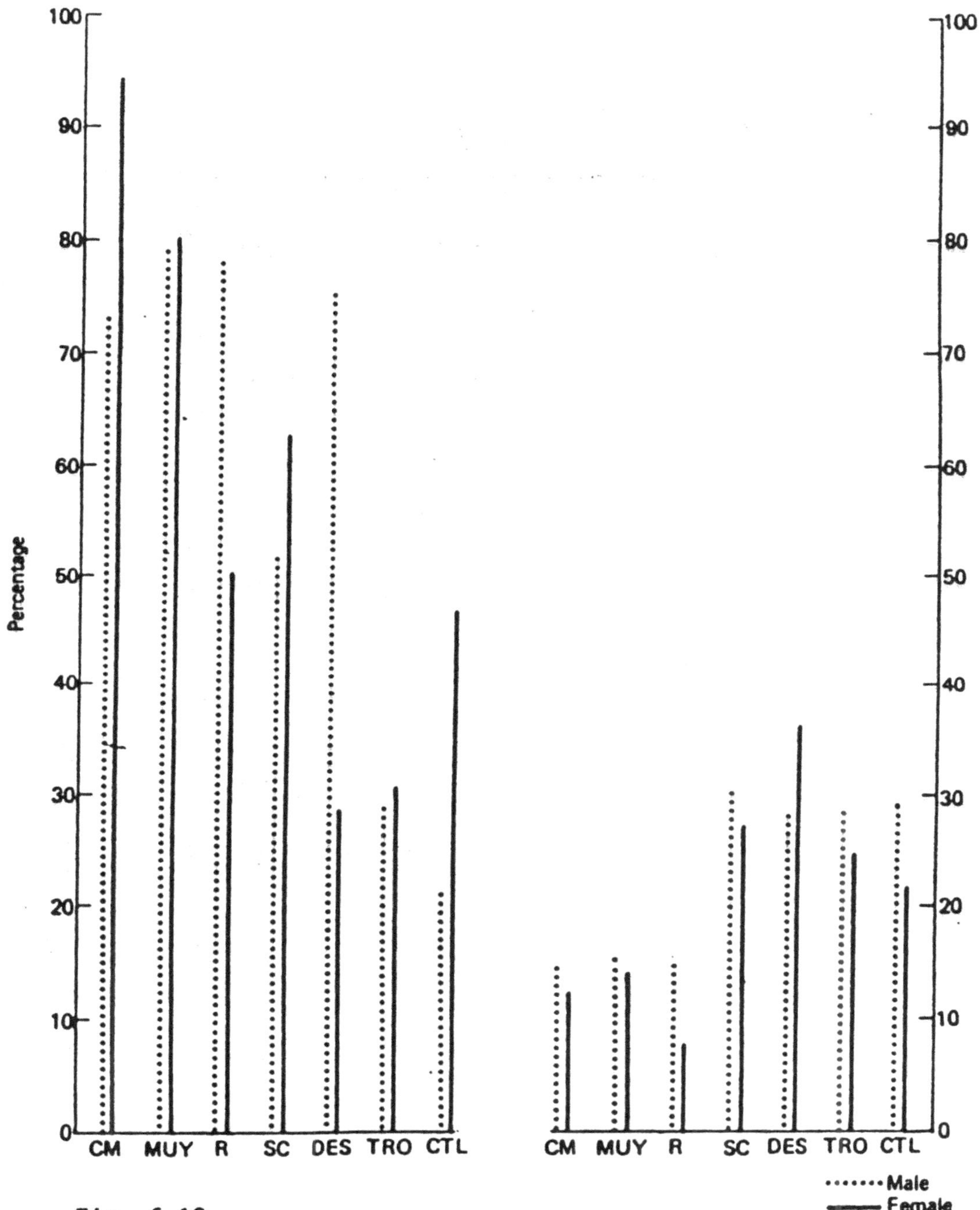

Fig. 6-12

Frequency of individuals with pre-mortem tooth loss and TMJ arthritis as a proportion of all those individuals who have either. Note: those only with incisors missing have been excluded

Fig. 6-13

Individuals with pre-mortem tooth loss as a proportion of all crania in this sub-survey (see note Fig. 6-12)

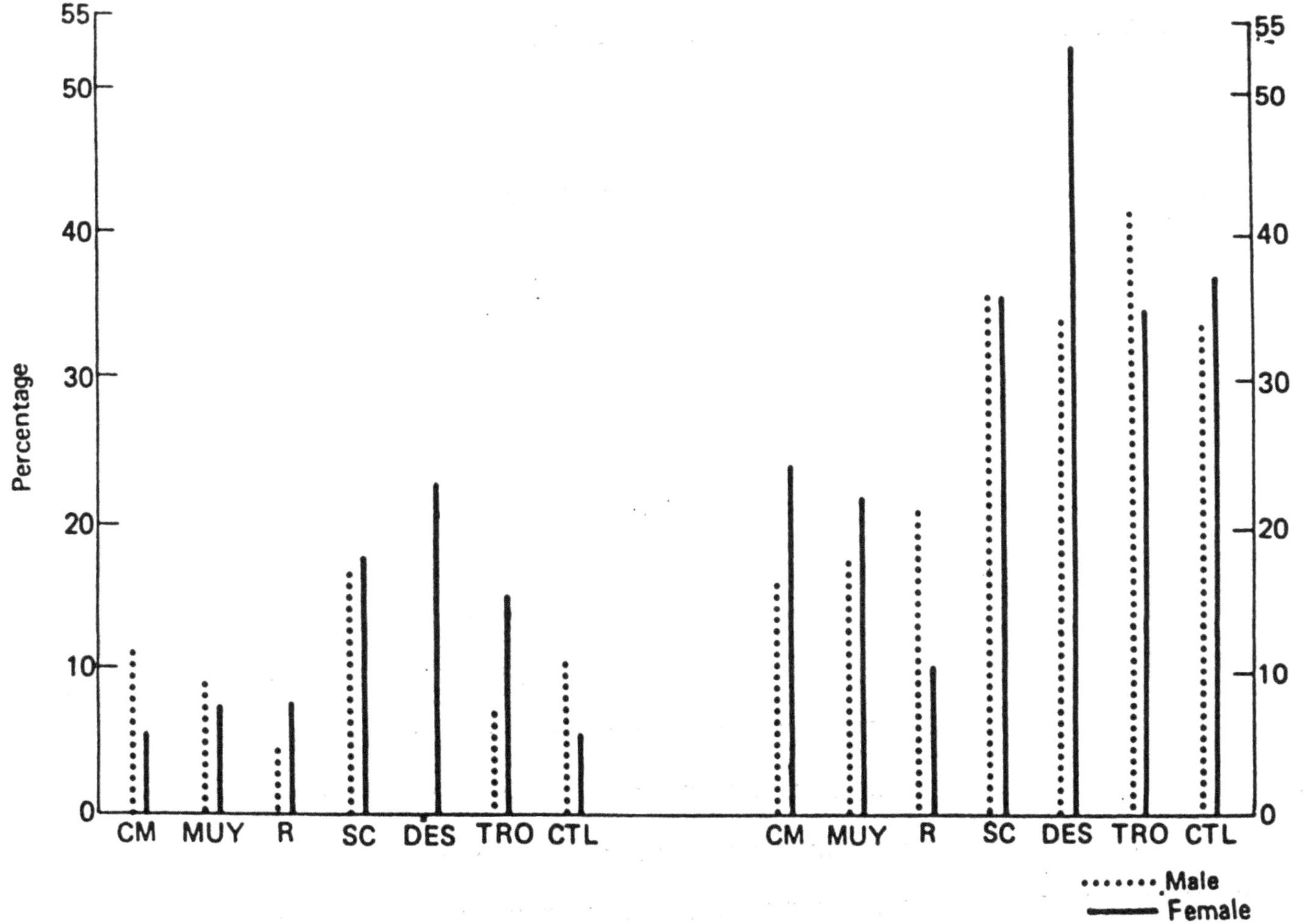

Fig. 6-14A Frequency of young adults with pre-mortem tooth loss only (see note Fig. 6-12)

Fig. 6-14B Old adults with pre-mortem tooth loss only (see note Fig 6-12)

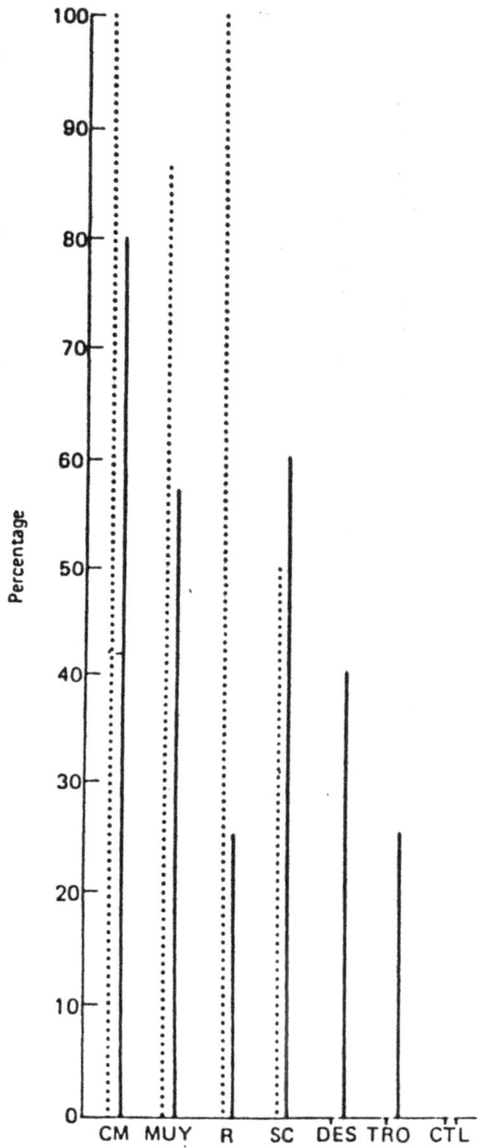

Fig. 6-15A

Young adults with more than one tooth missing and have TMJ arthritis (see note Fig 6-12)

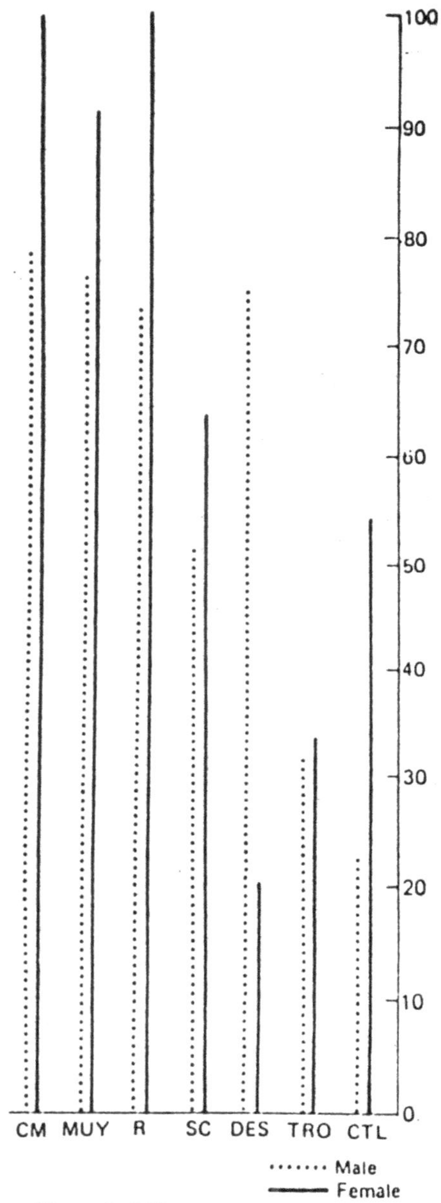

Fig. 6-15B

Old adults with more than one tooth missing and have TMJ arthritis (see note Fig 6-12

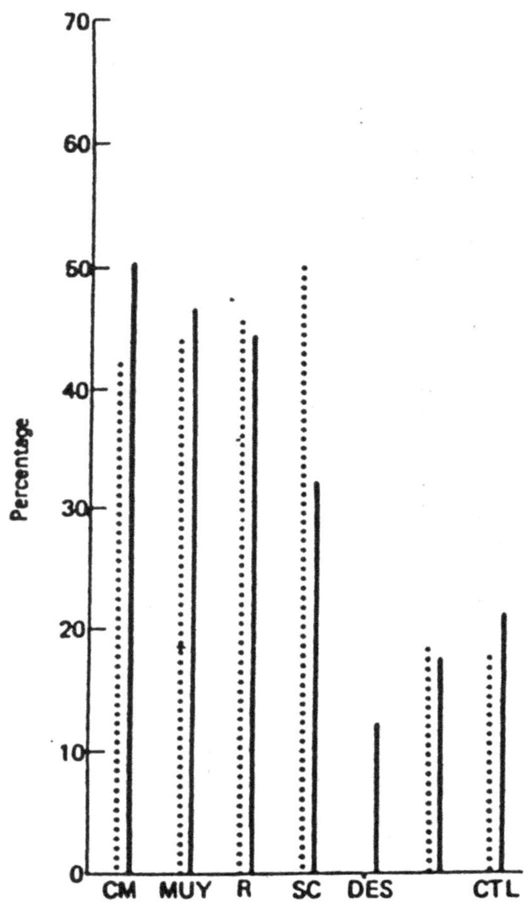

Fig. 6-16A

Young adults with TMJ arthritis but retaining all teeth (see note Fig. 6-12)

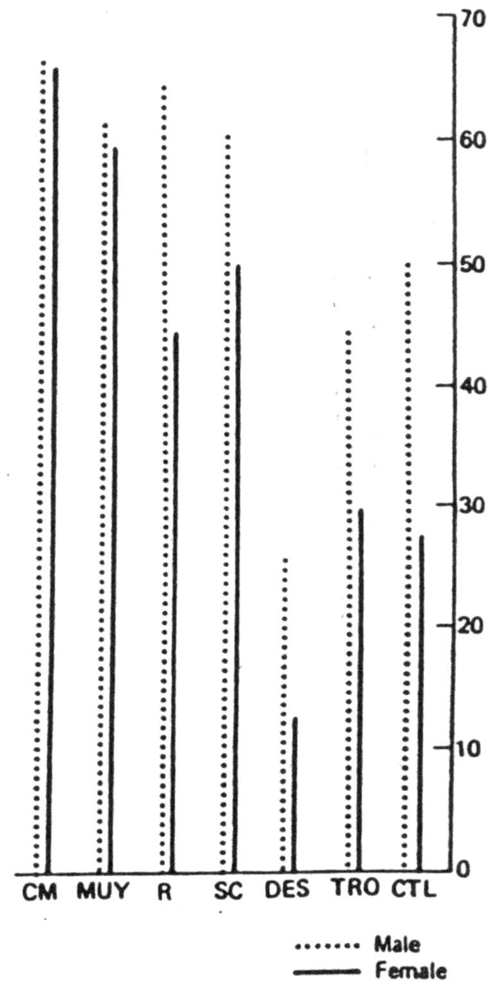

Fig. 6-16B

Old adults with TMJ arthritis but retaining all teeth (see note Fig. 6-12)

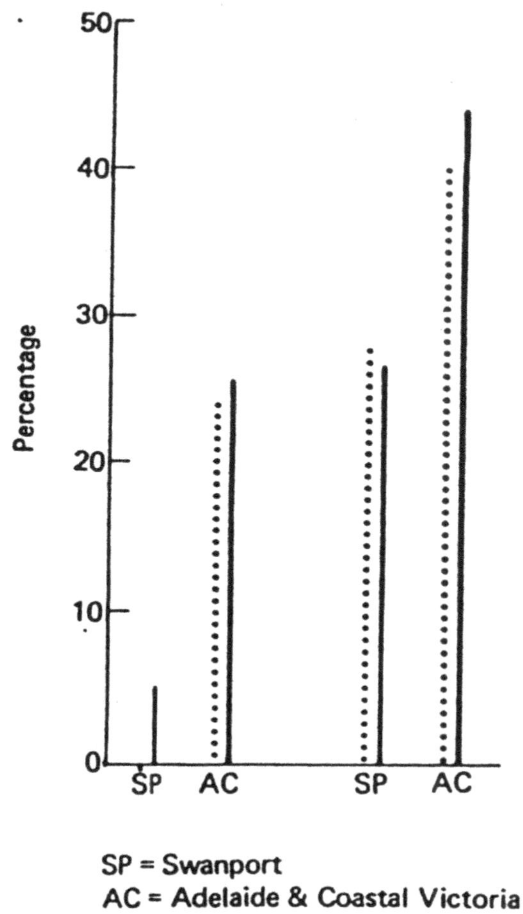

Fig. 6-17 Tooth loss in young and old adults. A comparison between Swanport on the Lower Murray River, and coastal groups (Adelaide and coastal Victoria)

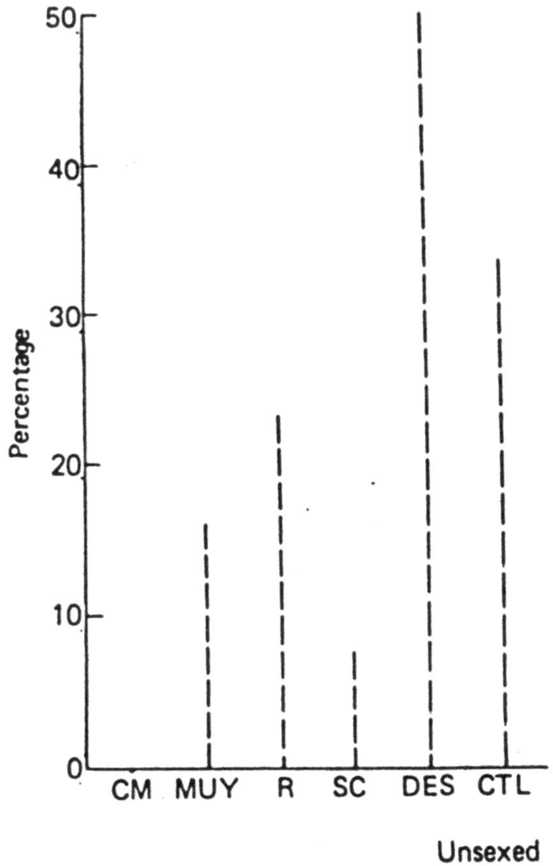

Fig. 7-1 Leg fractures as a percentage of all post-cranial fractures (humerus, ulna, radius, femur, tibia and fibula)

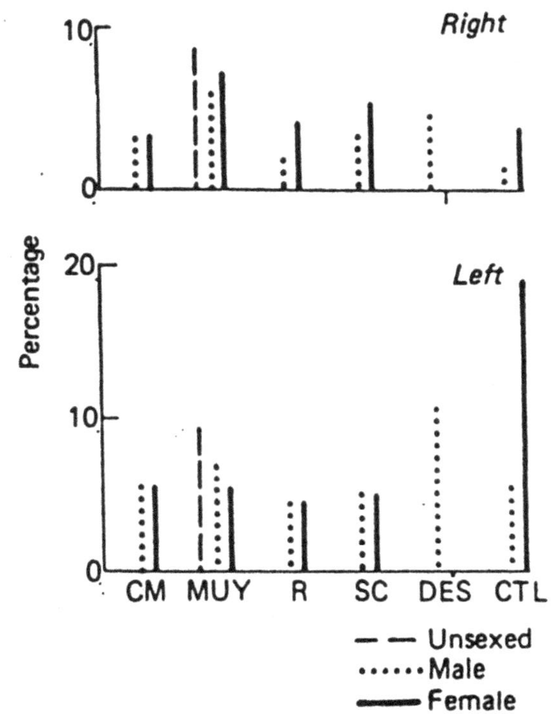

Fig. 7-2 Parrying fractures, by area

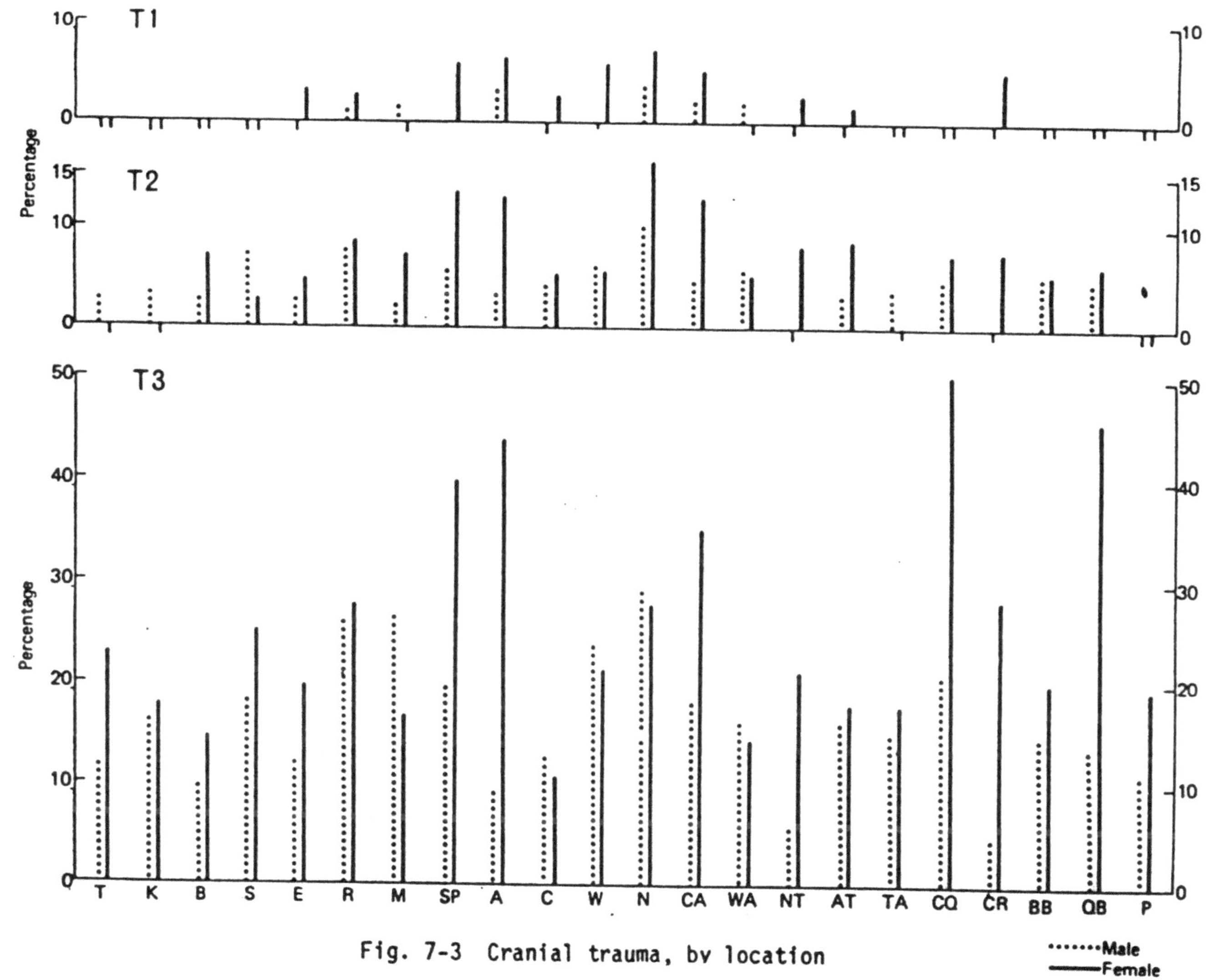

Fig. 7-3 Cranial trauma, by location

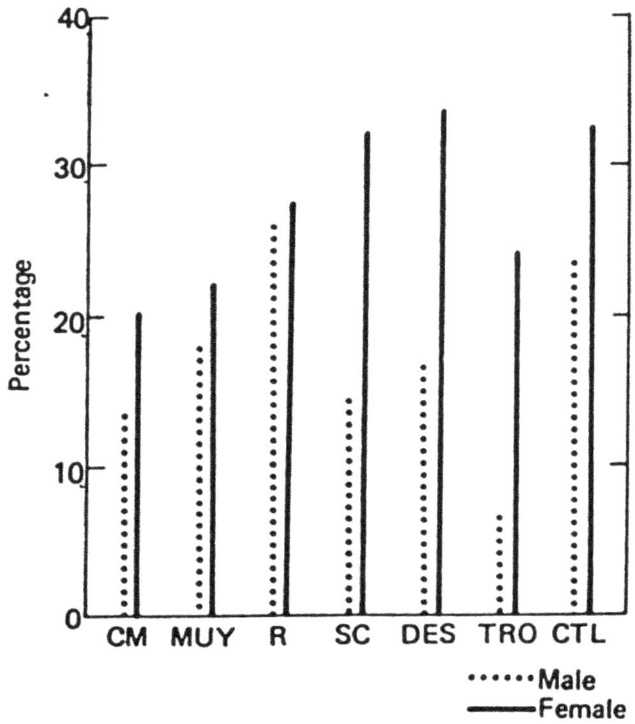

Fig. 7-4 Cranial trauma (T1 only), by area

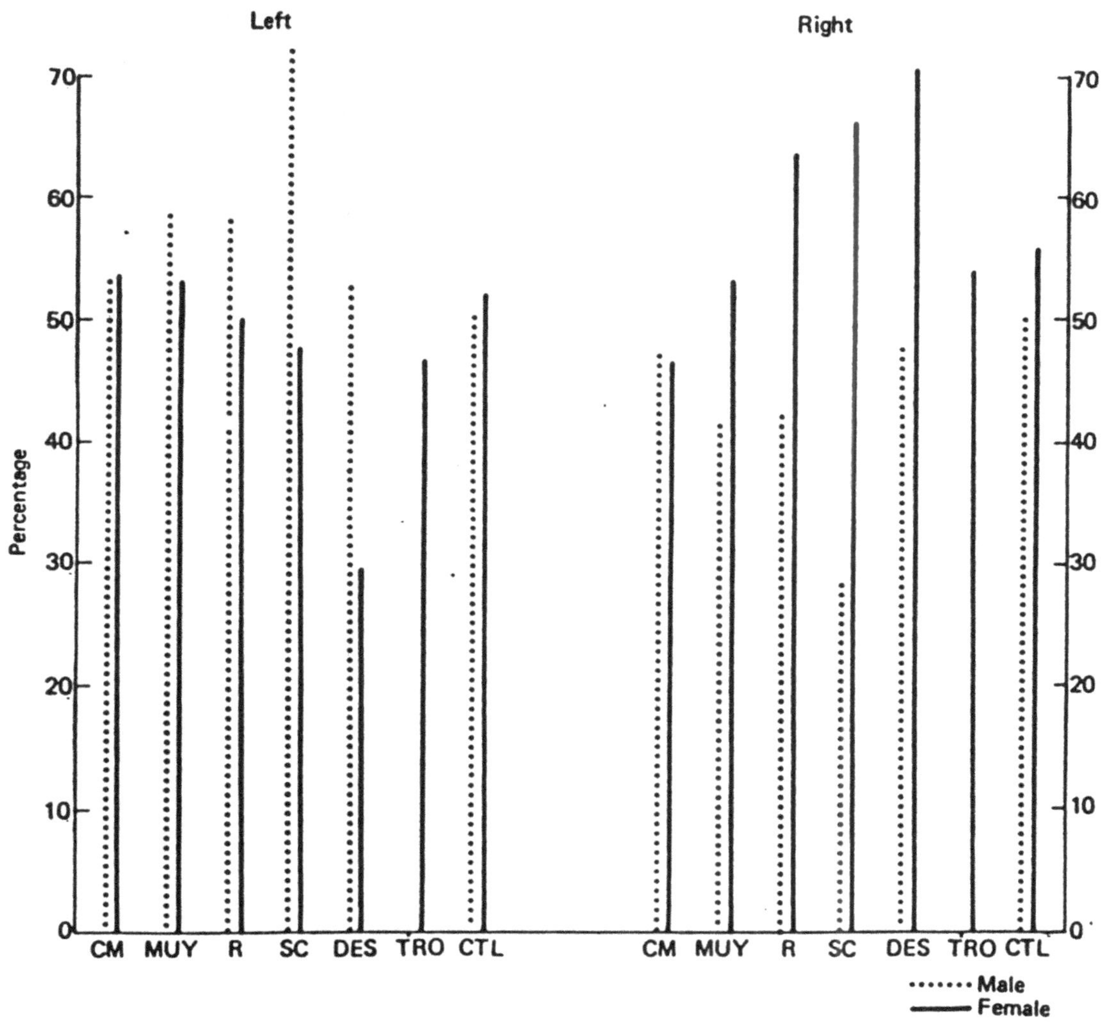

Fig. 7-5 Distribution of parietal trauma. Frequency of left and right parietals as a percentage of all parietals examined, irrespective of side

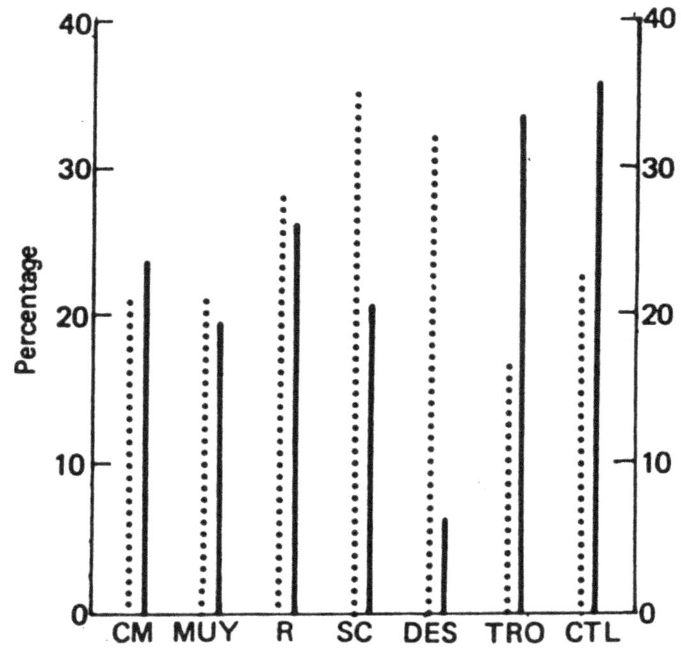

Fig. 7-6 Frontal trauma as a percentage of all cranial trauma

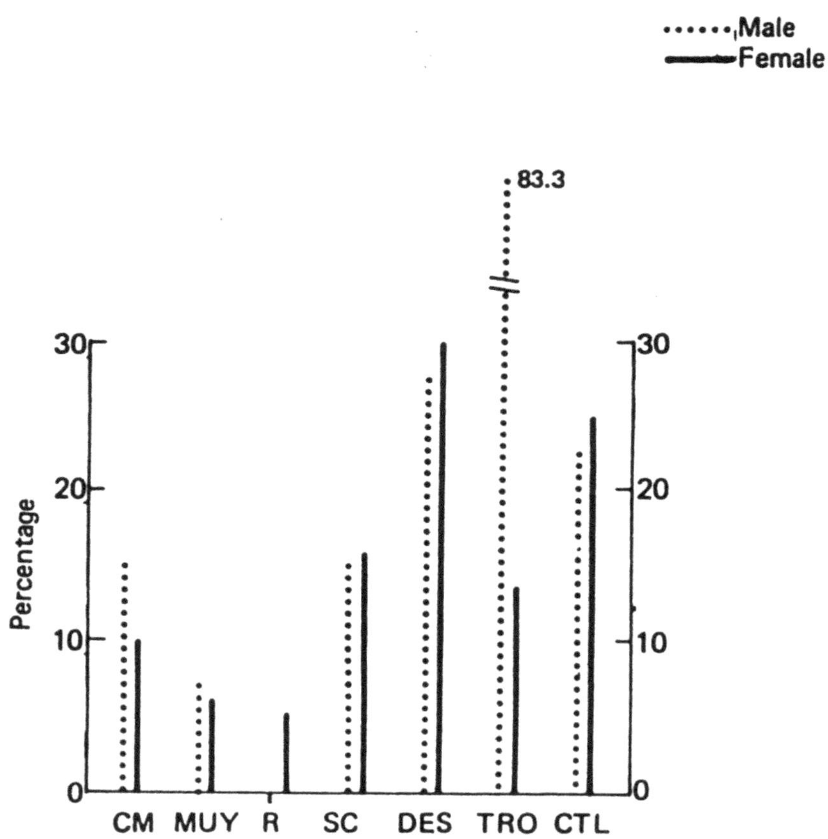

Fig. 7-7 Occipital trauma as a percentage of all cranial traumas

TABLES

Table 1-1

Adult post-cranial remains. Sample size from each location and area

Post-cranial samples: 10 = Humerus, 11 = Ulna, 12 = Radius,
20 = Femur, 21 = Tibia, 22 = Fibula

Location		LEFT							RIGHT						
		10	11	12	20	21	22	T	10	11	12	20	21	22	T
T	♂	40	2	2	2	7	3	56	42	1	2	2	3	3	53
	♀	42	1	-	-	5	-	48	44	-	2	1	1	-	48
	U	31	57	54	-	-	-	142	16	55	37	-	-	-	108
	T	113	60	56	2	12	3	246	102	56	41	3	4	3	209
K	♂	2	3	2	35	3	2	47	1	3	-	21	6	3	34
	♀	5	5	5	11	8	3	37	5	2	4	18	6	4	39
	U	-	-	-	20	81	38	139	-	-	-	16	84	34	134
	T	7	8	7	66	92	43	223	6	5	4	55	96	41	207
B	♂	47	23	22	47	37	25	201	54	28	30	33	35	25	205
	♀	18	12	11	20	16	14	91	22	8	12	20	20	11	93
	U	12	18	21	12	31	18	112	12	25	21	14	28	18	118
	T	77	53	54	79	84	57	404	88	61	63	67	83	54	416
S	♂	42	31	24	31	35	29	192	43	28	35	32	33	28	189
	♀	25	19	24	22	25	17	132	24	18	20	21	23	16	122
	U	5	-	3	-	-	1	15	7	-	4	-	-	1	17
	T	72	50	51	53	60	47	339	74	46	59	53	56	45	328
E	♂	9	-	-	8	2	-	19	13	-	-	10	1	-	24
	♀	6	-	-	5	1	-	12	4	-	-	6	-	-	10
	U	4	15	14	-	20	3	57	3	16	13	-	15	2	56
	T	19	15	14	13	23	3	88	20	16	13	16	16	2	90
R	♂	34	44	64	52	52	44	290	35	48	62	54	52	44	295
	♀	21	22	30	33	26	31	163	23	23	32	29	25	29	161
	U	73	13	16	9	20	12	143	76	15	19	5	24	14	153
	T	128	79	110	94	98	87	596	134	86	113	88	101	87	609
M	♂	79	-	-	100	-	-	179	83	1	-	99	-	-	183
	♀	34	-	-	33	3	-	70	31	1	1	32	-	-	65
	U	37	144	118	45	160	89	593	38	131	117	40	159	93	578
	T	150	144	118	178	163	89	842	152	133	118	171	159	93	826
SP	♂	10	3	4	13	28	6	64	15	8	6	14	42	5	90
	♀	25	16	15	30	33	11	130	19	16	16	27	31	11	120
	U	89	66	74	74	42	-	345	88	70	47	74	38	-	317
	T	124	85	93	117	103	17	539	122	94	69	115	111	16	527
A	♂	14	12	14	13	18	12	83	14	12	12	10	14	12	74
	♀	5	4	5	4	3	4	25	4	3	4	4	4	1	20
	U	-	-	-	1	-	-	1	-	-	-	1	2	-	3
	T	19	16	19	18	21	16	109	18	15	16	15	20	13	97

Table 1-1 (continued)

Location		LEFT 10	11	12	20	21	22	T	RIGHT 10	11	12	20	21	22	T
C	♂	8	6	7	8	9	7	45	7	8	7	9	8	7	46
	♀	1	1	1	1	5	1	10	-	-	-	1	6	-	7
	U	-	-	-	-	-	-	-	-	-	-	-	-	-	-
	T	9	7	8	9	14	8	55	7	8	7	10	14	7	53
W	♂	5	4	4	3	12	3	31	5	5	4	4	6	4	28
	♀	6	8	4	4	11	5	38	7	8	7	4	10	4	40
	U	-	-	-	-	1	-	1	-	-	-	-	-	-	-
	T	11	12	8	7	24	8	70	12	13	11	8	16	8	68
N	♂	12	15	13	19	16	11	86	14	14	16	16	23	9	93
	♀	10	10	8	8	10	4	50	10	10	9	9	6	6	50
	U	4	2	2	2	3	3	16	3	3	3	3	6	3	21
	T	26	27	23	29	29	18	152	27	27	28	28	35	18	164
CA	♂	11	9	11	6	14	7	58	9	11	6	7	14	6	53
	♀	4	4	4	3	3	3	21	5	4	5	4	4	2	24
	U	1	-	-	-	-	-	1	-	-	-	-	1	-	1
	T	16	13	15	9	17	10	80	14	15	11	11	19	8	78
WA	♂	5	6	8	6	9	8	42	7	6	7	5	18	6	49
	♀	1	1	1	1	-	1	5	1	1	1	1	8	1	13
	U	-	-	-	-	-	-	-	-	-	-	-	-	-	-
	T	6	7	9	7	9	9	47	8	7	8	6	26	7	62
NT	♂	3	-	1	2	2	1	9	3	-	1	3	4	-	11
	♀	1	1	1	1	1	1	6	1	1	1	1	1	1	6
	U	-	-	-	-	-	-	-	-	-	-	-	-	-	-
	T	4	1	2	3	3	2	15	4	1	2	4	5	1	17
AT	♂	4	1	2	3	3	2	15	4	1	2	4	5	1	17
	♀	1	1	1	1	1	1	6	1	1	1	1	1	1	6
	U	-	-	-	-	-	-	-	-	-	-	-	-	-	-
	T	5	2	3	4	4	3	21	5	2	3	5	6	2	23
TA	♂														
	♀	No post-cranial remains							No post-cranial remains						
	U														
	T														
CQ	♂	13	19	12	12	10	11	67	13	12	9	18	15	10	77
	♀	13	12	9	13	13	10	70	13	12	11	14	13	13	76
	U	1	-	-	8	4	1	14	2	-	-	7	2	-	11
	T	27	31	21	33	27	22	151	28	24	20	39	30	23	164
CR	♂	1	1	1	1	1	1	6	1	3	1	1	1	1	8
	♀	3	2	2	3	3	2	15	2	2	2	3	2	2	13
	U	2	2	2	1	2	3	12	1	2	1	1	3	3	11
	T	6	5	5	5	6	6	33	4	7	4	5	6	6	32

Table 1-1 (continued)

Location		LEFT 10	11	12	20	21	22	T	RIGHT 10	11	12	20	21	22	T
BB	♂	29	26	30	24	19	24	152	34	30	28	28	24	19	163
	♀	3	3	2	3	1	2	14	2	2	1	4	-	3	12
	U	11	12	12	9	10	5	59	11	12	12	8	11	5	59
	T	43	41	44	36	30	31	225	47	44	41	40	35	27	234
QB	♂	42	35	42	36	29	35	219	47	42	37	46	38	29	239
	♀	16	15	11	16	14	12	84	15	14	12	18	13	16	88
	U	12	12	12	17	14	6	73	13	12	12	15	13	5	70
	T	70	62	65	69	57	53	376	75	68	61	79	64	50	397
P	♂														
	♀	No post-cranial remains							No post-cranial remains						
	U														
	T														
AREAS															
CM	♂	100	57	46	124	78	56	464	111	60	56	98	77	56	458
	♀	54	36	40	58	50	34	272	55	28	36	66	49	31	265
	U	22	34	39	34	133	61	323	23	42	39	38	128	56	326
	T	176	127	128	216	261	151	1059	189	130	131	202	254	143	1049
MUY	♂	253	102	115	278	137	103	988	271	108	120	253	132	103	987
	♀	151	59	70	124	84	65	553	153	52	71	128	75	60	539
	U	163	248	227	88	313	162	1201	153	243	212	83	311	163	1165
	T	567	409	412	490	534	330	2742	577	403	403	464	518	326	2691
SC	♂	32	21	25	34	55	25	192	36	28	25	33	64	24	210
	♀	31	21	21	35	41	16	165	23	19	20	32	41	12	147
	U	89	66	74	75	42	-	346	88	70	47	75	40	-	320
	T	152	108	120	144	138	41	703	147	117	92	140	145	36	677
DES	♂	23	19	21	18	24	18	123	21	21	18	19	29	16	124
	♀	4	5	4	4	4	4	25	6	5	6	5	6	3	31
	U	1	-	-	-	-	-	1	-	-	-	-	1	-	1
	T	28	24	25	22	28	22	149	27	26	24	24	36	19	156
TRO	♂	4	1	2	3	3	2	15	4	3	2	4	5	1	19
	♀	4	3	3	4	4	3	21	3	3	3	4	3	3	19
	U	2	2	2	1	2	3	12	1	2	1	1	3	3	11
	T	10	6	7	8	9	8	48	8	8	6	9	11	7	49
CTL	♂	56	53	59	56	49	49	322	64	60	56	65	69	41	355
	♀	27	26	20	25	24	17	139	26	25	22	28	26	23	150
	U	16	14	14	19	17	9	89	16	15	15	18	19	8	91
	T	99	93	93	100	90	75	550	106	100	93	111	114	72	596

Table 1-2

Sub-adult cranial and post-cranial remains.
Sample size from each area

		POST-CRANIAL							CRANIA			
Bone	Code	10	11	12	20	21	22	T	1	2	3	T
CM	L	17	7	9	12	12	11	68	7	19	29	55
	R	20	7	10	12	9	10	68				
	T	37	14	19	24	21	21	136				
MUY	L	29	9	14	17	21	16	106	18	32	44	94
	R	31	11	16	22	15	13	108				
	T	60	20	30	39	36	29	214				
R	L	5	1	3	4	5	4	22	4	5	7	16
	R	5	1	4	6	5	2	23				
	T	10	2	7	10	10	6	45				
SC	L	12	3	6	2	3		26	6	12	21	33
	R	9	3	5	3	5		25				
	T	21	6	11	5	8		25				
DES	L	2	2	2		3	2	11	5	7	9	21
	R	2	1	1	1	2	1	8				
	T	4	3	3	1	5	3	19				
TRO	L								2	4	10	16
	R	No post-cranial remains										
	T											
CTL	L	12	13	13	11	11	7	67	9	13	10	32
	R	12	13	13	8	10	7	63				
	T	24	26	26	19	21	14	130				

Cranial age: 1 = <6 years, 2 = 6-11 years, 3 = 12-20 years

Bone Code:
10 = Humerus, 11 = Ulna, 12 = Radius, 20 = Femur, 21 = Tibia, 22 = Fibula

Area Codes:
CM - Central Murray, MUY - Murray Valley, R - Rufus River, SC - South Coast, DES - Desert, TRO - Tropics, CTL - East Coast

Table 1-3

Adult cranial sample size by location and area

		LOCATION								AREAS				
Age		6	7	T			6	7	T			6	7	T
T	M	17	25	42	N	M	16	43	59	CM	M	92	152	244
	F	13	13	26		F	22	19	41		F	98	51	149
	T	30	38	68		T	38	62	100		T	190	203	393
K	M	8	22	30	CA	M	18	75	93	MUY	M	180	285	465
	F	9	7	16		F	16	24	40		F	190	109	299
	T	17	29	46		T	34	99	133		T	370	394	764
B	M	13	28	41	WA	M	7	46	53	SC	M	42	96	138
	F	17	11	28		F	11	10	21		F	57	66	123
	T	30	39	69		T	18	56	74		T	99	162	261
S	M	24	30	54	NT	M	20	33	53	DES	M	24	106	130
	F	32	8	40		F	16	21	37		F	23	28	51
	T	56	38	94		T	36	54	90		T	47	134	181
E	M	47	72	1·19	AT	M	39	55	94	TRO	M	36	56	92
	F	40	25	65		F	29	32	61		F	27	34	61
	T	87	97	184		T	68	87	155		T	63	90	153
R	M	48	74	122	TA	M	9	17	26	CTL	M	30	103	133
	F	53	30	83		F	25	10	35		F	45	38	83
	T	101	104	205		T	34	27	61		T	75	141	216
M	M	23	34	57	CQ	M	7	37	44					
	F	26	15	41		F	16	12	28					
	T	49	49	98		T	23	49	72					
SP	M	16	41	57	CR	M	2	7	9					
	F	21	32	53		F	7	9	16					
	T	37	73	110		T	9	16	25					
A	M	5	28	33	BB	M	7	14	21					
	F	7	25	32		F	3	2	5					
	T	12	53	65		T	10	16	26					
C	M	21	27	48	QB	M	14	51	65					
	F	29	9	38		F	19	14	33					
	T	50	36	86		T	33	65	98					
W	M	4	13	17	P	M	7	21	28					
	F	13	6	19		F	5	10	15					
	T	17	19	36		T	12	31	43					

Crania age: 6 = Young Adults, 7 = Old Adults,
M = Males, F = Females, T = totals

Location Codes:
T - Tocumwal, K - Kow Swamp (modern), B - Baratta, S - Swan Hill, E - Euston, R - Rufus River, M - Murray River, SP - Swanport, A - Adelaide, C - Coastal Victoria, W - West Central Victoria, N - Coastal New South Wales, CA - Central Australia, WA - Western Australia, NT - Northern Territory (northern half), AT - All Northern Territory, TA - Tasmania, CQ - Coastal Queensland, CR - Cape York and Cairns, BB - Broadbeach, P - Pleistocene sample.

Area Codes:
CM - Central Murray, MUY - Murray Valley, R - Rufus River, SC - South Coast, DES - Desert, TRO - Tropics, CTL - East Coast

Table 1-4

Fossil	Date	(Reference)	Comment
WLH 3	28-32,000BP	(1)	Excluded from the survey£
WLH 1	25-28,000BP	(2)	Excluded from the survey*
Talgai	12-18,000BP?	(3)	Excluded from the survey*
Bourkes Bridge	12-17,000BP?	(4)	See Chapters 4 and 7
Murrabit	8-14,000BP?	(5)	See Chapter 7
Keilor	13,000BP?	(6)	See Chapter 6, TMJ section
Nitchie	6,800BP	(7)	No pathology$
Green Gully	6,500BP	(8)	Excluded from the survey+
Cossack	<6,500BP	(9)	No pathology
Mt Cameron West	4-6,000BP	(10)	No pathology
Mossgiel	>4,600BP	(11)	Excluded from the survey@

£ The pathology on this individual is described in Bowler and Thorne 1976 and Webb 1989.

* These individuals are too fragmented for comment

$ See Barker 1975

\+ See Macintosh 1967

@ See Macintosh 1967 and Freedman 1985

References

(1) Bowler and Thorne 1976, Webb 1989

(2) Bowler *et al.* 1970, Thorne 1975, Webb 1989

(3) Mulvaney 1975

(4) Not accurately dated, A.G.Thorne pers. comm.

(5) Not accurately dated but morphologically similar to individuals in the Kow Swamp and Coobool Creek populations (approx. 10-12,000BP), Thorne 1975, Brown 1982

(6) Mulvaney 1975

(7) Barker 1975

(8) Macintosh 1967

(9) Freedman and Lofgren 1979

(10) Wallace and Doran 1976

(11) Macintosh 1967 and Freedman 1985

Table 2-1

Frequencies of cribra orbitalia for adults, by area

		CATEGORIES 1-3			CATEGORY 4 (%)		
		n	n̄	%	Young Adult	Old Adult	Combined
CM	M	246	73	29.7	4.3	18.4	13.0
	F	149	60	40.3	6.1	5.9	6.0
	T	395	133	33.7			
MUY	M	465	110	23.7	6.1	17.9	13.3
	F	299	103	34.4	7.9	6.4	7.4
	T	764	213	27.9			
R	M	122	29	23.7	8.3	18.9	14.8
	F	83	21	25.3	11.3	13.3	12.1
	T	205	50	24.4			
SC	M	138	21	15.2	2.4	13.5	10.1
	F	123	27	22.0	10.5	12.5	11.4
	T	261	48	18.4			
DES	M	130	10	7.7	16.7	12.3	13.1
	F	51	5	9.8	4.3	3.6	3.9
	T	181	15	8.3			
TRO	M	92	10	10.9	2.8	10.7	7.6
	F	61	11	18.0	14.8	5.9	9.8
	T	153	21	13.7			
CTL	M	133	21	15.8	3.3	11.7	10.5
	F	83	21	25.3	4.4	13.2	8.4
	T	216	42	19.4			

M = Males, F = Females, T = Total
n = Sample size
n̄ = Affected individuals

Area Codes:
CM - Central Murray, MUY - Murray Valley, R - Rufus River, SC - South Coast, DES - Desert, TRO - Tropics, CTL - East Coast

Table 2-2

Chi-square analysis for population totals (males plus females), by area

	CM	MUY	R	SC	DES	TRO	CTL
MUY	(4.17) <0.05	-					
R	(5.48) <0.02	n.s.	-				
SC	(21.46) <0.001	(11.69) <0.001	n.s.	-			
DES	(41.89) <0.001	(30.68) <0.001	(17.8) <0.001	(7.3) <0.01	-		
TRO	(21.71) <0.001	(13.44) <0.001	(6.27) <0.02	n.s.	n.s.	-	
CTL	(13.83) <0.001	(6.22) <0.02	n.s.	n.s.	(9.97) <0.005	n.s.	-

Area Codes:
CM - Central Murray, MUY - Murray Valley, R - Rufus River,
SC - South Coast, DES - Desert, TRO - Tropics, CTL - East Coast

Table 2-3

Chi-square analysis for males and females, by area

Areas		CM ♂	CM ♀	MUY ♂	MUY ♀	R ♂	R ♀	SC ♂	SC ♀	DES ♂	DES ♀	TRO ♂	TRO ♀	CTL ♂	CTL ♀
MUY	♂	n.s.		-											
	♀		n.s.		-										
R	♂	n.s.		n.s.		-									
	♀		(4.42) <0.05		n.s.		-								
SC	♂	(10.43) <0.005		(4.75) <0.05		n.s.		-							
	♀		(10.39) <0.005		(6.39) <0.02		n.s.		-						
DES	♂	(20.95) <0.001		(16.08) <0.001		(12.44) <0.001		n.s.		-					
	♀		(16.07) <0.001		(12.4) <0.001		(5.48) <0.02		n.s.		-				
TRO	♂	(12.78) <0.001		(7.43) <0.01		(5.86) <0.02		n.s.		n.s.		-			
	♀		(9.56) <0.01		(6.31) <0.02		n.s.		n.s.		n.s.		-		
CTL	♂	(8.92) <0.005		n.s.		n.s.		n.s.		(4.15) <0.05		n.s.		-	
	♀		(8.04) <0.005		(4.62) <0.05		n.s.		n.s.		n.s.		n.s.		-

Table 2-4

Frequencies of cribra orbitalia for adults, categories 1-3 inclusive, by location

Location		n	$\bar{n}$	%	Location		n	$\bar{n}$	%
T	M	40	7	17.5	N	M	51	13	25.5
	F	26	8	30.7		F	41	15	36.6
K	M	24	12	50.0	CA	M	79	10	12.7
	F	15	7	46.6		F	38	4	10.5
B	M	39	11	28.2	WA	M	47	3	6.4
	F	27	10	37.0		F	18	2	11.1
S	M	49	14	28.5	NT	M	49	8	16.3
	F	36	15	41.7		F	36	7	19.4
E	M	103	36	35.0	AT	M	84	14	16.7
	F	64	28	43.8		F	57	13	22.8
R	M	105	19	18.1	TA	M	25	2	8.0
	F	73	21	28.8		F	31	4	12.9
M	M	47	12	25.5	CQ	M	42	1	2.4
	F	39	14	35.9		F	24	5	20.8
SP	M	52	9	17.3	CR	M	36	2	5.6
	F	46	14	26.1		F	20	5	25.0
A	M	29	4	13.8	BB	M	18	5	27.8
	F	28	2	7.1		F	5	1	20.0
C	M	43	8	18.6	QB	M	60	6	10.0
	F	35	13	37.1		F	29	6	20.7
W	M	14	4	28.6	P	M	26	5	19.2
	F	18	7	38.9		F	15	2	13.3

M = Males, F = Females

Location Codes:
T - Tocumwal, K - Kow Swamp (modern), B - Baratta, S - Swan Hill, E - Euston, R - Rufus River, SP - Swanport, A - Adelaide, C - Coastal Victoria, W - West Central Victoria, N - Coastal New South Wales, CA - Central Australia, WA - Western Australia, NT - Northern Territory (northern half), AT - All Northern Territory, TA - Tasmania, CR - Cape York and Cairns, CQ - Coastal Queensland, BB - Broadbeach, QB - Coastal Queensland and Broadbeach combined, P - Pleistocene sample

Table 2-5

Frequency of cribra orbitalia for five age groups, by area.

	Age	n	$\bar{n}$	%
CM	1	7	5	71.4
	2	19	14	73.7
	3	28	13	46.4
	6	190	65	34.2
	7	203	68	33.5
MUY	1	18	15	83.3
	2	32	25	78.1
	3	44	20	45.5
	6	370	107	28.9
	7	394	98	24.9
R	1	4	4	100.0
	2	5	4	80.0
	3	7	2	28.6
	6	91	22	24.2
	7	97	18	20.7
SC	1	6	3	50.0
	2	12	4	33.3
	3	20	8	40.0
	6	92	33	35.9
	7	141	15	10.6
DES	1	5	1	20.0
	2	7	4	57.1
	3	8	1	12.5
	6	43	6	14.0
	7	118	10	8.5
TRO	1	2	2	100.0
	2	4	4	100.0
	3	10	2	20.0
	6	58	13	22.4
	7	82	8	9.8
CTL	1	9	2	22.2
	2	13	7	53.8
	3	10	4	40.0
	6	72	23	31.9
	7	122	19	15.6

Age Code: 1 = 0-5 years 2 = 6-11 years 3 = 12-20 years
6 = Young adults (<35-40) 7 = Old adults (>40)

Area Codes:
CM - Central Murray, MUY - Murray Valley, R - Rufus River, SC - South Coast, DES - Desert, TRO - Tropics, CTL - East Coast

Table 2-6

Frequencies of cribra orbitalia for various populations around the world

Location	n	Juveniles %	Adult male %	Adult female %	Combined adult %	Sex/age unknown %	Author
Europeans	612	-	-	-	-	19.0	Ahrens (1904)
Germans	470	-	-	-	-	.17.0	Ahrens (1904)
Germans	-	-	-	-	-	3.5	Welcker (1888)
Southern Zealand (Denmark)	200	-	-	-	-	47.0	Møller-Christensen (1961)
Northern Zealand (Denmark)	104	-	-	-	-	20.0	Møller-Christensen (1961)
Austria	100	-	18.9	27.7	23.0	-	Webb after Cybulski (1977)
Scotland	743	52.0	-	-	6.6	-	Moller-Christensen & Sandison (1963)
Old Egyptians	434	-	-	-	-	8.0	Welcker (1888)
Old Egypt	182	-	-	-	-	7.0	Oetteking (1909)
Sudanese Nubia	285	23.3	-	-	19.9	-	Carson <u>et</u> al. (1974)
Modern Egypt	-	-	-	-	-	11.0	Oetteking (1909)
Socotrans (Arab descent)	42	-	-	-	-	48.0	Koganei (1912)
East Sudan Negroes	-	-	-	-	-	35.0	Koganei (1912)
Fenno-Turks	-	-	-	-	-	4.0	Henschen (1961)
Southeast Mongols	-	-	-	-	-	9.0	Henschen (1961)
Mongols	70	-	-	-	-	37.0	Henschen (1961)
Chinese	-	-	-	-	-	10.0	Henschen (1961)
Japanese	259	-	-	-	-	19.0	Osawa (?)
Japanese	121	-	-	-	-	15.0	Adachi (1904)
Japanese	411	52.5	26.6	17.9	22.4	-	Webb after Akabori (1933)
Japanese	695	25-27	-	-	9-11	-	Koganei (1912)
Ainu	-	-	-	-	-	16.8	Koganei (1912)
Malayans	-	-	-	-	-	22.5	Koganei (1912)
Malayan Peninsula	51	-	21.6	28.6	23.5	-	Webb after Cybulski (1977)
Nahel Hever) Israel	35	93.3	12.5	75.0	50.0	-	Webb afterCybulski (1977)
Nahel Salem)							
Ein Gedi (Israel)	94	66.6	13.0	18.2	13.5	-	Webb after Cybulski (1977)
Polynesians	-	-	-	-	-	6.0	Henschen (1961)
Micronesians	-	-	-	-	-	7.0	Henschen (1961)
Papuans-Australians	-	-	-	-	-	30.0	Henschen (1961)
Hottentots	-	-	-	-	-	4.0	Henschen (1961)
Bantu	-	-	-	-	-	10.0	Henschen (1961)
African Negroes	-	-	-	-	-	21.0	Welcker (1888)
Old Peruvians	-	-	-	-	-	9.0	Koganei (1912)
Alaska (Eskimo)	90	54.6	11.0	17.3	15.2	-	Webb after Nathan & Haas (1966)
North American Indians	120	62.8	10.9	22.8	14.3	-	Webb after Nathan & Haas (1966)
British Columbian Indians	454	26.6	4.8	13.3	11.25	-	Webb after Cybulski (1977)
Late Woodland	44	13.6	-	-	-	-	Lallo <u>et al</u>. (1977)
Early Mississippian	93	31.2	-	-	-	-	Lallo <u>et al</u>. (1977)
Middle Mississippian	101	51.5	-	-	-	-	Lallo <u>et al</u>. (1977)
Eiden, Lorain, Ohio	31	51.6	-	-	-	-	Lallo <u>et al</u>. (1977)
Central Columbian Indians	60	42.9	12.9	27.3	18.9	-	Webb after Cybulski (1977)
South Columbian Indians	105	14.3	9.1	11.6	10.2	-	Webb after Cybulski (1977)

Table 3-1

Dental hypoplasia frequencies, by location

		Canine			Third Molar		
		n	$\bar{n}$	%	n	$\bar{n}$	%
T	M	33	15	45.5	33	8	24.2
	F	17	5	29.4	17	1	5.9
	T	50	20	40.0	50	9	18.0
K	M	13	5	38.5	18	5	27.8
	F	11	3	27.3	14	5	35.7
	T	24	8	33.3	32	10	31.3
B	M	30	9	30.0	32	5	15.6
	F	23	11	47.8	23	7	30.4
	T	53	20	37.7	55	12	21.8
S	M	50	18	36.0	50	7	14.0
	F	31	10	32.3	32	4	12.5
	T	81	28	34.6	82	11	13.4
E	M	89	49	55.1	89	31	34.8
	F	42	21	50.0	42	11	26.2
	T	131	70	53.4	131	42	32.1
M	M	42	18	42.9	42	9	21.4
	F	27	9	33.3	29	7	24.1
	T	69	27	39.1	71	16	22.5
SP	M	41	17	41.5	48	1	2.1
	F	34	9	26.5	40	1	2.5
	T	75	26	34.7	88	2	2.3
A	M	18	10	55.6	22	2	9.1
	F	16	7	43.8	19	1	5.3
	T	34	17	50.0	41	3	7.3
C	M	29	8	27.6	31	3	9.7
	F	20	9	45.0	21	2	9.5
	T	49	17	34.7	52	5	9.6
W	M	10	-	-	12	1	8.3
	F	10	1	10.0	12	3	25.0
	T	20	1	5.0	24	4	16.7
N	M	52	23	44.2	53	7	13.2
	F	30	10	33.3	36	6	16.7
	T	82	33	40.2	89	13	14.6
CA	M	51	19	37.3	51	4	7.8
	F	24	4	16.7	24	1	4.2
	T	75	23	30.7	75	5	6.7
WA	M	38	9	23.7	40	2	5.0
	F	15	1	6.7	18	0	.0
	T	53	10	18.9	58	2	3.4

Table 3-1 continued

NT	M	38	19	50.0	40	-	-
	F	23	7	30.4	25	-	-
	T	61	26	42.6	65	-	-
AT	M	70	33	47.1	73	5	6.8
	F	42	12	28.6	46	1	2.2
	T	112	45	40.2	119	6	5.0
TA	M	11	4	36.4	13	3	23.1
	F	16	5	31.3	17	1	5.9
	T	27	9	33.3	30	4	13.3
CQ	M	27	13	48.1	27	3	11.1
	F	16	8	50.0	16	1	6.3
	T	43	21	48.8	43	4	9.3
CR	M	35	16	45.7	35	8	22.9
	F	21	2	9.5	21	1	4.8
	T	56	18	32.1	56	9	16.1
BB	M	16	7	43.8	16	3	18.8
	F	9	1	11.1	9	-	-
	T	25	8	32.0	25	3	12.0
QB	M	44	27	61.4	44	6	13.6
	F	20	11	55.0	20	3	15.0
	T	64	38	59.4	64	9	14.1
P	M	19	5	26.3	19	-	-
	F	5	2	40.0	5	-	-
	T	24	7	29.2	24	-	-

M = Males, F = Females, T = Total

Location Codes:
T - Tocumwal, K - Kow Swamp (modern), B - Baratta, S- Swan Hill, E - Euston, R - Rufus River, SP - Swanport, A - Adelaide, C - Coastal Victoria, W - West Central Victoria, N - Coastal New South Wales, CA - Central Australia, WA - Western Australia, NT - Northern Territory (northern half), AT - All Northern Territory, TA - Tasmania, CR - Cape York and Cairns, CQ - Coastal Queensland, BB - Broadbeach, QB - Coastal Queensland and Broadbeach combined, P - Pleistocene sample

Table 3-2

Average frequencies of dental hypoplasia for all locations that have a sample size of <23 combined

	M/F	MALE		FEMALE		Total	
	n	$\bar{n}$	%	$\bar{n}$	%	$\bar{n}$	%
Canine	1029	291	40.0	86	28.5	377	36.6
Third Molar	1088	102	13.7	42	12.2	144	13.2

Table 3-3

The statistical significance between canine and third molar dental hypoplasia

		CANINE			THIRD MOLAR				
		n	$\bar{n}$	%	n	$\bar{n}$	%		p
CM	M	182	80	44.0	189	48	25.4	(14.13)	<0.001
	F	107	45	42.1	111	27	24.3	(7.74)	<0.01
	T	289	125	43.3	300	75	25.0	(21.87)	<0.001
MUY	M	359	138	38.4	364	74	20.3	(28.60)	<0.001
	F	223	70	31.4	227	46	20.3	(7.28)	<0.01
	T	582	208	35.7	591	120	20.3	(34.68)	<0.001
R	M	102	25	24.5	100	9	9.0	(8.68)	<0.005
	F	72	11	15.3	70	11	15.7	(0.01)	n.s.
	T	174	36	20.7	170	20	11.8	(5.03)	<0.025
SC	M	88	35	39.8	101	6	5.9	(31.69)	<0.001
	F	70	25	35.7	80	4	5.0	(22.58)	<0.001
	T	158	60	38.0	181	10	5.5	(54.21)	<0.001
DES	M	80	27	33.8	81	5	6.2	(19.22)	<0.001
	F	31	5	16.1	32	1	3.1	(3.09)	n.s.
	T	111	32	28.8	113	6	5.3	(21.99)	<0.001
TRO	M	73	35	47.9	75	8	10.7	(24.94)	<0.001
	F	44	9	20.5	46	1	2.2	(5.80)	<0.02
	T	117	44	37.6	121	9	7.4	(27.30)	<0.001
CTL	M	101	51	50.5	103	14	13.6	(12.88)	<0.001
	F	57	21	36.8	65	9	13.8	(5.22)	<0.025
	T	158	72	45.6	168	23	13.7	(40.07)	<0.001

M = Males, F = Females, T = Total

Area Codes:
CM - Central Murray, MUY - Murray Valley, R - Rufus River, SC - South Coast, DES - Desert, TRO - Tropics, CTL - East Coast

Table 3-4

Chi-square analysis for canine and third molar teeth, by area

CANINE

	CM	MUY	R	SC	DES	TRO	CTL
MUY	4.62 <0.05	-					
R	24.38 <0.001	13.88 <0.001	-				
SC	n.s.	n.s.	12.04 <0.001	-			
DES	7.00 <0.01	n.s.	n.s.	n.s.	-		
TRO	n.s.	n.s.	10.04 <0.005	n.s.	n.s.	-	
CTL	n.s.	5.11	23.35	n.s.	7.70	n.s.	-

THIRD MOLAR

	CM	MUY	R	SC	DES	TRO	CTL
MUY	n.s.	-					
R	11.79 <0.001	6.41 <0.02					
SC	29.43 <0.001	21.61 <0.001	4.37 <0.05	-			
DES	20.19 <0.001	14.52 <0.001	n.s.	n.s.	-		
TRO	16.65 <0.001	11.21 <0.001	n.s.	n.s.	n.s.	-	
CTL	8.82 <0.005	n.s.	n.s.	6.79 <0.01	5.13 <0.025	n.s.	-

Area Codes:
CM - Central Murray, MUY - Murray Valley, R - Rufus River,
SC - South Coast, DES - Desert, TRO - Tropics,
CTL - East Coast

Table 3-5

The statistical association between cribra orbitalia and dental hypoplasia

		CRIBRA ORBITALIA							
		DH Canine				DH Third Molar			
		Males		Females		Males		Females	
DENTAL HYPOPLASIA		n	$\bar{n}$	n	$\bar{n}$	n	$\bar{n}$	n	$\bar{n}$
CM	n	58	40	37	24	88	50	49	34
			n.s.		n.s.		n.s.		n.s.
	n	56	25	24	21	27	21	15	12
MUY	n	144	72	96	55	190	96	113	66
			n.s.		n.s.		n.s.		n.s.
	n	97	42	40	30	47	27	26	20
R	n	53	24	38	23	62	29	40	19
			n.s.		n.s.		n.s.		n.s.
	n	20	5	9	2	8	1	5	6
SC	n	43	10	31	14	74	21	50	26
			n.s.		n.s.		n.s.		n.s.
	n	23	12	13	12	3	3	2	2
DES	n	42	11	22	4	68	15(16.51)	27	5
			n.s.		n.s.		<0.001		n.s.
	n	21	6	4	2	1	6	2	0
TRO	n	22	7	22	6	42	12	30	6
			n.s.		n.s.		n.s.		n.s.
	n	25	5	6	1	5	0	1	0
CTL	n	37	13	28	5(10.64)	71	23	38	15(5.18)
			n.s.		<0.005		n.s.		<0.025
	n	37	16	10	13	8	5	5	5

n.s. Not significant

Table 4-1

The incidence of Harris lines among three age grades, by area

Area	%F	n	nL	nB	IM (H)
CM	21.5	62	475	289	1.64
MUY	38.0	211	840	556	1.51
R	54.4	49	118	90	1.31
SC	19.9	36	575	181	3.18
DES	28.9	13	67	43	1.56
CTL	57.7	94	131	163	0.80

	H1	(nL)	H2	(nL)	H3	(nL)
CM	0.66	(190)	0.65	(188)	0.34	(97)
MUY	0.59	(330)	0.59	(329)	0.33	(181)
R	0.41	(37)	0.59	(53)	0.31	(28)
SC	0.86	(155)	1.85	(335)	0.47	(85)
DES	0.72	(31)	0.44	(19)	0.40	(17)
CTL	0.49	(79)	0.23	(37)	0.09	(15)

%F = Percentage of line free bones
n = Number of bones without lines
nL = Total number of lines in area
nB = Total number of bones in area with lines
IM = Index of Morbidity for area regardless of age

Table 4-2

Chi-square analysis of numbers of individuals with Harris lines, by area

	CM	MUY	R	SC	DES
MUY	n.s.	-			
R	n.s.	n.s.	-		
SC	(34.51) <0.001	(55.13) <0.001	(30.15) <0.001	-	
DES	n.s.	n.s.	n.s.	(11.49) <0.001	-
CTL	(26.75) <0.001	(24.13) <0.001	(7.22) <0.01	(95.35) <0.001	(8.86) <0.005

Area Code:
CM - Central Murray, MUY - Murray valley, R - Rufus River, SC - South Coast, DES - Desert, TRO - Tropics, CTL - East Coast

Table 4-3

The incidence of Harris lines among three sub-adult age grades, by location

	%F	n	nL	nB	IM (H)	H1	(nL)	H2	(nL)	H3	(nL)
T	28.6	4	16	14	1.14	0.71	10	0.14	2	0.29	4
K	30.9	30	129	97	1.33	0.71	69	0.31	30	0.31	30
B	25.5	24	164	94	1.73	0.61	57	0.85	80	0.28	27
S	32.0	24	149	75	1.99	0.65	49	0.93	70	0.40	30
E	34.8	8	33	23	1.44	0.65	15	0.35	8	0.44	10
R	54.4	49	118	90	1.31	0.41	37	0.59	53	0.31	28
M	44.2	72	231	163	1.42	0.57	93	0.53	86	0.32	52
SP	14.8	18	432	122	3.54	0.95	116	2.14	261	0.45	55
A	33.3	8	54	24	2.25	0.58	14	1.21	29	0.46	11
G	25.0	4	46	16	2.88	0.69	11	1.81	29	0.38	6
C	21.1	4	43	19	2.26	0.74	14	0.84	16	0.68	13
W	27.6	8	59	29	2.03	0.41	12	1.04	30	0.59	17
N	51.3	20	39	39	1.00	0.64	25	0.28	11	0.08	3
CA	34.4	11	48	32	1.50	0.66	21	0.40	13	0.44	14
WA	39.3	11	40	28	1.43	0.71	20	0.54	15	0.18	5
CQ	47.9	23	45	48	0.94	0.55	26	0.29	14	0.10	5
BB	71.0	42	26	59	0.44	0.31	18	0.05	3	0.09	5

%F = Percentage of line free bones
n = Number of bones without lines
nL = Total number of lines in location
nB = Total number of bones in location with lines
IM = Index of Morbidity for location regardless of age

Location Codes:
T - Tocumwal, K - Kow Swamp (modern), B - Baratta,
S - Swan Hill, E - Euston, R - Rufus River, SP - Swanport,
A - Adelaide, C- Coastal Victoria, W - West Central Victoria,
N - Coastal New South Wales, CA - Central Australia,
WA - Western Australia, CQ - Coastal Queensland,
BB - Broadbeach

Table 4-4

Distribution of Harris lines for males and females, in three age grades for six locations. Individuals with more than four lines are shown also

Location		H1	H2	H3	H	nB	n>4	%
Baratta (B)	M	0.76	0.74	0.21	1.61	34	5	14.7
	F	0.56	0.94	0.22	1.72	18	2	11.1
Swan Hill (S)	M	0.82	1.18	0.36	2.36	28	6	21.4
	F	0.80	1.20	0.80	2.80	20	4	20.0
Rufus River (R)	M	0.34	0.82	0.32	1,48	50	5	10.0
	F	0.25	0.38	0.33	0.96	24	2	8.3
Swanport (SP)	M	0.98	2.45	0.69	4.12	42	17	40.5
	F	0.48	1.41	1.19	1.10	31	9	29.0
Adelaide (A)	M	0.86	1.21	0.36	2.43	14	4	28.6
	F	0.25	-	-	0.25	4	-	-
New South Wales	M	0.52	0.17	-	0.69	23	-	-
coast (N)	F	0.17	-	-	0.17	6	-	-

M = Males, F = Females

nB = Sample size
n>4 = Individuals with more than four lines
% = Percentage of individuals having more than four lines

Table 5-1

Post-cranial non-specific infection, by area

UPPER LIMB

Humerus	Left n	Left n̄	Left %	Right n	Right n̄	Right %
CM	176	15	8.5	189	18	9.5
MUY	567	33	5.8	577	38	6.6
R	128	4	3.1	134	4	3.0
SC	152	1	0.7	147	2	1.4
DES	28	2	7.1	27	-	-
TRO*	10	1	10.0	8	1	12.5
CTL	99	1	1.0	106	2	1.9
Ulna						
CM	127	8	6.3	130	8	6.2
MUY	409	27	6.6	403	24	6.0
R	78	2	2.6	84	5	6.0
SC	108	1	.9	117	2	1.7
DES	24	1	4.2	26	1	3.8
TRO*	6	-	-	8	1	12.5
CTL	93	-	-	100	1	1.0
Radius						
CM	128	7	5.5	131	7	5.3
MUY	412	25	6.1	403	26	6.5
R	110	3	2.7	113	5	4.4
SC	120	2	1.7	92	3	3.3
DES	25	2	8.0	24	-	-
TRO*	7	1	14.3	6	-	-
CTL	93	1	1.1	93	-	-

LOWER LIMB

Femur	Left n	Left n̄	Left %	Right n	Right n̄	Right %
CM	216	11	5.1	202	8	4.0
MUY	490	36	7.3	464	26	5.6
R	94	4	4.3	88	2	2.3
SC	144	2	1.4	140	4	2.9
DES	22	-	-	24	4	16.7
TRO*	8	-	-	9	-	-
CTL	100	5	5.0	111	2	1.8
Tibia						
CM	261	17	6.5	254	20	7.9
MUY	534	55	10.3	518	58	11.2
R	98	6	6.1	101	8	7.9
SC	138	4	2.9	145	4	2.8
DES	28	2	7.1	36	4	11.1
TRO*	9	-	-	11	2	18.2
CTL	90	5	5.6	114	7	6.1
Fibula						
CM	151	8	5.3	143	8	5.6
MUY	330	27	8.2	326	35	10.7
R	87	2	2.3	87	6	6.9
SC	41	-	-	36	1	2.8
DES	22	1	4.5	19	1	5.3
TRO*	8	1	12.5	7	1	14.3
CTL	75	-	-	72	1	1.4

*This Area not used in analysis because of small samples sizes

Area Codes:
CM - Central Murray, MUY - Murray River, R - Rufus River, SC - South Coast, DES - Desert, CTL - East Coast

Table 5-2

Highest frequencies of non-specific infection in each location by the bone displaying the highest incidence of infection

	n	$\bar{n}$	%	Bone
T	41	1	2.4	Radius
K	41	2	4.9	Fibula
B	86	8	9.3	Tibia
S	47	6	12.8	Ulna
E	21	2	9.5	Femur
R	100	8	8.0	Tibia
M	93	21	22.6	Fibula
SP	115	3	2.6	Femur
N	35	3	8.6	Tibia
CA	24	4	16.7	Tibia
WA	26	-	-	-
CQ	29	3	10.3	Tibia
BB	35	1	2.9	Tibia

Location Code:
T - Tocumwal, K - Kow Swamp (modern), B - Baratta, S - Swan Hill, E - Euston, R - Rufus River, M - Murray River, SP - Swanport, N - Coastal New South Wales, CA - Central Australia, WA - Western Australia, CQ - Coastal Queensland, BB- Broadbeach

Table 5-3

Chi-square analysis of non-specific infection, by area

	18/171 CM	58/460 MUY	8/93 R	3/89 SC	4/20 DES
MUY	n.s.	-			
R	n.s.	n.s.	-		
SC	n.s.	5.47 <0.025	n.s.	-	
DES	n.s.	n.s.	n.s.	6.03 <0.025	-
CTL	n.s.	n.s.	n.s.	n.s.	n.s.

Table 5-4

Treponemal infection on crania by area

	MALES			FEMALES		
	n	$\bar{n}$	%	n	$\bar{n}$	%
CM	247	18	7.3	151	14	9.3
MUY	469	26	5.5	302	23	7.6
R	122	6	4.9	83	7	8.4
SC	138	3	2.2	123	4	3.3
DES	132	6	4.5	51	3	5.9
TRO	92	3	3.3	62	3	4.8
CTL	133	3	2.3	86	1	1.2

Area Codes:

CM - Central Murray, MUY - Murray River, R - Rufus River, SC - South Coast, DES - Desert, TRO - Tropics, CTL - East Coast

Table 5-5

Treponemal infection on crania by location

	MALES			FEMALES		
	n	$\bar{n}$	%	n	$\bar{n}$	%
T	43	1	2.3	26	-	-
K	31	2	6.5	17	3	17.6*
B	41	3	7.3	28	4	14.3
S	56	9	16.1	40	3	7.5
E	119	4	3.4	67	5	7.5
R	122	6	4.9	83	7	8.4
M	57	1	1.8	42	2	4.8
SP	57	2	3.5	53	1	1.9
A	33	-	-	32	3	9.4
C	48	1	2.1	38	-	-
W	17	1	5.9*	19	4	21.1*
N	59	-	-	44	1	2.3
CA	94	6	6.4	40	2	5.0
WA	47	-	-	20	1	5.0
NT	53	3	5.7	38	1	2.6
TA	-	-	-	-	-	-
CQ	-	-	-	-	-	-
CR	39	-	-	24	2	8.3
BB	21	-	-	5	-	- *
QB	65	3	4.6	33	-	-

* Less than 20 in the sample

Location Codes:
T - Tocumwal, K - Kow Swamp (modern), B - Baratta, S - Swan Hill, E - Euston, R - Rufus River, M - Murray River, SP - Swanport, A - Adelaide, C - Coastal Victoria, W - West Central Victoria, N - Coastal New South Wales, CA - Central Australia, WA - Western Australia, NT - Northern Territory, TA - Tasmania, CQ - Coastal Queensland, CR - Cape York/Cairns, BB- Broadbeach, QB - Coastal Queensland and Broadbeach combined.

Table 5-6

Chi-square analysis of treponemal disease, by area

	* 32/366 CM	49/722 MUY	13/192 R	7/254 SC	9/174 DES	8/144 TRO
MUY	n.s.	-				
R	n.s.	n.s.	-			
SC	8.13 <0.005	5.13 <0.025	n.s.	-		
DES	n.s.	n.s.	n.s.	n.s.	-	
TRO	n.s.	n.s.	n.s.	n.s.	n.s.	-
CTL	9.93 <0.005	6.9 <0.01	5.61 <0.02	n.s.	n.s.	n.s.

* Pathological (n)/Non-Pathological (n)

Table 5-7

Comparison of non-specific and treponemal disease frequencies, by area

	Non-specific Infection %	**Treponemal Infection %**
CM	9.5	8.0
MUY	11.2	6.4
R	8.0	6.3
SC	3.3	2.7
DES	16.7	4.9
TRO	-	5.2
CTL	6.1	1.8

Area Codes:
CM - Central Murray, MUY - Murray River, R - Rufus River, SC - South Coast, DES - Desert, TRO - Tropics, CTL - East Coast

Table 5-8

Tibial bowing

	n	B	%	P	O	OM	
CM	523	11	2.1	-	1	-	8.43 (p=<0.005)
MUY	1052	39	3.7	4	8	1	4.42 (p=<0.05)
R	199	3	1.5	-	1	-	n.s.
SC	283	2	0.7	-	-	-	n.s.
DES	64	1	1.6	-	-	-	n.s.
TRO	21	1	4.8	-	1	-	n.s.
CTL	207	8	3.9	-	1	-	5.57 (p=<0.025)

B = Bowed tibia without infection
P = Bowed tibia with periostitis
O = Bowed tibia with osteitis
OM = Bowed tibia with osteomyelitis

(Chi-square comparison between MUY and SC = 6.74 (p=<0.01))

Area Codes:
CM - Central Murray, MUY - Murray River, R - Rufus River, SC - South Coast, DES - Desert, TRO - Tropics, CTL - East Coast

Table 6-1

Post-cranial samples, by area

ELBOW

		CM			MUY			SC			CTL		
		n	$\bar{n}$	%	n	$\bar{n}$	%	n	$\bar{n}$	%	n	$\bar{n}$	%
M	L	182	21	11.5	444	37	8.3	75	1	1.3	153	2	1.3
	R	216	44	20.4	493	80	16.2	86	7	8.1	167	7	4.2
F	L	110	9	8.2	256	17	6.6	27	-	-	70	-	-
	R	106	8	7.6	260	18	6.9	54	-	-	70	3	4.3

KNEE

		CM			MUY			SC			CTL		
		n	$\bar{n}$	%	n	$\bar{n}$	%	n	$\bar{n}$	%	n	$\bar{n}$	%
M	L	194	15	7.7	393	23	5.9	87	7	8.1	94	4	4.3
	R	173	12	6.9	380	29	7.6	96	4	4.2	122	5	4.1
F	L	100	1	1.0	191	4	2.1	70	-	-	48	4	8.3
	R	116	3	2.6	186	3	1.6	67	-	-	46	4	8.7

COMBINED ELBOW AND KNEE

		CM			MUY			SC			CTL		
		n	$\bar{n}$	%	n	$\bar{n}$	%	n	$\bar{n}$	%	n	$\bar{n}$	%
M	L	376	36	9.6	837	60	7.2	162	8	4.9	247	6	2.4
	R	389	56	14.4	873	109	12.5	182	11	6.0	289	12	4.2
F	L	210	10	4.8	447	21	4.7	97	-	-	118	4	3.4
	R	222	11	4.5	446	21	4.7	121	-	-	116	7	6.0

NOTE: The sample sizes for the Rufus River (R) and Desert (D) area samples can be seen on Table 6-4 under their respective location headings

M = Males, F = Females

Area Codes:
CM - Central Murray, MUY - Murray River, SC - South Coast, CTL - East Coast

Table 6-2

Chi-square analysis of post-cranial osteoarthritis

MALES

	Left Elbow					Right Elbow				
	21/161	37/407	7/133	1/74	3/38	44/172	80/413	15/134	7/79	5/31
	CM	MUY	R	SC	DES	CM	MUY	R	SC	DES
MUY	n.s.	-				n.s.	-			
R	4.26	n.s.	-			6.91	n.s.	-		
	<0.05					<0.01				
SC	7.07	4.63	n.s.	-		6.56	n.s.	n.s.	-	
	<0.01	<0.05				<0.02				
DES	n.s.	n.s.	n.s.	n.s.	-	n.s.	n.s.	n.s.	n.s.	-
CTL	13.61	9.20	n.s.	n.s.	4.65	21.36	15.79	4.20	n.s.	5.01
	<0.001	<0.005			<0.05	<0.001	<0.001	<0.05		<0.05

FEMALES

	Left Elbow					Right Elbow				
	9/101	17/239	2/68	-/27	-/12	8/98	18/242	-/76	-/54	-/17
	CM	MUY	R	SC	DES	CM	MUY	R	SC	DES
MUY	n.s.	-				n.s.	-			
R	n.s.	n.s.	-			6.00	5.56	-		
						<0.02	<0.02			
SC	n.s.	n.s.	n.s.	-		4.29	3.97	n.s.	-	
						<0.05	<0.05			
DES	n.s.	n.s.	n.s.	n.s	-	n.s.	n.s.	n.s.	n.s.	-
CTL	6.03	4.90	n.s.	n.s	n.s.	n.s.	n.s.	n.s.	n.s.	n.s.
	<0.02	<0.05								

MALES

	Left Knee					Right Knee				
	5/179	23/370	3/96	7/80	3/26	12/161	29/351	10/98	4/92	2/29
	CM	MUY	R	SC	DES	CM	MUY	R	SC	DES
MUY	n.s.	-				n.s.	-			
R	n.s.	n.s.	-			n.s.	n.s.	-		
SC	n.s.	n.s.	n.s.	-		n.s.	n.s.	n.s.	-	
DES	n.s.	n.s.	n.s.	n.s.	-	n.s.	n.s.	n.s.	n.s.	-
CTL	n.s.	n.s.	n.s.	n.s.	n.s.	n.s.	n.s.	n.s.	n.s.	n.s.

FEMALES

	Left Knee					Right Knee				
	1/99	4/87	2/53	-/70	-/7	3/113	3/183	-/49	-/67	-/9
	CM	MUY	R	SC	DES	CM	MUY	R	SC	DES
MUY	n.s.	-				n.s.	-			
R	n.s.	n.s.	-			n.s.	n.s.	-		
SC	n.s.	n.s.	n.s.	-		n.s.	n.s.	n.s.	-	
DES	n.s.	n.s.	n.s.	n.s.	-	n.s.	n.s.	n.s.	n.s.	-
CTL	5.34	4.62	n.s.	6.04	n.s.	n.s.	6.32	4.45	6.04	n.s.
	<0.025	<0.005		<0.02			<0.02	<0.05	<0.02	

n.s. Not statistically significant

Area Codes:
CM - Central Murray, MUY - Murray River, R - Rufus River, SC - South Coast, DES - Desert, CTL - East Coast

Table 6-3

Knee samples, by location

	MALES						FEMALES					
	Left			Right			Left			Right		
	n	n̄	%	n	n̄	%	n	n̄	%	n	n̄	%
K	38	7	18.4	27	5	18.5	19	-	-	24	-	-
B	87	6	6.9	71	3	4.2	32	1	3.1	38	2	5.3
S	59	2	3.4	64	3	4.7	43	-	-	48	1	2.1
E	10	-	-	11	1	9.1	6	-	-	6	-	-
R	99	3	3.0	108	10	9.3	55	2	3.6	49	-	-
M	100	5	5.0	99	7	7.1	36	1	2.8	31	-	-
SP	41	5	12.2	56	3	5.4	60	-	-	56	-	-
A	29	-	-	23	-	-	4	-	-	5	-	-
C	17	2	11.8	17	1	5.9	6	-	-	6	-	-
W	15	-	-	10	1	10.0	15	-	-	14	-	-
N	32	2	6.3	39	2	5.1	18	-	-	15	-	-
CA	29	3	10.3	31	2	6.5	7	-	-	9	-	-
WA	14	1	7.1	24	-	-	1	-	-	9	-	-
CQ	20	2	10.0	32	3	9.4	26	4	15.4	27	4	14.8
QB	62	2	3.2	83	3	3.6	30	4	13.3	31	4	12.9
BB	42	-	-	51	-	-	4	-	-	4	-	-

POPULATION TOTALS
(Male + Female + Unsexed)

	Left			Right		
	n	n̄	%	n	n̄	%
K	155	10	6.5	148	7	4.7
B	157	9	5.7	146	8	5.5
S	102	2	2.0	112	4	3.6
E	37	-	-	39	2	5.1
R	183	7	3.8	187	13	7.0
M	332	13	3.9	322	17	5.3
SP	177	10	5.6	187	7	3.7
A	34	-	-	29	-	-
C	23	2	8.7	23	1	4.3
W	30	-	-	24	1	4.2
N	53	2	3.8	56	2	3.6
CA	36	3	8.3	40	2	5.0
WA	15	1	6.7	32	-	-
CQ	53	6	11.3	63	7	11.1
QB	99	6	6.1	118	7	5.9
BB	46	-	-	55	-	-

Sample not available or too small from T, TA, NT, ANT, CR and P

Location Codes:
T - Tocumwal, B - Baratta, S - Swan Hill, E - Euston, R - Rufus River, M - Murray River, SP - Swanport, A - Adelaide, C - Coastal Victoria, W - West Central Victoria, N - Coastal New South Wales, CA - Central Australia, WA - Western Australia, NT - Northern Territory (northern half), AT - All Northern Territory, TA - Tasmania, CQ - Coastal Queensland, CR - Cape York and Cairns, BB - Broadbeach, P - Pleistocene sample.

Table 6-4

Elbow samples, by location

	MALES						FEMALES					
	Left			Right			Left			Right		
	n	$\bar{n}$	%	n	$\bar{n}$	%	n	$\bar{n}$	%	n	$\bar{n}$	%
T	44	7	15.9	45	10	22.2	43	3	7.0	45	6	13.3
B	90	11	12.2	107	18	16.8	40	6	15.0	39	4	10.3
S	83	10	12.0	96	23	24.0	65	3	4.6	63	4	6.3
E	9	-	-	13	3	23.1	5	-	-	4	-	-
R	140	7	5.0	149	15	10.1	70	2	2.9	76	-	-
M	78	9	11.5	83	11	13.3	33	3	9.1	33	4	12.1
SP	17	-	-	29	4	13.8	15	-	-	48	-	-
A	37	1	2.7	35	1	2.9	9	-	-	6	-	-
C	21	-	-	22	2	9.1	3	-	-	-	-	-
W	13	1	7.7	14	-	-	18	-	-	22	-	-
N	37	-	-	44	5	11.4	28	-	-	29	-	-
CA	41	3	7.3	36	5	13.9	12	-	-	17	-	-
WA	19	1	5.3	21	2	9.5	3	-	-	3	-	-
CQ	31	2	6.5	31	1	3.2	34	-	-	36	3	8.3
QB	116	2	1.7	123	2	1.6	42	-	-	41	3	7.3
BB	85	-	-	92	-	-	8	-	-	5	-	-

POPULATION TOTALS
(Male + Female + Unsexed)

	Left			Right		
	n	$\bar{n}$	%	n	$\bar{n}$	%
T	227	22	9.7	194	34	17.5
B	174	19	10.9	198	28	14.1
S	152	14	9.2	165	28	17.0
E	47	2	4.3	49	13	26.5
R	307	13	4.2	330	21	6.4
M	404	55	13.6	396	54	13.6
SP	250	6	2.4	235	8	3.4
A	46	1	2.2	41	1	2.4
C	24	-	-	22	2	9.1
W	31	1	3.2	36	-	-
N	70	-	-	79	5	6.3
CA	53	3	5.7	53	5	9.4
WA	22	1	4.5	24	2	8.3
CQ	66	2	3.0	69	4	5.8
QB	159	2	1.3	166	5	3.0
BB	93	-	-	97	-	-

Samples not available or too small from K, TA, NT, ANT, CR and P

Location Codes:
T - Tocumwal, B - Baratta, S - Swan Hill, E - Euston, R - Rufus River, M - Murray River, SP - Swanport, A - Adelaide, C - Coastal Victoria, W - West Central Victoria, N - Coastal New South Wales, CA - Central Australia, WA - Western Australia, NT - Northern Territory (northern half), AT - All Northern Territory, TA - Tasmania, CQ - Coastal Queensland, CR - Cape York and Cairns, BB - Broadbeach, P - Pleistocene sample.

Table 6-5

Frequency of post-cranial osteophytosis, by area

		MALES						FEMALES					
		Left			Right			Left			Right		
		n	$\bar{n}$	%	n	$\bar{n}$	%	n	$\bar{n}$	%	n	$\bar{n}$	%
	10	99	7	7.1	111	21	18.9	54	5	9.3	55	3	5.5
C	11	57	9	15.8	60	8	13.3	36	4	11.1	28	2	7.1
M	12	49	3	6.1	56	3	5.4	40	3	7.5	36	4	11.1
	20	124	13	10.5	98	13	13.3	58	1	1.7	66	2	3.0
	21	78	1	1.3	77	6	7.8	50	-	-	49	-	-
	10	34	1	2.9	35	3	8.6	21	-	-	23	-	-
	11	43	1	2.3	46	4	8.7	22	1	4.5	23	-	-
R	12	64	2	3.1	62	4	6.5	30	1	3.3	32	-	-
	20	52	3	5.8	54	2	3.7	33	2	6.1	29	-	-
	21	52	-	-	52	2	3.8	26	-	-	25	-	-
	10	32	-	-	36	1	2.8	31	-	-	23	-	-
S	11	21	-	-	28	6	21.4	21	-	-	19	-	-
C	12	25	1	4.0	25	-	-	21	-	-	20	-	-
	20	34	2	5.9	33	1	3.0	35	-	-	32	-	-
	21	55	5	9.1	64	3	4.7	41	-	-	41	-	-
	10	23	3	13.0	21	2	9.5	4	-	-	6	-	-
D	11	19	1	5.3	19	2	10.5	5	-	-	5	-	-
E	12	21	-	-	15	3	20.0	4	-	-	6	-	-
S	20	18	2	11.1	19	1	5.3	4	-	-	5	-	-
	21	24	1	4.2	28	1	3.6	4	-	-	6	-	-
	10	56	2	3.6	64	2	3.1	27	-	-	26	1	3.8
C	11	53	-	-	60	1	1.7	26	-	-	24	1	4.2
T	12	59	-	-	56	1	1.8	20	-	-	21	1	4.8
L	20	56	2	3.6	65	2	3.1	25	2	8.0	28	2	7.1
	21	49	2	4.1	66	3	4.5	24	2	8.3	26	2	7.7

Bone Code: 10 = Humerus, 11 = Ulna, 12 = Radius,
20 = Femur, 21 = Tibia,

Area Codes:
CM - Central Murray, MUY - Murray River, R - Rufus River,
SC - South Coast, DES - Desert, CTL - East Coast

Table 6-6

Frequencies of osteoarthritis of the mandibular condyle (MAC), by location

Location		Age	n	n̄	%1	n̄	%2	Tn	T%
T	M	6	5	-		-	-	-	-
		7	14	1	7.1	2	14.3	3	21.4
		T	19	1	5.3	2	10.5	3	15.8
T	F	6	9	-	-	-	-	-	-
		7	6	1	16.7	1	16.7	2	33.3
		T	15	1	6.7	1	6.7	2	13.3
K	M	6	5	-	-	3	60.0	3	60.0
		7	13	5	38.5	5	38.5	10	76.9
		T	18	5	27.8	8	44.4	13	72.2
K	F	6	4	1	25.0	-	-	1	25.0
		7	3	-	-	2	66.7	2	66.7
		T	7	1	14.3	2	28.6	3	42.9
B	M	6	9	1	11.1	-	-	1	11.1
		7	18	3	16.7	3	16.7	6	33.3
		T	27	4	14.8	3	11.1	7	25.9
B	F	6	12	1	8.3	-	-	1	8.3
		7	6	1	16.7	2	33.3	3	50.0
		T	18	2	11.1	2	11.1	4	22.2
S	M	6	21	3	14.3	8	38.1	11	52.4
		7	22	7	31.8	4	18.2	11	50.0
		T	43	10	23.3	12	27.9	22	51.2
S	F	6	20	6	30.0	1	5.0	7	35.0
		7	6	3	50.0	1	16.7	4	66.7
		T	26	9	34.6	2	7.7	11	42.3
E	M	6	34	3	8.8	2	5.9	5	14.7
		7	48	6	12.5	3	6.3	9	18.8
		T	82	9	11.0	5	6.1	14	17.1
E	F	6	22	6	27.3	1	4.5	7	31.8
		7	13	1	7.7	-	-	1	7.7
		T	35	7	20.0	1	2.9	8	22.9
R	M	6	33	5	15.2	2	6.1	7	21.2
		7	48	12	25.0	5	10.4	17	35.4
		T	81	17	21.0	7	8.6	24	29.6
R	F	6	42	7	16.7	2	4.8	9	21.4
		7	22	4	18.2	-	-	4	18.2
		T	64	11	17.2	2	3.1	13	20.3
M	M	6	12	1	8.3	1	8.3	2	16.7
		7	18	5	27.8	3	16.7	8	44.4
		T	30	6	20.0	4	13.3	10	33.3
M	F	6	11	1	9.1	-	-	1	9.1
		7	8	1	12.5	1	12.5	2	25.0
		T	19	2	10.5	1	5.3	3	15.8
SP	M	6	7	2	28.6	-	-	2	28.6
		7	16	2	12.5	5	31.3	7	43.8
		T	23	4	17.4	5	21.7	9	39.1
SP	F	6	9	1	11.1	-	-	1	11.1
		7	17	4	23.5	5	29.4	9	52.9
		T	26	5	19.2	5	19.2	10	38.5

Table 6-6 continued

A	M	6	3	-	-	-	-	-	-
		7	17	3	17.6	1	5.9	4	23.5
		T	20	3	15.0	4	20.0	7	35.0
A	F	6	4	-	-	-	-	-	-
		7	10	3	30.0	-	-	3	30.0
		T	14	3	21.4	-	-	3	21.4
C	M	6	11	1	9.1	-	-	1	9.1
		7	11	3	27.3	1	9.1	4	36.4
		T	23	4	17.4	1	4.3	5	21.7
C	F	6	18	2	11.1	-	-	2	11.1
		7	4	1	25.0	-	-	1	25.0
		T	22	3	13.6	-	-	3	13.6
W	M	6	3	-	-	-	-	-	-
		7	9	3	33.3	1	11.1	4	44.4
		T	12	3	25.0	1	8.3	4	33.3
W	F	6	12	1	8.3	1	8.3	2	16.7
		7	3	-	-	-	-	-	-
		T	15	1	6.7	1	6.7	2	13.3
N	M	6	11	-	-	-	-	-	-
		7	26	1	3.8	3	11.5	4	15.4
		T	37	1	2.7	3	8.1	4	10.8
N	F	6	10	1	10.0	1	10.0	2	20.0
		7	14	2	14.3	-	-	2	14.3
		T	24	3	12.5	1	4.2	4	16.7
CA	M	6	9	1	11.1	-	-	1	11.1
		7	40	6	15.0	2	5.0	8	20.0
		T	49	7	14.3	2	4.1	9	18.4
CA	F	6	12	2	16.7	-	-	2	16.7
		7	18	3	16.7	2	11.1	5	27.8
		T	30	5	16.7	2	6.7	7	23.3
WA	M	6	4	-	-	-	-	-	-
		7	20	4	20.0	1	5.0	5	25.0
		T	24	4	16.7	1	4.2	5	20.8
WA	F	6	3	1	33.3	-	-	1	33.3
		7	8	-	-	-	-	-	-
		T	11	1	9.1	-	-	1	9.1
NT	M	6	12	-	-	-	-	-	-
		7	20	4	20.0	1	5.0	5	25.0
		T	34	4	11.8	1	2.9	5	14.7
NT	F	6	8	-	-	-	-	-	-
		7	15	1	6.7	3	20.0	4	26.7
		T	23	1	4.3	3	13.0	4	17.4
AT	M	6	30	3	10.0	1	3.3	4	13.3
		7	42	6	14.3	2	4.8	8	19.0
		T	72	9	12.5	3	4.2	12	16.7
AT	F	6	19	-	-	-	-	-	-
		7	25	3	12.0	6	24.0	9	36.0
		T	44	3	6.8	6	13.6	9	20.5
TA	M	6	2						
		7	4		Inadequate sample				
		T	6						
TA	F	6	2						
		7	3		Inadequate sample				
		T	5						

Table 6-6 continued

CQ	M	6	4	-	-	-	-	-	-
		7	21	6	28.6	-	-	6	28.6
		T	25	6	24.0	-	-	6	24.0
CQ	F	6	7	2	28.6	-	-	2	28.6
		7	7	1	14.3	2	28.6	3	42.9
		T	14	3	21.4	2	14.3	5	35.7
CR	M	6	12	-	-	2	16.7	2	16.7
		7	16	6	37.5	1	6.3	7	43.8
		T	28	6	21.4	3	10.7	9	32.1
CR	F	6	7	2	28.6	-	-	2	28.6
		7	9	2	22.2	4	44.4	6	66.7
		T	16	4	25.0	4	25.0	8	50.0
BB	M	6	6	3	50.0	-	-	3	50.0
		7	13	2	15.4	-	-	2	15.4
		T	19	5	26.3	-	-	5	26.3
BB	F	6	1						
		7	1		Inadequate sample				
		T	2						
QB	M	6	10	3	30.0	-	-	3	30.0
		7	34	8	23.5	-	-	8	23.5
		T	44	11	25.0	-	-	11	25.0
QB	F	6	8	2	25.0	-	-	2	25.0
		7	8	1	12.5	2	25.0	3	37.5
		T	16	3	18.8	2	12.5	5	31.3
P	M	6	4	2	50.0	-	-	2	50.0
		7	5	2	40.0	1	20.0	3	60.0
		T	9	4	44.4	1	11.1	5	55.6
P	F	6	1						
		7	1		Inadequate sample				
		T	2						

%1 = Percentage of individuals with category 1 arthritis
%2 = Percentage of individuals with category 2 arthritis
Tn = Total number of individuals affected
M = Males, F = Females, T = Totals

Location Codes:
T - Tocumwal, K - Kow Swamp (modern), B - Baratta, S - Swan Hill, E - Euston, R - Rufus River, M - Murray River, SP - Swanport, A - Adelaide, C - Coastal Victoria, W - West Central Victoria, N - Coastal New South Wales, CA - Central Australia, WA - Western Australia, NT - Northern Territory (northern half), AT - All Northern Territory, TA - Tasmania, CQ - Coastal Queensland, CR - Cape York and Cairns, BB - Broadbeach, QB - Coastal Queensland and Broadbeach combined, P - Pleistocene sample.

Table 6-7

Frequencies of osteoarthritis in the glenoid fossa (TMJ), by location

Location		Age	n	$\bar{n}$	%1	$\breve{n}$	%2	Tn	T%
T	M	6	17	8	47.1	3	17.6	11	64.7
		7	25	7	28.0	6	24.0	13	52.0
		T	42	15	35.7	9	21.4	24	57.1
T	F	6	13	1	7.7	-	-	1	7.7
		7	13	5	38.5	4	30.8	9	69.2
		T	26	6	23.1	4	15.4	10	38.5
K	M	6	8	4	50.0	2	25.0	6	75.0
		7	22	7	31.8	11	50.0	18	81.8
		T	30	11	36.7	13	43.3	24	80.0
K*	F	6	9	3	33.3	3	33.3	6	66.7
		7	7	2	28.6	4	57.1	6	85.7
		T	16	5	31.3	7	43.8	12	75.0
B	M	6	12	3	25.0	1	8.3	4	33.3
		7	28	9	32.1	8	28.6	17	60.7
		T	40	12	30.0	9	22.5	21	52.5
B	F	6	17	6	35.3	-	-	6	35.3
		7	11	6	54.5	4	36.4	10	90.9
		T	28	12	42.9	4	14.3	16	57.1
S	M	6	24	7	29.2	8	33.3	15	62.5
		7	30	13	43.3	11	36.7	24	80.0
		T	54	20	37.0	19	35.2	39	72.2
S	F	6	32	15	46.9	2	6.3	17	53.1
		7	8	3	37.5	2	25.0	5	62.5
		T	40	18	45.0	4	10.0	22	55.0
E	M	6	47	16	34.0	3	6.4	19	40.4
		7	71	29	40.8	15	21.1	44	62.0
		T	118	45	38.1	18	15.3	63	53.4
E	F	6	40	16	40.0	5	12.5	21	52.5
		7	25	14	56.0	3	12.0	17	68.0
		T	65	30	46.2	8	12.3	38	58.5
R	M	6	47	21	44.7	2	4.3	23	48.9
		7	74	31	41.9	18	24.3	49	66.2
		T	121	52	43.0	20	16.5	72	59.5
R	F	6	53	23	43.4	-	-	23	43.4
		7	30	13	43.3	2	6.7	15	50.0
		T	83	36	43.4	2	2.4	38	45.8
M	M	6	23	3	13.0	6	26.1	9	39.1
		7	34	16	47.1	4	11.8	20	58.8
		T	57	19	33.3	10	17.5	29	50.9
M	F	6	26	9	34.6	6	23.1	15	57.7
		7	15	8	53.3	3	20.0	11	73.3
		T	41	17	41.5	9	22.0	26	63.4
SP	M	6	16	6	37.5	1	6.3	7	43.8
		7	41	7	17.1	18	43.9	25	61.0
		T	57	13	22.8	19	33.3	32	56.1
SP	F	6	21	6	28.6	2	9.5	8	38.1
		7	30	9	30.0	8	26.7	17	56.7
		T	51	15	29.4	10	19.6	25	49.0

Table 6-7 continued

A	M	6	5	-	-	-	-	-	-
		7	28	15	53.6	3	10.7	18	64.3
		T	33	15	45.5	3	9.1	18	54.5
A	F	6	7	-	-	-	-	-	-
		7	25	15	60.0	-	-	15	60.0
		T	32	15	46.9	-	-	15	46.9
C	M	6	21	10	47.6	3	14.3	13	61.9
		7	27	11	40.7	3	11.1	14	51.9
		T	48	21	43.8	6	12.5	27	56.3
C	F	6	29	13	44.8	-	-	13	44.8
		7	9	3	33.3	1	11.1	4	44.4
		T	38	16	42.1	1	2.6	17	44.7
W*	M	6	4	1	25.0	-	-	1	25.0
		7	13	4	30.8	2	15.4	6	46.2
		T	17	5	29.4	2	11.8	7	41.2
W*	F	6	13	5	38.5	1	7.7	6	46.2
		7	6	3	50.0	-	-	3	50.0
		T	19	8	42.1	1	5.3	9	47.4
N	M	6	16	4	25.0	-	-	4	25.0
		7	43	17	39.5	3	7.0	20	46.5
		T	59	21	35.6	3	5.1	24	40.7
N	F	6	22	4	18.2	3	13.6	7	31.8
		7	19	6	31.6	-	-	6	31.6
		T	41	10	24.4	3	7.3	13	31.7
CA	M	6	17	1	5.9	-	-	1	5.9
		7	71	26	36.6	4	5.6	30	42.3
		T	88	27	30.7	4	4.5	31	35.2
CA	F	6	16	2	12.5	-	-	2	12.5
		7	23	6	26.1	3	13.0	9	39.1
		T	39	8	20.5	3	7.7	11	28.2
WA	M	6	7	-	-	-	-	-	-
		7	46	17	37.0	3	6.5	20	43.5
		T	53	17	32.1	3	5.7	20	37.7
WA*	F	6	10	2	20.0	-	-	2	20.0
		7	10	2	20.0	-	-	2	20.0
		T	20	4	20.0	-	-	4	20.0
NT	M	6	19	3	15.8	-	-	3	15.8
		7	33	10	30.3	1	3.0	11	33.3
		T	52	13	25.0	1	1.9	14	26.9
NT	F	6	16	2	12.5	-	-	2	12.5
		7	21	3	14.3	3	14.3	6	28.6
		T	37	5	13.5	3	8.1	8	21.6
AT	M	6	38	12	31.6	2	5.3	14	36.8
		7	55	20	36.4	3	5.5	23	41.8
		T	93	32	34.4	5	5.4	37	39.8
AT	F	6	29	4	13.8	-	-	4	13.8
		7	31	6	19.4	5	16.1	11	35.5
		T	60	10	16.7	5	8.3	15	25.0
TA	M	6	9	2	22.2	2	22.2	4	44.4
		7	17	6	35.3	-	-	6	35.3
		T	26	8	30.8	2	7.7	10	38.5
TA	F	6	24	10	41.7	-	-	10	41.7
		7	10	3	30.0	-	-	3	30.0
		T	34	13	38.2	-	-	13	38.2

Table 6-7 continued

CQ	M	6	7	-	-	-	-	-	-
		7	37	14	37.8	1	2.7	15	40.5
		T	44	14	31.8	1	2.3	15	34.1
CQ	F	6	16	1	6.3	1	6.3	2	12.5
		7	12	4	33.3	1	8.3	5	41.7
		T	28	5	17.9	2	7.1	7	25.0
CR	M	6	27	3	11.1	-	-	3	11.1
		7	37	6	16.2	3	8.1	9	24.3
		T	64	9	14.1	3	4.7	12	18.8
CR	F	6	17	4	23.5	-	-	4	23.5
		7	22	3	13.6	6	27.3	9	40.9
		T	39	7	17.9	6	15.4	13	33.3
BB*	M	6	4	1	25.0	-	-	1	25.0
		7	11	4	36.4	-	-	4	36.4
		T	15	5	33.3	-	-	5	33.3
BB*	F	6	3	1	33.3	-	-	1	33.3
		7	2	1	50.0	-	-	1	50.0
		T	5	2	40.0	-	-	2	40.0
QB	M	6	11	1	9.1	-	-	1	9.1
		7	48	18	37.5	1	2.1	19	39.6
		T	59	19	32.2	1	1.7	20	33.9
QB	F	6	19	2	10.5	1	5.3	3	15.8
		7	14	5	35.7	1	7.1	6	42.9
		T	33	7	21.2	2	6.1	9	27.3
P	M	6	6	4	66.7	-	-	4	66.7
		7	19	5	26.3	3	15.8	8	42.1
		T	25	9	36.0	3	12.0	12	48.0
P*	F	6	5	-	-	-	-	-	-
		7	10	6	60.0	1	10.0	7	70.0
		T	15	6	40.0	1	6.7	7	46.7

* = A sample less than 25
%1 = Percentage of those with category 1 arthritis
%2 = Percentage of those with category 2 arthritis
Tn = Total number of individuals affected
M = Males, F = Females, T = Totals

Location Codes:
T - Tocumwal, K - Kow Swamp (modern), B - Baratta, S - Swan Hill, E - Euston, R - Rufus River, M - Murray River, SP - Swanport, A - Adelaide, C - Coastal Victoria, W - West Central Victoria, N - Coastal New South Wales, CA - Central Australia, WA - Western Australia, NT - Northern Territory (northern half), AT - All Northern Territory, TA - Tasmania, CQ - Coastal Queensland, CR - Cape York and Cairns, BB - Broadbeach, QB - Coastal Queensland and Broadbeach combined, P - Pleistocene sample.

Table 6-8 A-D

Chi-square analysis for Figure 6-10A (A and B) and Figure 6-11A (C and D)

A

TMJ Males

	* 147/244 CM	272/465 MUY	72/12 R	77/138 SC	44/130 DES	29/92 TRO
MUY	n.s.	-				
R	n.s.	n.s.	-			
SC	n.s.	n.s.	n.s.	-		
DES	24.49 <0.001	24.79 <0.001	16.6 <0.001	13.02 <0.001	-	
TRO	22.84 <0.001	22.5 <0.001	16.41 <0.001	13.09 <0.001	n.s.	-
CTL	22.19 <0.001	22.5 <0.001	14.86 <0.001	11.42 <0.001	n.s.	n.s.

B

TMJ Females

	88/149 CM	162/299 MUY	38/83 R	57/123 SC	13/51 DES	16/61 TRO
MUY	n.s.	-				
R	n.s.	n.s.	-			
SC	<0.05 4.88	n.s.	n.s.	-		
DES	17.13 <0.001	14.35 <0.001	5.52 <0.02	6.52 <0.02	-	
TRO	18.66 <0.001	15.83 <0.001	5.74 <0.02	6.89 <0.01	n.s.	-
CTL	20.99 <0.001	18.23 <0.001	5.83 <0.02	7.24 <0.01	n.s.	n.s.

C

MAC Males

	* 56/170 CM	93/300 MUY	24/81 R	18/65 SC	12/77 DES	11/50 TRO
MUY	n.s.	-				
R	n.s.	n.s.	-			
SC	n.s.	n.s.	n.s.	-		
DES	5.31 <0.025	4.59 <0.05	n.s.	n.s.	-	
TRO	n.s.	n.s.	n.s.	n.s.	n.s.	-
CTL	6.35 <0.02	5.61 <0.02	n.s.	n.s.	n.s.	n.s.

D

MAC Females

	26/86 CM	44/184 MUY	13/64 R	15/61 SC	8/35 DES	8/32 TRO
MUY	n.s.	-				
R	n.s.	n.s.	-			
SC	n.s.	n.s.	n.s.	-		
DES	n.s.	n.s.	n.s.	n.s.	-	
TRO	n.s.	n.s.	n.s.	n.s.	n.s.	-
CTL	n.s.	n.s.	n.s.	n.s.	n.s.	n.s.

* Pathological (n)/Non-Pathological

Table 6-9

Numbers of individuals with and without TMJ osteoarthritis and/or tooth loss

	YOUNG ADULTS				OLD ADULTS			
	A	B	C	D	A	B	C	D
CM	43	-	31	9	40	5	79	18
	43	1	44	4	13	-	25	12
MUY	82	2	65	13	84	11	135	36
	90	6	78	8	33	2	48	21
R	24	-	20	2	20	4	36	11
	27	3	21	1	15	-	12	3
SC	15	3	15	3	22	15	34	16
	32	4	15	6	20	8	20	14
DES	13	-	-	-	29	5	10	15
	15	3	2	2	7	7	1	2
TRO	22	2	5	-	14	13	11	6
	19	3	4	1	12	6	5	3
CTL	14	2	3	-	22	17	22	5
	26	2	7	-	16	6	6	7

A = Individuals without TMJ arthritis and without tooth loss*
B = Individuals without TMJ arthritis but with tooth loss
C = Individuals with TMJ arthritis but without tooth loss
D = Individuals with TMJ arthritis and tooth loss

* One or more teeth missing but including those with avulsed central and/or lateral incisors

Area Codes:
CM - Central Murray, MUY - Murray River, R - Rufus River,
SC - South Coast, DES - Desert, CTL - East Coast

Table 6-10

Young and old adults with missing teeth and TMJ osteoarthritis

	MALES				FEMALES			
	Young Adults		Old Adults		Young Adults		Old Adults	
	$\bar{n}$	%	$\bar{n}$	%	$\bar{n}$	%	$\bar{n}$	%
CM	9	100.0	18	78.3	4	80.0	12	100.0
MUY	13	86.7	36	76.6	8	57.1	21	91.3
R	2	100.0	11	73.3	1	25.0	3	100.0
SC	3	50.0	16	51.6	6	60.0	14	63.6
DES	-	-	15	75.0	2	40.0	2	20.2
TRO	-	-	61	31.6	1	25.0	3	33.3
CTL	-	-	5	22.7	-	-	7	53.9

% = Percentage having both conditions of all those affected by either

Table 6-11

Young and old adults with missing teeth but without TMJ osteoarthritis

	MALES				FEMALES			
	Young Adults		Old Adults		Young Adults		Old Adults	
	n	%	n	%	n	%	n	%
CM	83	10.8	142	16.2	92	5.4	50	24.0
MUY	164	9.2	266	17.7	184	7.7	104	22.1
R	49	2.0	71	21.2	52	7.7	30	10.0
SC	36	16.7	87	35.6	55	17.5	62	35.5
DES	13	-	59	33.9	22	22.7	17	52.9
TRO	29	6.9	46	41.3	27	14.8	26	34.6
CTL	19	10.5	66	33.3	35	5.7	35	37.1

n = Total sample size

Table 6-12

Young and old adults with TMJ osteoarthritis and retaining their teeth

	MALES				FEMALES			
	Young Adults		Old Adults		Young Adults		Old Adults	
	$\bar{n}$	%	$\bar{n}$	%	$\bar{n}$	%	$\bar{n}$	%
CM	31	41.9	79	66.4	44	50.6	25	65.8
MUY	65	49.2	135	61.6	78	46.4	48	59.3
R	20	45.5	36	64.3	21	43.8	12	44.4
SC	15	50.0	34	60.7	15	31.9	20	50.0
DES		-	10	25.6	2	11.8	1	12.5
TRO	52	18.5	11	44.4	4	17.4	5	29.4
CTL	2	17.7	22	50.0	2	21.2	6	27.3

% = Percentage having both conditions of all those affected by either

Area Codes:
CM - Central Murray, MUY - Murray River, R - Rufus River, SC - South Coast, DES - Desert, TRO - Tropics, CTL - East Coast

Table 7-1

Distribution of fractures among the six major long bones, by area

		LEFT				RIGHT			
		n	n̄	%	Bone	n	n̄	%	Bone
CM	M	100	1	1.0	Humerus	111	-	-	Humerus
	F	54	1	1.9	Humerus	55	-	-	Humerus
	M	57	4(3)	7.1	Ulna*	60	4(2)	6.7	Ulna*
	F	36	(2)	5.6	Ulna*	28	1(1)	7.8	Ulna*
	M	49	-	-	Radius	55	1	1.8	Radius
	F	40	-	-	Radius	36	-	-	Radius
	M	124	-	-	Femur	98	-	-	Femur
	F	58	-	-	Femur	66	-	-	Femur
	M	78	-	-	Tibia	77	-	-	Tibia
	F	50	-	-	Tibia	49	-	-	Tibia
	M	56	-	-	Fibula	56	-	-	Fibula
	F	34	-	-	Fibula	31	-	-	Fibula
MUR	M	252	3	1.2	Humerus	271	1	0.4	Humerus
	F	151	3	2.0	Humerus	153	2	1.3	Humerus
	M	102	7(5)	6.9	Ulna*	108	7(3)	6.5	Ulna*
	F	59	3(2)	5.1	Ulna*	52	3(2)	5.7	Ulna*
	M	115	1	0.9	Radius	119	1	0.8	Radius
	F	70	1	1.4	Radius	71	-	-	Radius
	M	277	2	0.8	Femur	253	1	0.4	Femur
	F	124	-	-	Femur	128	-	-	Femur
	M	137	-	-	Tibia	132	-	-	Tibia
	F	84	1	1.2	Tibia	75	-	-	Tibia
	M	103	1	1.0	Fibula	103	-	-	Fibula
	F	65	1	1.5	Fibula	60	-	-	Fibula
R	M	34	-	-	Humerus	35	1	2.9	Humerus
	F	21	-	-	Humerus	23	1	4.3	Humerus
	M	43	3(2)	7.0	Ulna*	46	(1)	2.2	Ulna*
	F	22	(1)	4.5	Ulna*	23	(1)	4.3	Ulna*
	M	64	1	1.6	Radius	62	-	-	Radius
	F	30	1	3.3	Radius	32	-	-	Radius
	M	51	-	-	Femur	54	-	-	Femur
	F	33	-	-	Femur	29	-	-	Femur
	M	52	-	-	Tibia	52	-	-	Tibia
	F	26	1	3.8	Tibia	25	-	-	Tibia
	M	44	1	2.3	Fibula	44	-	-	Fibula
	F	31	1	3.2	Fibula	29	-	-	Fibula
SC	M	32	-	-	Humerus	36	1	2.8	Humerus
	F	31	-	-	Humerus	23	-	-	Humerus
	M	21	(1)	4.8	Ulna*	28	3(1)	10.7	Ulna*
	F	21	4(1)	19.1	Ulna*	19	1	5.3	Ulna*
	M	25	-	-	Radius	25	-	-	Radius
	F	21	2	9.6	Radius	20	-	-	Radius
	M	34	-	-	Femur	33	1	3.0	Femur
	F	35	-	-	Femur	32	-	-	Femur
	M	55	-	-	Tibia	64	-	-	Tibia
	F	41	-	-	Tibia	41	-	-	Tibia
	M	25	-	-	Fibula	24	-	-	Fibula
	F	16	-	-	Fibula	12	-	-	Fibula

Table 7-1 continued

DES	M	23	2	8.7	Humerus	21	-	-	Humerus
	F	3	-	-	Humerus	6	-	-	Humerus
	M	19	(2)	10.5	Ulna*	21	1	4.8	Ulna*
	F	5	-	-	Ulna*	5	-	-	Ulna*
	M	21	1	4.8	Radius	18	-	-	Radius
	F	4	-	-	Radius	6	-	-	Radius
	M	18	-	-	Femur	19	-	-	Femur
	F	4	-	-	Femur	5	1	20.0	Femur
	M	24	1	4.2	Tibia	29	-	-	Tibia
	F	4	-	-	Tibia	6	-	-	Tibia
	M	18	2	11.1	Fibula	16	1	6.3	Fibula
	F	4	-	-	Fibula	3	-	-	Fibula
CTL	M	56	1	1.8	Humerus	63	-	-	Humerus
	F	27	-	-	Humerus	26	-	-	Humerus
	M	53	5(3)	9.5	Ulna*	60	2(1)	3.4	Ulna*
	F	26	(5)	19.2	Ulna*	25	2(1)	4.0	Ulna*
	M	59	1	1.7	Radius	56	1	1.8	Radius
	F	20	-	-	Radius	22	2	9.1	Radius
	M	56	-	-	Femur	65	-	-	Femur
	F	25	1	4.0	Femur	27	-	-	Femur
	M	49	1	2.0	Tibia	69	-	-	Tibia
	F	24	2	8.2	Tibia	26	1	3.8	Tibia
	M	49	2	4.1	Fibula	41	1	2.4	Fibula
	F	17	1	5.9	Fibula	23	-	-	Fibula

* This frequency is calculated minus parrying fractures
(n) Number of parrying fractures excluded

M = Male, F = Female,

Area Codes:
CM - Central Murray, MUY - Murray River, R - Rufus River, SC - South Coast, DES - Desert, TRO - Tropics, CTL - East Coast

Table 7-2A

Unified and malunified fractures of the six major post-cranial bones, by area

LEFT

		10		11		12		20		21		22		(1)	(2)
		+ U	m	U	m	U	m	U	m	U	m	U	m	A	L
*CM	M	-	1	3	1	-	-	-	-	-	-	-	-	5	-
	F	-	1	2	-	-	-	-	-	-	-	-	-	3	-
MUY	M	-	3	6	1	1	-	1	-	-	-	1	-	11	2
	F	-	3	3	-	1	-	-	-	-	1	-	-	7	1
*R	M	-	-	3	-	1	-	-	-	-	-	1	-	4	1
	F	-	-	1	-	1	-	1	-	-	1	-	-	2	2
SC	M	-	-	-	1	-	-	-	-	-	-	-	-	1	-
	F	-	-	3	1	-	-	-	-	-	-	-	-	4	-
DES	M	1	2	2	-	1	-	-	-	1	-	1	1	6	3
	F	-	-	-	-	-	-	-	-	-	-	-	-	-	-
CTL	M	1	-	4	1	1	-	-	-	-	1	1	1	7	3
	F	-	-	5	-	-	-	-	1	1	1	1	-	5	4
	Total	2	8	12	4	4	-	1	1	2	3	4	2	40	13

RIGHT

		10		11		12		20		21		22		(1)	(2)
		+ U	m	U	m	U	m	U	m	U	m	U	m	A	L
*CM	M	-	-	3	1	-	1	-	-	-	-	-	-	5	-
	F	-	-	-	-	-	-	-	-	-	-	-	-	-	-
MUY	M	-	1	6	1	-	1	1	1	-	-	-	-	9	1
	F	-	2	2	1	-	-	-	-	-	-	-	-	5	1
*R	M	-	1	1	-	-	1	-	1	-	-	-	-	2	1
	F	-	1	-	1	-	-	-	-	-	-	-	-	2	-
SC	M	-	1	3	-	-	-	-	1	-	-	-	-	4	1
	F	-	-	-	-	1	-	-	-	-	-	-	-	1	-
DES	M	-	-	-	1	-	-	-	-	-	-	1	-	1	1
	F	-	-	-	-	-	-	-	1	-	-	-	-	-	1
CTL	M	-	-	1	-	1	1	-	-	-	-	1	-	3	1
	F	-	-	-	-	1	2	-	-	-	1	-	-	3	1
	Total	-	4	12	3	3	4	1	4	-	1	2	-	26	7

* - These totals omitted from calculation because of their incorporation in the MUY totals
\+ U - unified bones, m - malunified bones
M - Males, F - Females
(1) A - arm, total fractured, (2) L - leg, total fractures

Area Codes:
CM - Central Murray, MUY - Murray Valley, R - Rufus River,
SC - South Coast, DES - Desert, CTL - East Coast

Table 7-2B

Combined post-cranial fractures for each bone type for Australia as a whole

	LEFT			RIGHT		
	UN	MAL	%	UN	MAL	%
Humerus	2	8	24.4	-	4	15.4
Ulna	23	4	65.9	12	3	57.7
Radius	4	-	9.7	3	4	26.9
Femur	1	1	15.4	1	3	57.1
Tibia	2	3	38.5	-	1	14.3
Fibula	4	2	46.1	2	-	28.6
Total	36	18		18	15	

	Left	Right	Both
Arm	41	26	67
Leg	13	7	21
Total	54	33	88
Arm %	75.9	78.8	77.0

UNIFIED FRACTURES

	Left %	Right %
Arm	70.0	57.7
Leg	53.9	42.9

UN = Unified fractures
MAL = Malunified fractures

Table 7-3

The distribution of multiple cranial trauma, by location

			T1		T2		T3		T4	
		n	n̄	%	n̄	%	n̄	%	n̄	%
T	M	43	5	11.6	1	2.3	-	-	-	-
	F	26	6	23.1	-	-	-	-	-	-
K	M	31	5	16.1	1	3.2	-	-	-	-
	F	17	3	17.6	-	-	-	-	-	-
B	M	41	4	9.8	1	2.4	-	-	-	-
	F	28	4	14.3	2	7.1	-	-	-	-
S	M	56	10	17.9	4	7.1	-	-	-	-
	F	40	10	25.0	1	2.5	-	-	-	-
E	M	119	14	11.8	3	2.5	-	-	-	-
	F	66	13	19.7	3	4.5	2	3.0	-	-
R	M	122	32	26.2	9	7.4	1	.8	-	-
	F	83	23	27.7	7	8.4	2	2.4	-	-
M	M	57	15	26.3	1	1.8	1	1.8	1	1.8
	F	42	7	16.7	3	7.1	-	-	-	-
SP	M	57	11	19.3	3	5.3	-	-	-	-
	F	53	21	39.6	7	13.2	3	5.7	-	-
A	M	33	3	9.1	1	3.0	1	3.0	-	-
	F	32	14	43.8	4	12.5	2	6.3	1	3.1
C	M	48	6	12.5	2	4.2	-	-	-	-
	F	38	4	10.5	2	5.3	1	2.6	-	-
W	M	17	4	23.5	1	5.9	-	-	-	-
	F	19	4	21.1	1	5.3	1	5.3	-	-
N	M	59	17	28.8	6	10.2	2	3.4	-	-
	F	44	12	27.3	7	15.9	3	6.8	1	2.3
CA	M	94	17	18.1	4	4.3	2	2.1	2	2.1
	F	40	14	35.0	5	12.5	2	5.0	-	-
WA	M	54	9	16.7	3	5.6	1	1.9	-	-
	F	21	4	19.0	1	4.8	-	-	-	-
NT	M	53	3	5.7	-	-	-	-	-	-
	F	38	8	21.1	3	7.9	1	2.6	-	-
AT	M	94	15	16.0	3	3.2	-	-	-	-
	F	62	14	22.6	5	8.1	1	1.6	-	-
TA	M	27	4	14.8	1	3.7	-	-	-	-
	F	35	6	17.1	-	-	-	-	-	-
CQ	M	44	9	20.5	2	4.5	-	-	-	-
	F	28	14	50.0	2	7.1	-	-	-	-
CR	M	64	3	4.7	-	-	-	-	-	-
	F	39	11	28.2	3	7.7	2	5.1	-	-
BB	M	21	3	14.3	1	4.8	-	-	-	-
	F	5	1	20.0	-	-	-	-	-	-
QB	M	65	12	18.5	3	4.6	-	-	-	-
	F	33	15	45.5	2	6.1	-	-	-	-
P	M	28	3	10.7	-	-	-	-	-	-
	F	16	3	18.8	-	-	-	-	-	-

T1 = Crania having at least one cranial bone displaying trauma
T2 = Crania having a second bone affected by trauma
T3 = Crania having a third bone affected by trauma
T4 = Crania having a fourth bone affected by trauma
M = Males, F = Females

Location Codes:

T - Tocumwal, K - Kow Swamp (modern), B - Baratta, S - Swan Hill, E - Euston, R - Rufus River, M - Murray River, SP - Swanport, A - Adelaide, C - Coastal Victoria, W - West Central Victoria, N - Coastal New South Wales, CA - Central Australia, WA - Western Australia, NT - Northern Territory (northern half), AT - All Northern Territory, TA - Tasmania, CQ - Coastal Queensland, CR - Cape York and Cairns, BB - Broadbeach, P - Pleistocene sample.

Table 7-4

Frequency of trauma on the frontal parietal and occipital bones

	n	$\bar{n}$	%	P	P%	PL	PR	PL%	PR%	F	F%	O	O%
CM	247	33	13.4	32	97.0	17	15	53.1	46.9	7	21.2	5	15.2
	151	30	19.9	30	100.0	16	14	53.3	46.7	7	23.3	3	10.0
MUY	469	85	18.1	84	98.8	49	35	58.3	41.7	18	21.2	6	7.1
	302	66	21.9	64	97.0	34	34	53.1	53.1	13	19.7	4	6.1
R	122	32	26.2	31	96.9	18	13	58.1	41.9	9	28.1	-	-
	83	23	27.7	22	95.7	11	14	50.0	63.6	6	26.1	1	4.3
SC	138	20	14.5	18	90.0	13	5	72.2	27.8	7	35.0	2	10.0
	123	39	31.7	38	97.4	18	25	47.4	65.8	8	20.5	6	15.4
DES	132	22	16.7	19	86.4	10	9	52.6	47.4	7	31.8	6	27.3
	51	17	33.3	17	100.0	5	12	29.4	70.6	1	5.9	5	29.4
TRO	92	6	6.5	-	-	-	-	-	-	1	16.7	5	83.3
	62	15	24.2	15	100.0	7	8	46.7	53.3	5	33.3	2	13.3
CTL	133	31	23.3	30	96.8	15	15	50.0	50.0	7	22.6	7	22.6
	86	28	32.6	27	96.4	14	15	51.9	55.6	10	35.7	7	25.0

P = individuals with parietal trauma
PL = individuals with left parietal trauma
PR = individuals with right parietal trauma
PL% = percentage of left parietal trauma
PR% = percentage of right parietal trauma
F = individuals with frontal trauma
O = individuals with occipital trauma

Note:
PL and PR will not always equal P because some individuals have trauma on both left and right parietals and P indicates only the presence of trauma to this bone irrespective of side.

Area Codes:
CM - Central Murray, MUY - Murray River, R - Rufus River, SC - South Coast, DES - Desert, TRO - Tropics, CTL - East Coast

APPENDICES

APPENDIX 1

KEY TO PATHOLOGY RECORDING SHEETS

RECORD 1

CSN — Collection Series Number. The accession number allocated to particular bone by the institute in which it is kept.

Collection . — This number corresponds to the institution in which the skeletal remains were held.

Age — 1 = <6 years, 2 = 6-11 yrs, 3 = 12-20 yrs, 4 = sub-adult, 5 = adult, 6 = young adult, 7 = old adult.

Sex — 1 = male, 2 = female, 0 = not sexed

Origin — Only boxes 41 and 42 used for population code

Bone —
1 = Frontal, 2 = Parietal, 3 = Temporal
4 = Occipital, 5 = Zygomatic, 6 = Maxilla
7 = Mandible, 8 = Scapula, 9 = Clavicle
10 = Humerus, 11 = Ulna, 12 = Radius
13 = Cervical vertebrae, 14 = Thoracic vertebrae
15 = Lumbar vertebrae, 16 = Sacrum, 17 = Ilium
18 = Ischium, 19 = Pubis, 20 = Femur, 21 = Tibia
22 = Fibula

L/R — 1 = left, 2 = right hand side

Pathology Box — 1 = non-pathological, 2 = pathological

Harris Lines — Two boxes for each age grade e.g. <5 yrs, 5-10 yrs and >11 years. Two boxes for each of these grades allows for up to 99 lines to be recorded

C.O. — Cribra orbitalia. Left and right columns for left and right orbits. 1 = category 1,
2 = category 2, 3 = category 3, 4 = category 4, 9 = not observable

"Trach"	Same as for cribra orbitalia but the number in either box indicates the number of lesions found. 9 = not observable. (This lesion has not been used in this study but is the subject of a separate publication)
Dental Hypoplasia	Eight boxes, one for each tooth: 1 = tooth having lines or grooves 2 = tooth having pits 9 = not observable (missing data) Only the canine and third molar were inspected for this condition
Dental Pathology	Three boxes for the right (66-68 and 74-76) and three (70-72 and 78-80) for left side of the maxilla and mandible. Boxes 65, 69, 72 and 77 not used.

Box 2 = Caries	
None	-
Single	1
(2-4 teeth) Polycaries 1	2
(5-8 teeth) Polycaries 2	3
Not observable	9
Box 3 = Alveolar Pathology	
None	-
Single	1
Multiple	2
Not observable	9
Box 4 = Edentulism (missing teeth)	
None	-
1 tooth lost	1
2-4 teeth lost	2
5-8 teeth lost	3
Full edentulism	4
Incisors missing	5
Incisor + toothloss	6
Not observable	9

RECORD 2

CSN	As for Record 1
Arthritis	Boxes 10-27 are for the proximal end of the bone and boxes 28-45 are for the distal end. Boxes 10-18 and 28-36 are for a left, and boxes 19-27 and 37-45 for a right. Each of the nine boxes represents one of a number of symptoms arising from arthritic disease:

Box '1' = Osteoarthritis
Box '2' = Vertebral oseophytosis
Box '3' = Ankylosing spondylitis
Box '4' = Traumatic arthritis
Box '5' = Infectious or Pyogenic arthritis

Box '6' = Osteophytosis: 1 = minor
2 = major

Box '7' = Erosion:

1 = <1cm2 medial facet or condyle
2 = <1cm2 lateral facet or condyle
3 = >1cm2 medial facet or condyle
4 = >1cm2 lateral facet or condyle
5 = <1cm2 on both
6 = >1cm2 on both
7 = severe erosion on either f/c
8 = erosion on head of femur, humerus, radius,and fibula (proximal and distal ends)
9 = not observable

Box '8' = Eburnation:

1 = <1/2 medial articular facet or condyle
2 = <1/2 lateral articular facet or condyle
3 = >1/2 medial articular facet or condyle
4 = >1/2 lateral articular facet or condyle
5 = >1/2 both articular facet or condyle
6 = >1/2 both art. facet or condyle
7 = eburnation on head of femur, humerus, radius and fibula (proximal and distal ends)
9 = Not observable

Box '9' = Ankylosing between adjacent long bones 1 = present

Arthritis T.M.J. One box each for right and left articular fossae (48-51), and one box for right and left mandibular condyles (50-51). Only two forms of arthritic wear recorded:

1 = minor wear and/or erosion
2 = major wear and/or erosion

Wear and flattening of the articular surface are synonymous. Arthritic attrition of in the T.M.J. area will be noted whether the mandibular (articular) fossa and/or the articular tubercle are involved.

9 = not observable

Arthritis M.A.C. Mandibular condyle (as for T.M.J.)

RECORD 3

CSN As for Record 1.

Trauma Two boxes for a left bone and two for a right.
These accommodate 19 different types of trauma:

1 = Parrying fracture (distal one third of ulna and sometimes the radius when accompanying a parrying fracture of the ulna).
2 = Transverse and oblique fractures (across the bone).
3 = Comminuted fracture (bone shattered at point of break).
4 = Greenstick fracture (partial break).
5 = Impacted fracture ("squashed" head of femur or humerus).
6 = Colles fracture (distal end of radius).
7 = Pott's fracture (distal fibula and medial malleolus).
8 = Cranial dents (small oval depressions <2cm at greatest diameter). The first box (10 or 12 etc) will denote the amount of dents present in that particular bone e.g. 28 will mean two dents are present. A prefix of 8 denotes >7 dents.
9 = Depressed fracture >2cm etc, skull or zygomatic arch.
10 = Monteggia fracture (proximal half of the ulna shaft).
11 = Transcervical fracture (neck of femur only).
12 = Linear or split fracture, cranium only.
13 = Gauging, on cranium only.
14 = Amputation (six major long bones only).
15 = Dislocation (shoulder and hip joints only).
16 = Trephination.
17 = Piercing wound.
18 = Atrophy.
19 = Pathological fracture.

Mend	Fracture healing: 1 = unified, 2 = malunified 3 = lateral bend, 4 = medial bend
Mult.	Denotes whether fracture is multiple or not 1 = single, 2 = multiple
Position	Position of fracture or trauma along shaft 1 = Prox. 1/3, 2 = Mid 1/3, 3 = Distal 1/3
N-Sp Inf.	Non-specific infection: Box 1 = Periostitis 1 = present Box 2 = Osteo-periostitis 1 = present Box 3 = Osteomyelitis 1 = present Box 4 = Bowing: 1 = antero-posterior 2 = anteverted 3 = lateral 4 = medial
Exostosis	Box 5. Code 1 = present

CR (Cranium) — Treponemal disease on the cranium only. Terminology follows that of Hackett (1976)

Contiguous Series:

Box 1 = 1 = Serpiginous cavitation
2 = Nodular cavitation
3 = Caries sicca plus stellate cracks
4 = All three of the above
5 = Nodular scars
9 = not observable.

Discrete Series:

Box 2 = 1 = Circumvallate cavitation
2 = Radial scars
9 = not observable.

Box 3 = 1 = Bone shrinkage, only around nasal aperture, usually with periostitis.
2 = Naso-maxillary destruction. This can extend from nasion to prosthion and includes shrinkage around the nasal aperture, signs of sub-glabella collapse ('saddle nose') and sub-nasal bone resorption.
3 = '1' and '2' plus involvement of the palate

Prox/Dist. — 1 = Proximal, 2 = distal, 3 = both ends

Position — 1 = 1/3, 2 = 2/3, 3 = Middle 1/3, 4 = All over

Comment Box — This box is used for a code corresponding to a standard comment (i.e. '1' might refer to the poor condition of the bone). Not used in this study.

APPENDIX 2

SAMPLE COMPOSITION AND PATHOLOGY CODES

Locations

Computer code	Figure abbrev.	Sample Origins
1	T	Tocumwal, Numurkah, Cobram.
2	T	Echuca, Moama, Barmah, Nathalia, Tulla Creek and Tulla Station.
3	T	Deniliquin.
4	K	Kerang, Murrabit.
5	K	Loddon River, Barham, Barr Creek and Koondrook.
6	K	Kow Swamp (modern), Cohuna, Gumbower.
7	P	Kow Swamp (Pleistocene).
8	B	Baratta, Coobool Crossing or Creek (modern), Nacurrie.
9	P	Coobool Crossing (Pleistocene).
10	S	Swan Hill, Lake Boga, Vinifera, Nyah and Nyah West, Poon Boon, Piangle, Kyalite.
11	E	Balranald. Note: this sample is incorporated in the Euston sample.
12	E	Euston, Robinvale, Lake Benanee.
13	R	Mildura, Merbein, Dareton, Wentworth, Red Cliffs.
14	R	Rufus River, Lake Victoria, Lindsay Creek, Chowilla.
15*	RK	Roonka.

16	SP	Swanport, Murray Bridge, Tailem Bend, Lake Alexandrina.
17	A	Adelaide. (Fulham, Sydenham, Zoo, Torrens, Glenelg, Port Adelaide, St Peters, Henley Beach Reedbeds, Mt Pleasant, Parafield Gardens)
18	CR	Cape York.
19	CR	Cairns.
20	CQ	Coastal Queensland (from Cardwell to Queensland/NSW border).
21	BB	Broadbeach Collection.
-	QB	Combined CQ and BB samples
22*	KI	Keppel Islands.
23	N	North Coast of New South Wales.
24	N	Metropolitan Sydney.
25	N	South Coast of New South Wales.
26	C	South Eastern Coast of Victoria, including Metropolitan Melbourne.
27	C	South Western Coast of Victoria.
28	W	West Central Victoria (the Mallee or Wimmera), to within 50 km of the Murray River.
29	TA	Tasmania
30	WA	South Western, Western Australia.
31	WA	West Central Western Australia (Pilbara region).
32	WA	North Western, Western Australia (Kimberley region).
33	NT	Western Northern Territory.
34	NT	Eastern Northern Territory.
37	CA	Central Australia. This group represents all areas roughly lying within and near the 10

		inch rainfall isohyet. These include arid areas in Western Australia, the Northern Territory, Queensland, South Australia and New South Wales.
38	G	Goolwa, South Australia.
51	M	A general code used for any material from the Murray Valley area but lacking exact provenance.

* These groups do not appear in figures and tables and are not discussed in the text.

Areas

Area	Population Code
CM (Central Murray)	4, 5, 6, 8, 10, 11, 12
MUY (Murray Valley)	1, 2, 3, 4, 5, 6, 8, 9, 10, 11, 12, 13, 14, 51
RUF (Rufus River)	13, 14
SC (South Coast)	16, 17, 26, 27
DES (Desert Regions)	31, 32, 37
TRO (Tropical Regions)	18, 19, 33, 34
CTL (Eastern Coast)	20, 21, 23, 24, 25, 30

www.ingramcontent.com/pod-product-compliance
Ingram Content Group UK Ltd.
Pitfield, Milton Keynes, MK11 3LW, UK
UKHW061845041026
15591UKWH00015B/918

9 780860 546276